Glycoproteins and Glycolipids in Disease Processes

Earl F. Walborg, Jr., Editor
The University of
Texas System Cancer Center

A symposium sponsored by the ACS Division of Carbohydrate Chemistry at the 175th Meeting of the American Chemical Society, Anaheim, March 14–15, 1978.

ACS SYMPOSIUM SERIES **80**

AMERICAN CHEMICAL SOCIETY
WASHINGTON, D. C. 1978

Library of Congress CIP Data

Glycoproteins and glycolipids in disease processes.
(ACS symposium series; 80 ISSN 0097-6156)

Includes bibliographies and index.

1. Glycoproteins—Congresses. 2. Glycolipids—Congresses. 3. Tumor antigens—Congresses. 4. Carbohydrate metabolism disorders—Genetic aspects—Congresses.

I. Walborg, Earl F. II. American Chemical Society. Division of Carbohydrate Chemistry. III. Series: American Chemical Society. ACS symposium series; 80.

QP552.G59G6 616.3'99 78-11920
ISBN 0-8412-0452-7 ASCMC8 80 1–480 1978

PRINTED IN THE UNITED STATES OF AMERICA

ACS Symposium Series

FOREWORD

The ACS SYMPOSIUM SERIES was founded in 1974 to provide a medium for publishing symposia quickly in book form. The format of the SERIES parallels that of the continuing ADVANCES IN CHEMISTRY SERIES except that in order to save time the papers are not typeset but are reproduced as they are submitted by the authors in camera-ready form. As a further means of saving time, the papers are not edited or reviewed except by the symposium chairman, who becomes editor of the book. Papers published in the ACS SYMPOSIUM SERIES are original contributions not published elsewhere in whole or major part and include reports of research as well as reviews since symposia may embrace both types of presentation.

CONTENTS

PREFACE

It is fitting that the proceedings of this symposium on "Glycoproteins and Glycolipids in Disease Processes" should be dedicated to the life and scientific contributions of my friend and fellow scientist, Ward W. Pigman. Much of Ward's professional effort was devoted to the interchange of scientific information, both by his editorial skills and his support of national and international symposia on glycoconjugate chemistry. This symposium was born in Ward's mind in 1976 during deliberations of the Division of Carbohydrate Chemistry of the American Chemical Society concerning symposia topics for the coming years. Ward felt that sufficient progress had been made concerning the relationship of glycoconjugates to a variety of disease processes to warrant a symposium devoted to this topic. At that early date, the Carbohydrate Chemistry Division made a commitment to sponsor a symposium on some aspect of glycoprotein research and designated Ward as chairman with responsibility to formulate the program. Early in 1977, Ward informed me of his plans for the symposium and asked if I would serve as co-chairman. I gladly agreed and we set about the task of preparing a tentative program and contacting prospective speakers. On September 28, 1977, during the Fourth International Symposium on Glycoconjugates being held at Woods Hole, Ward and I finalized the symposium program. Upon completion of this task Ward expressed to me his personal satisfaction that the years of basic research on glycoconjugates was finding increasing relevance to understanding a variety of disease processes. Ward's sudden death on September 30, 1977, was a shock to the many participants of the Wood's Hole meeting, many of whom were his long-time friends. In a way it was fitting that his last days found him engaged in an international meeting on glycoconjugates, an endeavor he had vigorously promoted and fostered for many years. Many of the participants in the symposium on "Glycoproteins and Glycolipids in Disease Processes" were present at the Wood's Hole meeting and by common consent it was decided that the planned symposium and this monograph should be dedicated to Ward Pigman's life and contributions to glycoconjugate research.

The efforts of many persons and agencies were responsible for the success of the symposium and the publication of this monograph. First, I should like to express my appreciation to all of the participants for their contributions to the symposium and to the monograph. The support of the Division of Carbohydrate Chemistry, particularly the efforts of Derek H. Ball and Gary D. McGinnis, is gratefully acknowledged. Grants from the National Institute of General Medical Sciences (GM 25226) and the National Cancer Institute (CA 24513), which provided travel support for participants in the symposium, are gratefully acknowledged. Particular gratitude is extended to Edward Hampp (N.I.G.M.S.) and Mary A. Fink (N.C.I.) for their efforts in making this symposium a reality. On numerous occasions I have been encouraged by Gladys Pigman. Able assistance was obtained from Beverly Sea, Carey Osborn, Reita Maughan, and Pat McKirahan, staff at the Research Division—Science Park, The University of Texas System Cancer Center.

Science Park—Research Division
The University of Texas System Cancer Center
July 20, 1978

EARL F. WALBORG, JR.

To Ward W. Pigman

DEDICATION TO WARD W. PIGMAN

It is appropriate that this Symposium on "Glycoproteins and Glycolipids in Disease Processes" should be dedicated to the memory of my friend and colleague, Ward W. Pigman, who passed away on September 30, 1977, during his attendance at the 4th International Meeting on Glycoconjugates being held at Woods Hole, Massachusetts. Ward convinced the Carbohydrate Division of the American Chemical Society that a symposium on this subject warranted consideration for the spring meeting of the ACS to be held in Anaheim, California. Through his efforts and those of Earl F. Walborg, Jr., this symposium has become a reality. Barely two hours before Ward's death, I had the opportunity to discuss his plans for this symposium, and I realized how deeply he felt about it. In a way, this symposium represents the ultimate goal of the long scientific road he and quite a few others among us have traveled, a road through carbohydrate chemistry that has led us from the use of analytical and physical chemistry in the study of reaction mechanisms to synthetic chemistry and finally to the elucidation of the structure of more and more complex carbohydrates, including glycosaminoglycans of connective tissue and glycoproteins from mucous secretions and the surface of the cancer cell.

I feel that I have a deep understanding of Ward's research interests because both of us followed quite similar pathways, becoming interested in connective tissue polysaccharides in the late forties. In 1930 at the age of 20, he joined the National Bureau of Standards, where he was active until 1943. During this time, he obtained his B.A. and M.A. degrees at George Washington University and his Ph.D. degree at the University of Maryland. At the Bureau he was associated with Horace Isbell, one of the leading American carbohydrate chemists. A series of papers reflecting Isbell's interests in the physical and organic chemistry of carbohydrates resulted from this collaboration. Soon a change in Ward's interests was noticeable, a turn to more biological problems with a study of glycosidases. This turning point was marked by a paper published in collaboration with H. Isbell and B. Helferich and by Ward's obtaining a Lalor Fellowship to spend one year in Leipzig, a stay that was unfortunately interrupted by the start of World War II. He returned to the National Bureau of Standards, and then during the last two years of the war he became a group leader at Corn Products, where most of his efforts were directed toward starch chemistry. Following his brief association with Corn Products, he moved to the Institute for Paper Chemistry at Appleton, Wisconsin, an institute for basic research supported by the industry, where he also served as a group leader. During the Wisconsin period, Ward's interests included pectins, cellulose, carbohydrate-degrading enzymes, and tannins.

In 1949 Ward moved into the academic community, accepting a position at the University of Alabama Dental School where he was Associate Professor until 1960, Professor and then Dean of the Graduate School from 1963 to 1968. He remained associated with this institution as a Visiting Professor until his death. In 1960 Ward assumed the additional burden of a second academic position as Professor and Chairman of the Department of Biological Chemistry at New York Medical College. These moves to academic positions in a dental and medical school correspond to a final shift in Ward's research interests. After completing a series of papers on the reaction between amino acids and carbohydrates, the so-called Browning reaction, initiated at Appleton, papers appeared on dental plaque and saliva, and then the beginning of the work for which Ward is now best known: first, investigations on hyaluronic acid degradation and then an outstanding series of papers on submaxillary mucins. Before this work had been started, the contribution of Ward to carbohydrate chemistry had already been recognized by his selection as the recipient of the Hudson Award by the Carbohydrate Chemistry Division of the American Chemical Society in 1959. Other awards followed, such as medals of the Société de Biochimie de France and of the University of Milan.

Enumerated below are a few of Ward's contributions to glycoconjugate chemistry.

(a) Elucidation of the mechanism of degradation of hyaluronic acid by ascorbic acid and free radicals.

(b) Purification of submaxillary glycoproteins and elucidation of many aspects of their peptide and saccharide structure.

(c) Elucidation of the β-elimination mechanism for cleavage of the glycopeptide bond between 2-acetamido-2-deoxy-D-galactose and serine or threonine, the predominant glycopeptide linkage of the mucin glycoproteins.

(d) Suggestion that the peptide moiety of the mucin glycoproteins is composed of repeating sequences of 28 amino acids. Although present evidence suggests a more complex arrangement, this hypothesis will continue to deserve attention.

This symposium and the resulting monograph bear the mark and vision of Ward W. Pigman, and therefore are fitting memorials to his life and scientific contributions to glycoconjugate chemistry.

Harvard Medical School
and Massachusetts General Hospital
Boston, Massachusetts
August 21, 1978

ROGER JEANLOZ

Structure and Metabolism of Glycoproteins and Glycolipids

Introduction

ROGER W. JEANLOZ

Laboratory for Carbohydrate Research, Harvard Medical School and Massachusetts General Hospital, Boston, MA 02114

To some chemist members of the Division of Carbohydrate Chemistry of the American Chemical Society, which is organizing this symposium, it may be unexpected that, for the next few days, we will be discussing compounds that encompass lipids and proteins, in addition to carbohydrates, and their relation to disease processes. It is within the life span of many of us that natural product chemists have realized that nature is not compartmentalized into such nice and clean-cut categories as carbohydrates, proteins, and lipids; in fact, very few carbohydrates exist that are devoid of a peptide or lipid component, as well as there are few proteins devoid of a carbohydrate component. The roles of the carbohydrate component, of glycoproteins and glycolipids that are located at the surface of the animal cell or present in the connective tissue that links the cells, have received increased recognition in the past decade and will be discussed during the next few days.

The subjects of the first session of this symposium will be the methods of structure identification presently being developed, as well as the biosynthesis and degradation of glycoproteins and glycolipids. Because glycoproteins and glycolipids of biological interest are generally available only in minute amounts, the methods of structure identification, which will be discussed by Dr. Walborg in his general presentation of current concepts of glycoprotein structure, are being developed at the micro- and even nano-gram level.

Dr. Schachter and associates will entertain us with the very active development of glycoprotein biosynthesis presently taking place. For many years, the concept of glycoprotein and glycolipid biosynthesis was dominated by the concept of elongation of the chains by one carbohydrate unit at a time, through sugar nucleotides; this concept, first developed for such homoglycans as glycogen and starch, was very ably demonstrated for glycoproteins and glycolipids by Roseman's group. Nearly twenty years ago, this unified concept of chain elongation was shown to be invalid for bacterial polysaccharides, and the importance of isoprenoid sugar phosphate as intermediates in polysaccharide synthesis was recognized. Ten years ago, Leloir's group, already known for its work on sugar nucleotides, established the role of isoprenoid sugar phosphates in glycoprotein biosynthesis. The past few years have witnessed the development of the role of these intermediates in the biosynthesis of the core component of *N*-glycoproteins, the role of D-glucose residues in the transfer to the peptide chain, etc. It is of more than passing interest that we are now faced with the existence of two processes: One is active in animal and plant *N*-glycoproteins and involves a rather complex mechanism, where the transfer of carbohydrate units to the peptide backbone takes place at the stage of large oligosaccharides synthesized through lipid intermediates and partially degraded before new elongation, the second is a very simple process where sugar units are transferred as single components from sugar nucleotides. Time will show whether these two processes coexist or are merely the expression of inadequate experimentation.

The glycolipids to be discussed by Dr. Sweeley and his associates will be restricted to those of eukaryotic organisms, which encompass solely glycoglycerolipids and glycosphingolipids. The diversity of their structures (for example α-D-galactopyranosyl residues are found only within the chain of glycolipids), and of their chain length (evidence for carbohydrate chains of complex composition having up to 50 units has been reported), their variation at the surface of the cells during the oncogenic process or as the result of specific genetic defects, their role in blood-group typing, have all been reasons for active investigation.

Finally, most of the recent chemical structure elucidations at the microgram level would not have been possible without the specific *endo*-glycosidases that are able to split off carbohydrate chains near the carbohydrate-peptide linkage and which will be discussed by Dr. Maley and his associates. I remember when, 25 years ago, Dr. Eylar was working in my laboratory with *exo*-glycosidases of *D. pneumoniae* on the structure of α_1-acid glycoprotein of human plasma, we were deploring the lack of an *endo*-glycosidase not realizing at the time that such enzyme, found later by Dr. Kobata and his group, was present in the extract that we were manipulating. It is the addition of such biochemical degradation methods to the more established chemical methods that has allowed the great progresses made recently in structural identification of glycoproteins and glycolipids and made possible the correlation of these structures with disease processes, the subject of this Symposium.

RECEIVED August 2, 1978.

Current Concepts of Glycoprotein Structure

EARL F. WALBORG, JR.

University of Texas System Cancer Center, Science Park—Research Division, Smithville, TX 78957

Glycoproteins are ubiquitous components of Nature, being found in the extra- and intracellular fluids, connective tissue and cellular membranes. They are involved in diverse physiological functions, including: ion transport, hormone storage, blood clotting, immunological protection, hormonal control, enzymatic catalysis, nutrition, surface protection, lubrication, structural support, transmission of hormonal stimuli, molecular recognition and cell-cell adhesion (1). Furthermore, the manifestation of many disease processes is accompanied by qualitative and/or quantitative alterations of cellular and extracellular glycoproteins. Our increasing appreciation of glycoproteins in normal and disease processes has resulted from recent advances in our understanding of the structure and properties of this diverse class of proteins. One of the last endeavors of Ward Pigman, in whose honor this symposium is held, was the editing of a new volume on the biochemistry of glycoproteins (2). Relying heavily on the monograph edited by Horowitz and Pigman (2) and on several other excellent reviews on this subject (1,3-6), I shall briefly examine current concepts of glycoprotein structure, including: the glycopeptide bond, structure of the heterosaccharide moieties, the information content of heterosaccharides, the distribution of the heterosaccharide moieties on the peptide chain, and the integration of glycoproteins into cellular membranes. For more adequate documentation of most of the concepts discussed herein, the reader should consult the comprehensive reviews previously cited. Hopefully, this discussion will provide a conceptual base for subsequent presentations concerning the role of these glyconjugates in disease processes.

For the purpose of this symposium I should like to propose that we use the more restrictive definition of glycoprotein, proposed by Gottschalk (6, p. 29); i.e., "a conjugated protein containing as a prosthetic group(s) one or more saccharide(s)..., lacking a serially repeating unit and bound covalently to the peptide chain." This definition excludes the proteoglycans which will not be emphasized at this symposium.

0-8412-0452-7/78/47-080-005$05.00/0

COVALENT LINKAGE OF SACCHARIDES TO THE PEPTIDE CHAIN

The covalent linkage of saccharides to the peptide chain represents a central aspect of glycoprotein structure. Figure 1 depicts the three major glycopeptide bonds: the glycosylamine linkage between 2-acetamido-2-deoxy-D-glucose and L-asparagine, the glycosidic linkage between 2-acetamido-2-deoxy-D-galactose and L-serine or L-threonine, and the glycosidic linkage between hydroxylysine and D-galactose. These first two linkages occur in a wide variety of glycoproteins, including those obtained from blood plasma, cellular membranes and connective tissue. The glycosidic linkage to serine or threonine is the predominant linkage present in the glycoproteins of mucous secretions. The β anomeric configuration of the glycosylamine linkage was established by synthesis; whereas the anomeric configuration of the glycosidic linkage to serine or threonine was determined by the use of glycosidases of known anomeric specificity. The glycosidic linkage between hydroxylysine and galactose was first noted in tropocollagen and glomerular basement membrane and is generally confined to fibrillar collagens. The β anomeric configuration of this linkage was established using glycosidases of known anomeric specificity. Other glycopeptide linkages have also been reported, e.g., Hallgren *et al*. demonstrated a linkage between L-fucose and L-threonine in certain urinary glycopeptides (7). Later in this symposium S. Steiner will report the isolation of some new cellular glycopeptides having this same linkage. A thioglycosidic linkage between L-cysteine and D-glucose has been proposed for glycopeptides isolated from the erythrocyte membrane and urine (8,9). Glycosidic linkages between galactose (10) or arabinose (11) and hydroxyproline are present in plant cell walls. A glycosidic linkage between mannose (12) or galactose (13) and serine or threonine has been proposed for glycoproteins of microbial origin and earthworm cuticle collagen, respectively.

STRUCTURE OF HETEROSACCHARIDES LINKED GLYCOSIDICALLY TO SERINE OR THREONINE

Of the three major glycopeptide linkages only the glycosidic linkage to serine or threonine is labile to cleavage by base. In this linkage the glycosidic bond is located β to a carbonyl function, thus facilitating cleavage of the glycosidic bond by a β-elimination reaction (1,3). Downs and Pigman (14) and others (15) explored this reaction and contributed to the development of methodology to isolate oligosaccharide moieties from a variety of mucous glycoproteins. Two types of oligosaccharides (simple and complex) are released from mucous glycoproteins by the β-elimination reaction. The structures of several glycopeptides representative of the simpler class containing 2 to 6 monosaccharide units are presented in Figure 2. Examination of the structures of the simpler class of heterosaccharides linked glycosidically to

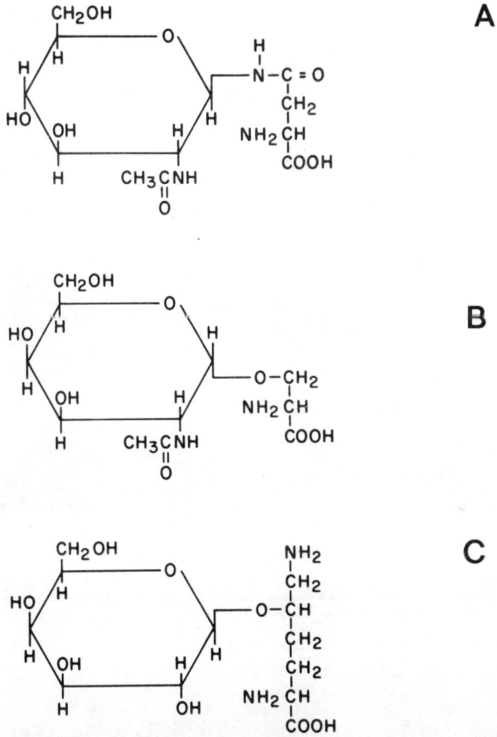

Advances in Protein Chemistry

Figure 1. Structure of the three major glycopeptide bonds: N-*glycosyl linkage to asparagine (A), glycosidic linkage to serine or threonine (B), and glycosidic linkage to hydroxylysine (C) (1)*

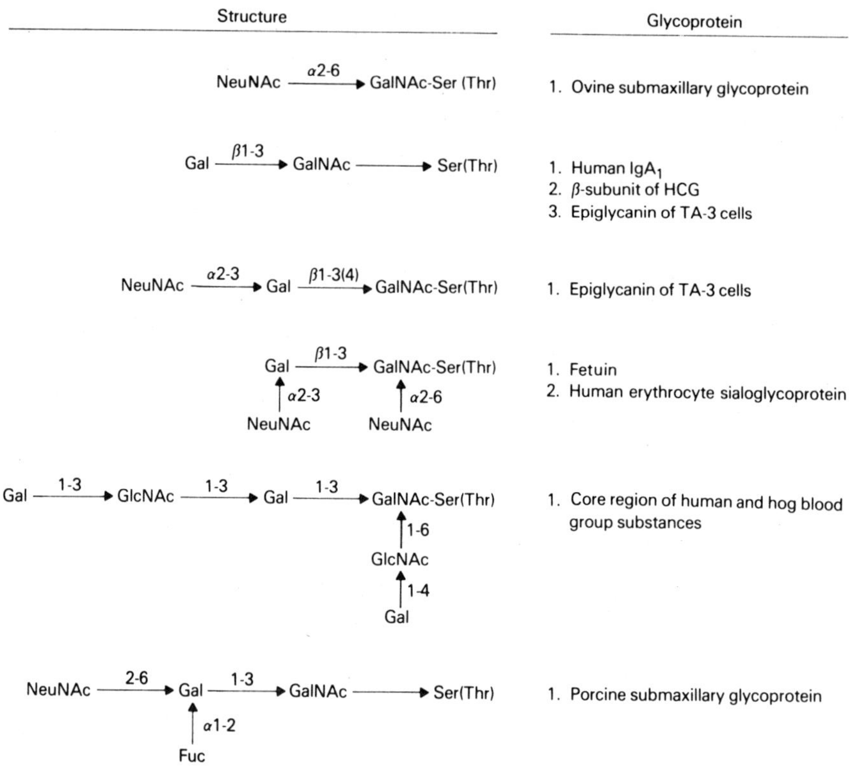

Figure 2. Proposed structures of glycopeptides of simpler type linked glycosidically to serine or threonine. Adapted from Ref. 3. See Ref. 3 for references concerning individual glycopeptides.

serine or threonine reveals certain structural generalizations; e.g., except in the case of the simplest glycopeptide from ovine submaxillary glycoprotein, galactose is always linked β(1→3) to the 2-acetamido-2-deoxy-D-galactose residue involved in the glycopeptide linkage; thus establishing a core structure consisting of D-galactosyl-2-acetamido-2-deoxy-D-galactosyl-L-serine as a characteristic heterosaccharide sequence in the mucous glycoproteins. L-fucose and sialic acid residues are always located at the nonreducing periphery of this class of glycopeptides. As the structures of more glycopeptides of this class are elucidated, it is apparent that biosynthesis places certain restrictions on the structure of the heterosaccharide moieties. The biosynthetic basis of these restrictions will be discussed in a subsequent lecture by H. Schachter.

More complex heterosaccharides may also be linked glycosidically to threonine or serine, as exemplified by the heterosaccharide determinant of the blood group substances (Figure 3). The elucidation of the structure of these heterosaccharide determinants represents a milestone in carbohydrate chemistry (17). A definitive correlation between gene activity and heterosaccharide structure was established, genetic variation being achieved through specific glycosyltransferases involved in the biosynthesis of the heterosaccharide determinants (18). The H gene codes for a fucosyltransferase which adds L-fucose to a peripheral D-galactose residue of the heterosaccharide precursor. A and B blood group specificities arise from the addition of a 2-acetamido-2-deoxy-D-galactose or a D-galactose residue to the nonreducing terminal D-galactose residue of the H specific heterosaccharide. The Lewis or Le gene codes for another fucosyltransferase which adds fucose to a peripheral 2-acetamido-2-deoxy-D-glucose residue of the heterosaccharide precursor. In the absence of the H gene this gives rise to the Le^a specific determinant, while the presence of both H and Le genes yields the interaction product responsible for the Le^b specificity.

INFORMATION CONTENT OF HETEROSACCHARIDES

The composite heterosaccharide determinant of blood group substances (Figure 3) illustrates the information content of heterosaccharides, a concept which intrigued the later Professor Richard Winzler (19). Unlike peptides in which information content is based only on the number of different monomeric units and their sequence, heterosaccharide information resides in the number of different monomeric units, the sequence of these units, branching of the monomeric sequences, and the position and anomeric configuration of the glycosidic linkages; thus per unit mass heterosaccharides contain considerable information. The implication of this concept for biological processes involving glycoproteins has yet to be fully realized. Detracting from this concept is the apparent infidelity of enzyme systems involved in heterosaccharide

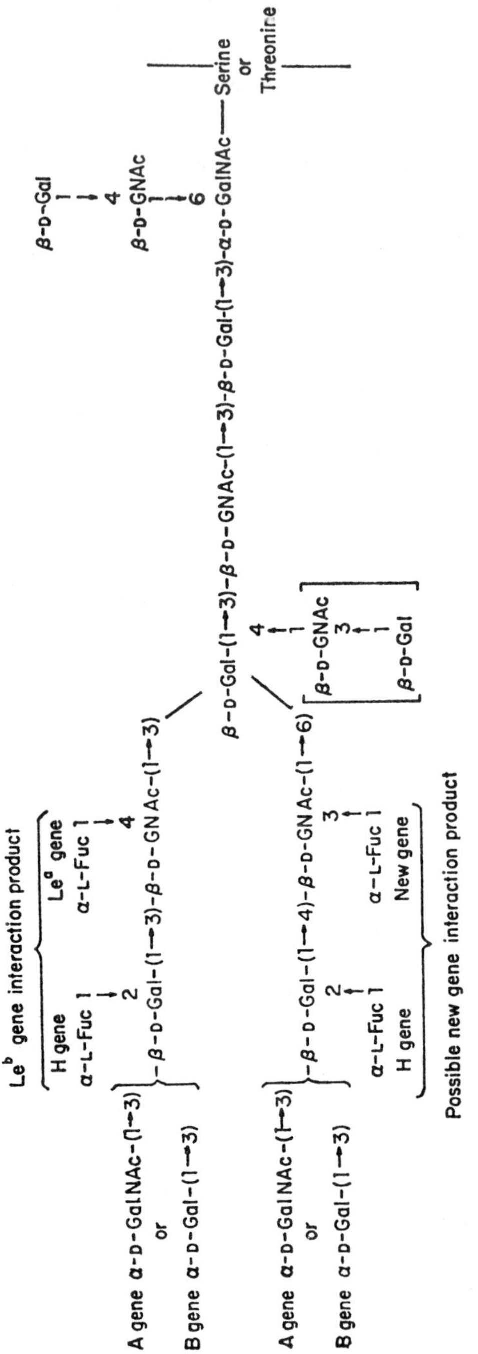

Journal of Immunology

Figure 3. Composite heterosaccharide showing relationships of blood group determinants (16)

biosynthesis. A major problem of researchers working on glycoprotein structure has been the microheterogeneity (20) of oligosaccharide moieties, a property which can be explained on the basis of incomplete addition of monosaccharide residues to the nonreducing periphery of the heterosaccharide. The following quote by Alfred Gottschalk is pertinent to this discussion: "One may be tempted to compare the biosynthesis of proteins with flawlessly tailored suits, a flawlessness essential for their function. To continue the comparison... heteropolysaccharides may be related to ready-made suits; their fit is not as accurate as that of the tailored garment; but they will do fine for the purposes for which they are used," (6, Part A, p. 29). It should be noted that in certain cases it is not possible to distinguish between incomplete biosynthesis and partial degradation by glycosidases as the causal factors responsible for the microheterogeneity of glycoproteins.

STRUCTURE OF HETEROSACCHARIDES LINKED N-GLYCOSICALLY TO ASPARAGINE

As the structures of more heterosaccharides linked N-glycosically to asparagine are elucidated, certain common structural features emerge. Figure 4 shows the structure of a glycopeptide from hen egg ovalbumin, an example of a simpler class of glycopeptides which contains only 2-acetamido-2-deoxy-D-glucose and D-mannose. Many glycopeptides of this type contain a core heterosaccharide sequence composed of a mannotriose unit linked to di-N-acetylchitobiose. The mannose residue adjacent to the di-N-acetylchitobiose unit is commonly attached by a $\beta(1\rightarrow4)$ glycosidic bond; while other more peripheral mannose residues are linked by α glycosidic bonds. Ovalbumin glycopeptides, containing additional mannose and/or 2-acetamido-2-deoxy-D-glucose residues at the nonreducing periphery of the heterosaccharide have also been isolated, thus providing a classic example of microheterogeneity within the heterosaccharide moiety (20).

Figure 4 also shows the proposed structures of more complex heterosaccharide moieties linked to asparagine through a glycosylamine linkage, including glycopeptides from human immunoglobulins A_1 and E. A trimannosyl-di-N-acetylchitobiose core is a common, but not constant, feature of these glycopeptides. In addition, the heterosaccharide chains commonly terminate with the trisaccharide sialyl-galactosyl-2-acetamido-2-deoxy-D-glucose or partially completed sequences of this trisaccharide. Fucose is always present at the nonreducing periphery of the heterosaccharide. The proposed structures for the glycopeptides from human immunoglobulin E (Figure 4) and several other glycoproteins do not contain the trimannosyl-di-N-acetylchitobiose core. Although more confirmatory structural evidence would be desirable, it appears that core structures other than that seen in ovalbumin glycopeptide are possible. This observation has profound implications concerning the specificity and compartmentalization of

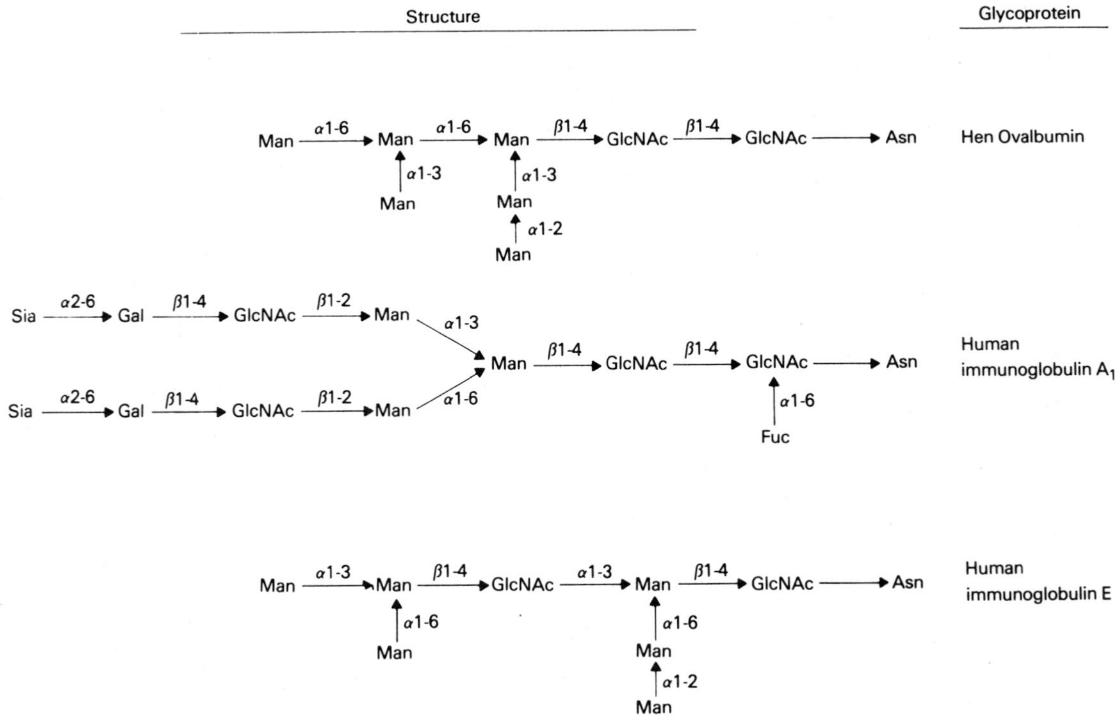

Figure 4. Proposed structures of several glycopeptides linked N*-glycosically to asparagine. Adapted from Ref. 3.*

enzymes involved in the biosynthesis of the heterosaccharide moieties of glycoproteins.

DISTRIBUTION OF THE HETEROSACCHARIDE MOIETIES ON THE PEPTIDE

The number of heterosaccharide moieties per peptide chain varies widely; e.g., each molecule of ovalbumin, fetuin, glycophorin and ovine submaxillary glycoprotein contain 1, 3, 16 and approximately 800 heterosaccharide moieties per molecule of glycoprotein, respectively (1,21). Figure 5, taken from Ward Pigman (2, p. 132), shows the distribution of oligosaccharide chains along the peptide moiety of bovine submaxillary glycoprotein. The high density of hydrophilic saccharide chains terminating in negatively charged sialic acid residues causes the mucous glycoproteins to assume an extended conformation. Gottschalk (22) demonstrated that removal of the sialic acid by neuraminidase digestion resulted in the collapse of mucous glycoprotein molecules to a more globular conformation; thus, in the case of the mucous glycoproteins, the heterosaccharide moieties serve as modulators of peptide conformation. On the other hand, the heterosaccharide determinants of proteins more sparsely substituted with heterosaccharide moieties do not appear to have a major role in the determination of protein conformation; and, in fact, protein conformation probably dictates the site(s) of glycosylation. Figure 6 shows the location of the three N-glycosically linked heterosaccharide determinants of porcine pancreatic ribonuclease (23). Using X-ray crystallographic data, it was determined that these three heterosaccharide moieties are located at the surface of the molecule and project into the solution environment (1). It is reasonable to assume that a similar situation will predominate in other nonmembrane-bound globular glycoproteins, which are sparsely substituted with heterosaccharide chains.

Several factors are involved in determining which amino acid residues of the peptide moiety are glycosylated. Only in the case of heterosaccharide moieties N-glycosically linked to asparagine has it been possible to implicate the amino acid sequence in the vicinity of the glycopeptide bond as a determinant of glycosylation, Figure 7 shows the amino acid sequence adjacent to the glycosylamine linkage for a variety of glycoproteins. The only common feature is the sequence Asn-X-Ser (Thr) at the site of glycosylation. It should be emphasized, however, that this sequence represents only a minimum requirement for glycosylation since other factors, such as steric hinderances imposed by the peptide chain, may prevent glycosylation. Amino acid sequence does not appear to be the major determinant for the glycosylation of serine, threonine or hydroxylysine; consequently, other determinants of glycosylation must be sought.

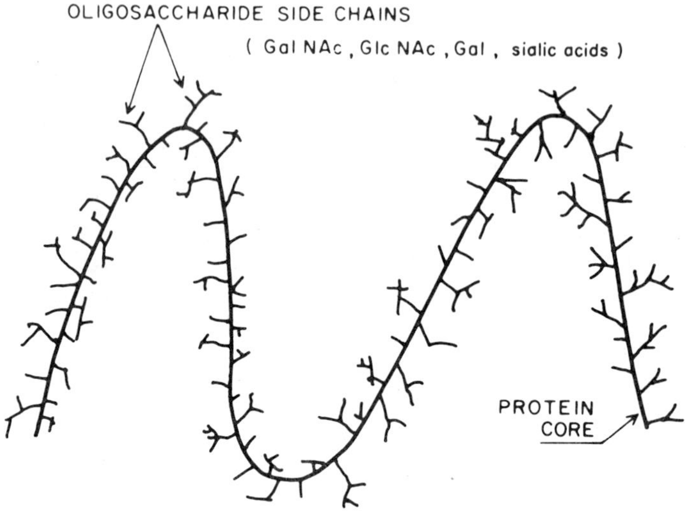

The Glycoconjugates

Figure 5. Structure of mucous glycoproteins (2)

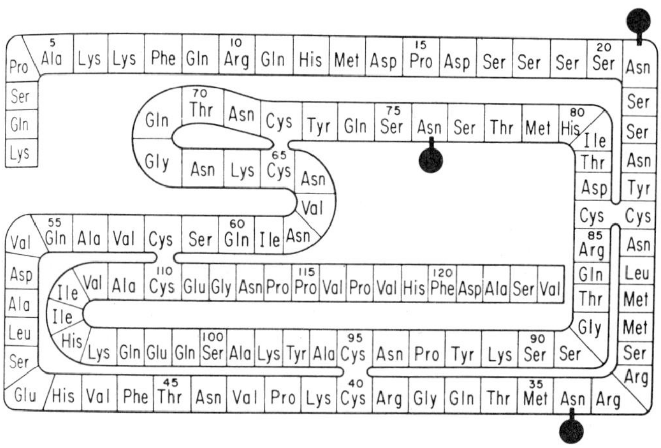

Figure 6. Primary structure of the protein moiety of porcine pancreatic ribonuclease. The filled circles represent the three heterosaccharide moieties (23).

Glycoprotein	Amino Acid Sequence
α_1-Acid glycoprotein	Glu-Glu-Tyr-Asn-Lys-Ser-Val
Rabbit IgG, H chain	Gln-Gln-Phe-Asn-Ser-Thr-Ile
Fibrinogen, γ chain	Gln-Val-Glu-Asn-Lys-Thr-Ser
Porcine ribonuclease	Ser-Ser-Ser-Asn-Ser-Ser-Asn
Lutenizing hormone	Gln-Pro-Ile-Asn-Ala-Thr-Leu
Thyroglobulin	Ala-Leu-Glu-Asn-Ala-Thr-Arg
Common sequence	Asn-x-Thr (Ser)

Figure 7. Amino acid sequences around asparagine-linked heterosaccharide moieties of several glycoproteins. See Ref. 1 for references to original literature concerning individual glycoproteins. Adapted from Ref. 1.

INTEGRATION OF GLYCOPROTEINS INTO CELLULAR MEMBRANES

Although early research focused on the more tractable soluble glycoproteins, increasing effort is now being devoted to membrane-bound species. A brief perusal of the research presented at this symposium indicates the current importance of membrane-bound glycoproteins; therefore, it is important to understand how these proteins are integrated into cellular membrane systems. Figure 8 shows a representation of the fluid mosaic model of plasma membrane structure (24), as envisioned by Garth Nicolson (25). The phospholipids are arranged in the classical lipid bilayer in which the acyl chains of the fatty acids form the hydrophobic domain and the charged polar groups interface with the cell exterior or the cytoplasm. Cholesterol and glycolipids, now shown in this model, are intercalated in the lipid bilayer. Membrane glycoproteins can be of two types: integral glycoproteins in which a portion of the peptide chain is associated with the hydrophobic region of the lipid bilayer, or peripheral glycoproteins which are not associated with the hydrophobic region of the membrane, but rather are associated with the outer membrane surface through ionic bonds. Peripheral proteins may be removed from the surface by extraction at high ionic strength, whereas agents which disrupt hydrophobic bonds are required to solubilize the integral membrane glycoproteins. One of the most attractive features of the fluid mosaic model is its ability to describe the dynamic aspects of membrane structure. Some integral glycoproteins can diffuse laterally within the membrane matrix or rotate about an axis normal to the membrane plane. Recently, it has become apparent that cytoskeletal elements are associated with certain membrane glycoproteins, thereby restricting their lateral mobility or inducing topographical rearrangement of integral proteins by active contractile processes (25).

A hallmark of the plasma membrane is the asymmetry of the heterosaccharide determinants; the saccharide moieties of both glycolipids and glycoproteins being located on the external face. Glycophorin, the major glycoprotein of the human erythrocyte membrane, has provided a model for investigating the integration of integral glycoproteins into cellular membranes. Its complete amino acid sequence and the sites of the heterosaccharide chain attachment have been determined and are shown in Figure 9. This transmembrane glycoprotein can be divided into three domains: the hydrophilic N-terminal sequence, a hydrophobic sequence which penetrates the membrane, and a cytoplasmic C-terminal sequence. The N-terminal sequence of 70 amino acids contains 22 residues of serine or threonine, 15 of which are substituted with glycosidically linked heterosaccharide moieties. There is also one heterosaccharide unit linked N-glycosically to asparagine-26. This hydrophilic sequence or receptor domain is oriented toward the exterior of the plasma membrane. The sequence encompassed by amino acid residues 71-92 is composed predominately of hydrophobic

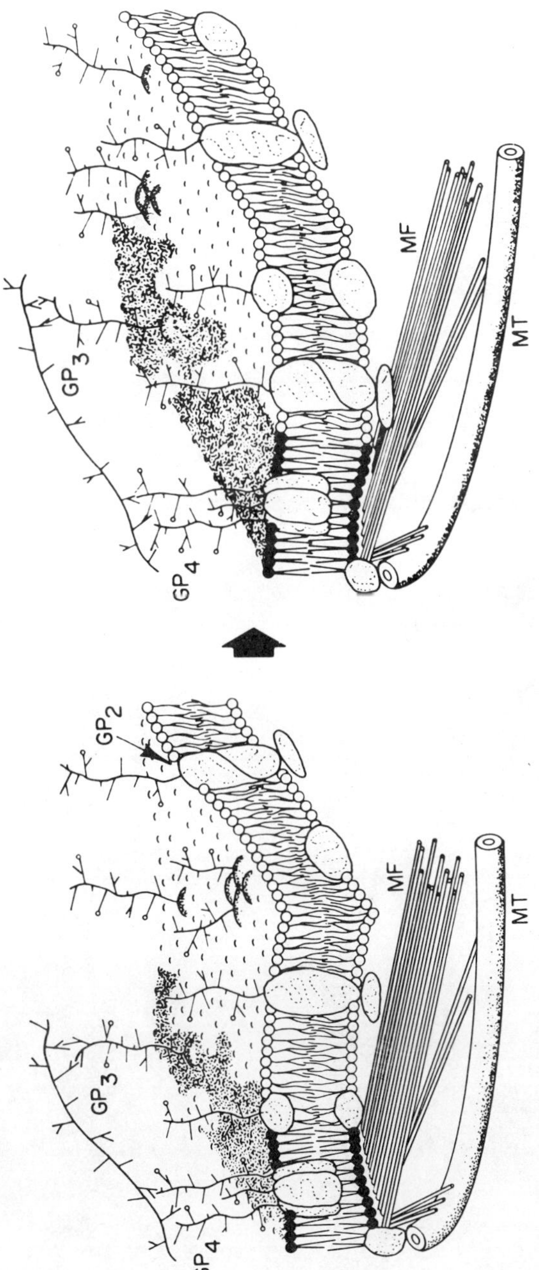

Biochimica et Biophysica Acta

Figure 8. Fluid mosaic model of membrane structure. This representation illustrates "transmembrane control over the distribution and mobility of cell-surface glycoproteins by peripheral and membrane-associated cytoskeleton components. In this example the mobility of integral glycoprotein complexes GP_3 *and* GP_4 *is controlled by outer surface peripheral components and also transmembrane linkages to membrane-associated cytoskeletal elements. In addition, complexes* GP_3 *and* GP_4 *could be sequestered into a specific lipid domain indicated by the shaded area while complex* GP_2 *exists in a free or unanchored state and is capable of lateral motion.* (MF) *microfilaments;* (MT) *microtubules"* (25).

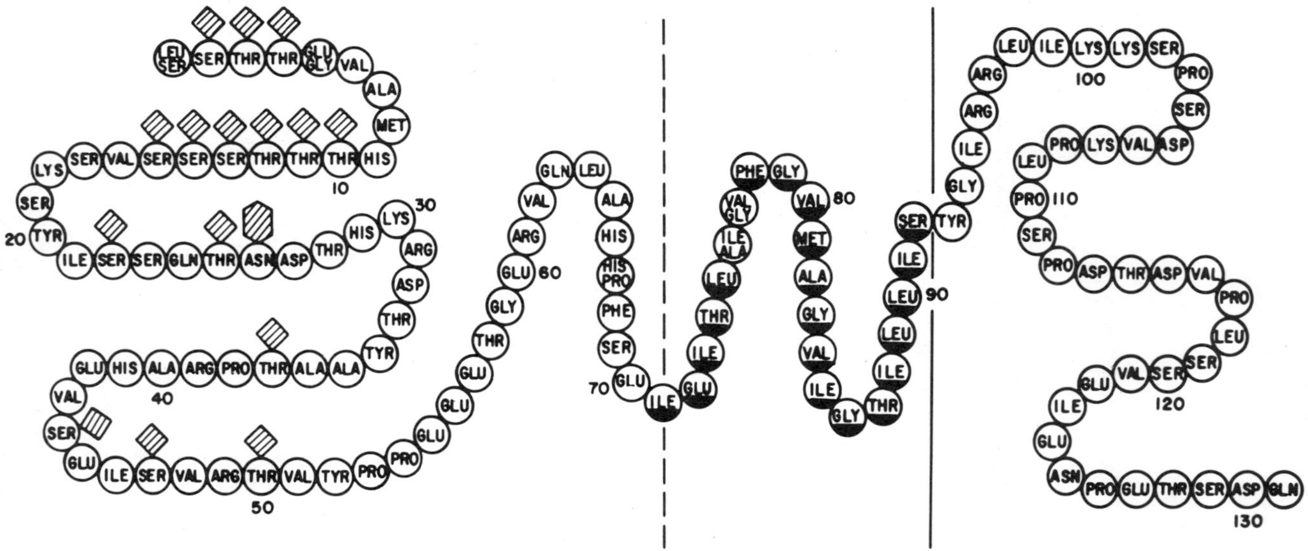

Figure 9. The amino acid sequence and domains of glycophorin. "The amino acids are arranged to simulate . . . the positions they might have if the glycophorin molecule was perpendicular to the lipid bilayer of the membrane. The limits of the bilayer are defined by the two vertical lines. The solid vertical line, which passes between residues 92 and 93, should be the approximate location of the polar groups of the inner half of the phospholipid bilayer" (21). The dotted vertical line indicates the approximate location of the polar groups of the outer half of the phospholipid bilayer. The shaded squares and hexagons represent the glycosidially linked and the N-glycosically linked heterosaccharide chains, respectively (21).

amino acids and is presumably intimately associated with the hydrophobic region of the lipid bilayer. This hydrophobic domain favors self-association, yielding multimeric units within the membrane matrix (21,26). The C-terminal sequence, rich in charged amino acids and proline, is exposed to the cytoplasm and free to interact with cytoplasmic elements. The functional properties of integral membrane glycoproteins probably depend not only on the structure of the monomer, but on their ability to form multimeric units by self-association and association with membrane lipid and on the topographical organization of these multimeric units within the membrane matrix (27).

Research on this prototype integral membrane glycoprotein has far-reaching implications for future research on plasma membrane function, particularly with respect to the reception and transmission of environmental stimuli, cell-cell contact, and membrane transport. Indeed, I believe this represents a glimpse into the future when we will be able to relate the structure and supramolecular organization of glycoproteins to normal and disease processes.

ACKNOWLEDGMENTS

The author gratefully acknowledges the National Cancer Institute (CA 11710; CA 18829), The Robert A. Welch Foundation (G-354), The Paul and Mary Haas Foundation and The George and Mary Josephine Hamman Foundation for their past and/or continuing support of his research on the biochemistry of glycoproteins.

LITERATURE CITED

1. Spiro, R. G. (1973), Advan. Protein Chem. 27, 349.
2. Horowitz, M. I., and Pigman, W., ed. (1977), The Glycoconjugates, New York, N.Y., Academic.
3. Kornfeld, R., and Kornfeld, S. (1976), Ann. Rev. Biochem. 45, 217.
4. Montreuil, J. (1975), Pure Appl. Chem. 42, 431.
5. Hughes, R. C. (1973), Prog. Biophys. Mol. Biol. 26, 189.
6. Gottschalk, A., ed. (1972), The Glycoproteins: Their Composition, Structure and Function, New York, N.Y., Elsevier.
7. Hallgren, P., Lundblad, A., and Svensson, S. (1975), J. Biol. Chem. 250, 5312.
8. Weiss, J. B., Lote, C. J., and Bobinski, H. (1971), Nature New Biol. 234, 25.
9. Lote, C. J., and Weiss, J. B. (1971), FEBS Letters 16, 81.
10. Miller, D. H., Lamport, D.T.A., and Miller, M. (1972), Science 176, 918.
11. Lamport, D.T.A. (1969), Biochemistry 8, 1155.
12. Rosenthal, A. L., and Nordin, J. H. (1975), J. Biol. Chem. 250, 5295.
13. Muir, L., and Lee, Y. C. (1970), J. Biol. Chem. 245, 502.

14. Downs, F., and Pigman, W. (1976), in Meth. Carbohydrate Chem., Vol. 7, Whistler, R. L., and BeMiller, J. N., eds., New York, N.Y., Academic, pp. 241-243.
15. Zinn, A. B., Plantner, J. J., and Carlson, D. M. (1977), in The Glycoconjugates, Vol. 1, Horowitz, M. I., and Pigman, W., eds., New York, N.Y., Academic, pp. 69-85.
16. Feizi, T., Kabat, E. A., Vicari, G., Anderson, B., and Marsh, W. L. (1971), J. Immunol. 106, 1588.
17. Kabat, E. A. (1969), in Blood and Tissue Antigens, Aminoff, D., ed., New York, N.Y., Academic, pp. 187-203.
18. Morgan, W.T.J. (1970), Ann. N.Y. Acad. Sci., 169, 118.
19. Winzler, R. J. (1970), Internatl. Rev. Cytol. 29, 112.
20. Montgomery, R. (1972), in Glycoproteins: Their Composition, Structure and Function, Part A, Gottschalk, A., ed., New York, N.Y., Elsevier, pp. 518-528.
21. Marchesi, V. T. (1978), in The Molecular Basis of Cell-Cell Interaction, Lerner, R. A., and Bergsma, D., eds., New York, N.Y., Alan R. Liss, pp. 127-138.
22. Gottschalk, A., and Thomas, M.A.W. (1961), Biochim. Biophys. Acta 46, 91.
23. Jackson, R.L. and Hirs, C.H.W. (1970), J. Biol. Chem 245, 637.
24. Singer, S. J., and Nicolson, G. L. (1972), Science 175, 720.
25. Nicolson, G. L. (1976), Biochim. Biophys. Acta 457, 57.
26. Segrest, J. P. (1977), in Mammalian Cell Membranes, Vol. 3, Jamieson, G. A., and Robinson, D. M., eds., Boston, Butterworths, pp. 1-26.
27. Nicolson, G. L. (1974), Internatl. Rev. Cytol. 39, 89.

RECEIVED August 3, 1978.

2

Biosynthesis and Catabolism of Glycoproteins

HARRY SCHACHTER, SAROJA NARASIMHAN, and JAMES R. WILSON

Research Institute, Hospital for Sick Children, Toronto, Canada

Glycoproteins are a complex group of macromolecules and their biosynthesis is, not unexpectedly, a rather complicated process. Although much has been learned about this process in the past ten years, there are many aspects which we do not understand. In the limited space permitted for this article, it will be possible only to outline the major steps in the biosynthesis of some of the glycoprotein types; more detailed reviews are available (Schachter and Rodén, 1973; Montreuil, 1975; Schachter, 1974a,b, 1977, 1978; Waechter and Lennarz, 1976).

I. Initiation of oligosaccharides of the asparagine-N-acetyl-D-glucosamine linkage type.

The largest group of animal glycoproteins are those which contain only or predominantly oligosaccharides that are linked to the polypeptide back-bone by an N-glycosidic linkage between N-acetyl-D-glucosamine (GlcNAc) and asparagine (Asn). Many secreted glycoproteins (α_1-acid glycoprotein and other plasma globulins, immunoglobulins, various gonadotrophins, ovalbumin, various enzymes, etc.) and membrane-bound glycoproteins (red cell membrane glycophorin, rhodopsin, the envelope glycoproteins of certain viruses such as vesicular stomatitis virus and Sindbis virus) contain Asn-GlcNAc linkage type oligosaccharides. It has become clear that the synthesis of the sugar-amino acid linkage pre-determines the general nature of the oligosaccharide that is subsequently assembled, i.e., that the synthesis of this linkage is an important control point. It is therefore important to understand the initiation process.

In recent years it has become apparent that many (but not all) Asn-GlcNAc oligosaccharides share a common "core" structure (Montreuil, 1975):

```
Man-α1,3 ↘
           Man-β1,4-GlcNAc-β1,4-GlcNAc-Asn
Man-α1,6 ↗
```

0-8412-0452-7/78/47-080-021$06.50/0

Further, the work of several laboratories (Leloir, Jeanloz, Hemming, Heath, Lennarz and others, see Waechter and Lennarz, 1976, and Schachter, 1978, for recent reviews) has provided strong evidence that this common core is pre-assembled as a lipid-linked oligosaccharide intermediate prior to incorporation into polypeptide. The general scheme for the initiation reaction is shown in Figure 1.

Soon after the role of polyprenol-linked sugars had been established in bacterial polysaccharide assembly, various workers suggested that a similar process might be occuring in higher cells. It was, however, the pioneering work of Leloir's group that first provided strong evidence for the transfer in mammalian tissues of glucose, mannose and N-acetylglucosamine from their respective sugar nucleotides to dolichol phosphate, a phosphorylated polyprenol (Fig. 1); the dolichol phosphate sugars formed are, respectively, GlcNAc-α-pyro-phosphate-dolichol, Man-β-monophosphate-dolichol and Glc-β-monophosphate-dolichol. The production of lipid intermediates has now been shown to occur in a variety of mammalian tissues (rat and pig liver, mouse myeloma tumor, bovine thyroid, hen oviduct, human lymphocytes, calf pancreas, etc).

Another major contribution from Leloir's group was the discovery of dolichol pyrophosphate oligosaccharide. It was noted that rat liver microsomes could effect the transfer of radioactive glucose from dolichol monophosphate glucose to an endogenous acceptor to form an acid-insoluble product which was at first thought to be protein. This product was subsequently shown to be a dolichol pyrophosphate oligosaccharide which was insoluble in water, in trichloroacetic acid and in chloroform-methanol (2:1, v/v) but was soluble in chloroform-methanol-water (10:10:3, v/v); the discovery of this solvent (Behrens _et al_., 1971) led the way to intense study of dolichol pyrophosphate oligosaccharides in many tissues.

The length of the oligosaccharide attached to dolichol varies with the tissue and with the conditions used for synthesis. The reports from Leloir's laboratory indicated the oligosaccharide to contain about 20 glycose units consisting of N-acetylglucosamine, mannose and glucose residues. With other tissues (hen oviduct and mouse myeloma microsomes, for example), the predominant dolichol pyrophosphate oligosaccharide appeared to lack glucose and contain two residues of GlcNAc and 5-7 residues of mannose, as follows:

$$(\text{Man-}\alpha\text{-})_n\text{-Man-}\beta 1,4\text{-GlcNAc-}\beta 1,4\text{-GlcNAc-pyrophosphate-dolichol}$$

The assembly of this structure is shown in Fig. 1. It is interesting to note the inversions of anomeric configuration in the formation of β-linked mannose from GDP-α-Man and α-linked mannose from Man-β-monophosphate-dolichol.

M.J. Spiro _et al_. (1976a,b) and R.G. Spiro _et al_ (1976) have carried out careful investigations on the formation of glucose-containing dolichol pyrophosphate oligosaccharides by several tissues. Working with thyroid slices rather than with microsomes, these workers demonstrated the formation of a dolichol pyrophosphate oligosaccharide with the tentative structure:

$$(\text{Man-}\alpha\text{-})_6\text{-(Glc)}_{1\text{-}2}\text{-(Man-}\alpha\text{-})_4\text{-Man-GlcNAc-GlcNAc-pyrophosphate-dolichol}$$

Evidence was also obtained for the formation by oviduct, kidney and thymus tissue slices of similar glucose-containing dolichol pyrophosphate oligosaccharides. It is interesting that microsome preparations from calf thyroid and hen oviduct do not produce these large dolichol pyrophosphate oligosaccharides but smaller molecules which contain only mannose and N-acetylglucosamine and lack glucose. However, microsomes from rat liver (Behrens et al., 1971; Parodi et al., 1972) and from calf pancreas (Herscovics et al., 1977a,b) were capable of synthesizing large glucose-containing dolichol pyrophosphate oligosaccharide molecules; the glucose is believed to be present in α-linked form in these molecules so that transfer of glucose from glucose-β-monophosphate-dolichol involves a single inversion of anomeric configuration.

The final step in the initiation process is the transfer of oligosaccharide from dolichol pyrophosphate oligosaccharide to an asparagine residue in the polypeptide chain (Fig. 1). Earlier work (Waechter and Lennarz, 1976) with rat liver, myeloma cells and hen oviduct established that microsome preparations catalyzed the transfer of labelled oligosaccharide from lipid intermediate to endogenous protein acceptors; however, it has not yet been established that an N-glycosidic linkage between asparagine and N-acetylglucosamine is synthesized in these reactions. The endogenous acceptors in these experiments were all tightly bound to membrane and could not be identified as known secretory proteins of the tissues under study; these findings initially suggested that the dolichol pathway might be involved only in the synthesis of membrane-bound glycoproteins. More recent work, however, has shown that glycosylation of ovalbumin is inhibited by tunicamycin, an antibiotic which blocks the synthesis of dolichol pyrophosphate N-acetylglucosamine (Struck and Lennarz, 1977); since synthesis of unglycosylated ovalbumin continued at almost normal rates in the presence of tunicamycin, it appears that ovalbumin, a secretory glycoprotein, is glycosylated by the dolichol pathway. Pless and Lennarz (1977) have obtained further evidence for the involvement of the dolichol pathway in secretory glycoprotein synthesis by showing the transfer of oligosaccharide from dolichol pyrophosphate oligosaccharide to denatured forms of three secretory proteins-ovalbumin, α-lactalbumin and ribonuclease A.

The substrate specificity of the transferase which effects incorporation of oligosaccharide into polypeptide is not understood. It has been recognized for some time that oligosaccharide is always attached to an asparagine in the tripeptide sequence-Asn-X-Ser (Thr)-where X can be almost any amino acid (Marshall, 1974). However the presence of this sequence does not necessarily result in glycosylation (Marshall, 1974; Pless and Lennarz, 1977); a protein may, for example, exist in both glycosylated and unglycosylated forms or may be fully glycosylated at one -Asn-X-Ser (Thr)- site but be completely unglycosylated at another such site. The work of Pless and Lennarz (1977) with exogenous acceptors indicates that the protein must be unfolded for glycosylation to occur. All three effective acceptor proteins (ovalbumin, α-lactalbumin and RNase A) contained the required tripeptide sequence whereas several proteins without this sequence were ineffective as acceptors; however, several denatured proteins with the required tripeptide sequence did not accept oligosaccharide and the exact requirements for glycosylation have not as yet been resolved.

Although dolichol pyrophosphate oligosaccharides which lack glucose and which carry relatively small oligosaccharide moieties can serve as oligosaccharide donors in the incorporation of oligosaccharide into protein, Turco *et al.* (1977) have made the important observation that glucose-containing dolichol pyrophosphate oligosaccharides are about 9 times more effective in oligosaccharide transfer to endogenous protein acceptors than the smaller glucose-free dolichol pyrophosphate oligosaccharides. This suggests that *in vivo* pathway involves primarily the large dolichol pyrophosphate oligosaccharides which contain N-acetylglucosamine, mannose and glucose.

II. Processing of protein-bound oligosaccharides of the Asn-GlcNAc linkage type.

Since oligosaccharides of the Asn-GlcNAc linkage type do not contain glucose and since the N-acetyllactosamine type structures (Montreuil, 1975) contain only 3 mannose residues, it is clear that incorporation of a large oligosaccharide containing 1-2 residues of glucose and 11 residues of mannose must be followed by removal of all the glucose residues and of some mannose residues. Several laboratories have recently reported evidence for such "oligosaccharide processing". Work with the enveloped viruses (vesicular stomatitis virus and Sindbis virus) has indicated that at early stages in the assembly of the envelope glycoproteins of these viruses, the protein-bound oligosaccharide is quite large and rich in mannose residues; subsequent processing of these envelope glycoproteins leads to removal of mannose (and probably glucose) residues (Hunt *et al.*, 1977; Robbins, 1977; Tabas *et al.*, 1978). The viral envelope glycoproteins are representative of membrane-bound glycoproteins; processing is, however, not confined to this class of glycoprotein, since biosynthesis of a secretory IgG by a mouse plasmacytoma cell line was also shown to involve oligosaccharide processing (Tabas *et al.*, 1978).

The initiation of Asn-GlcNAc oligosaccharides and subsequent processing are summarized in Fig. 2. It is postulated that all Asn-GlcNAc oligosaccharides may undergo initiation by transfer of a large oligosaccharide containing GlcNAc, Man and Glc (Fig. 2). This process occurs in the rough endoplasmic reticulum (see next

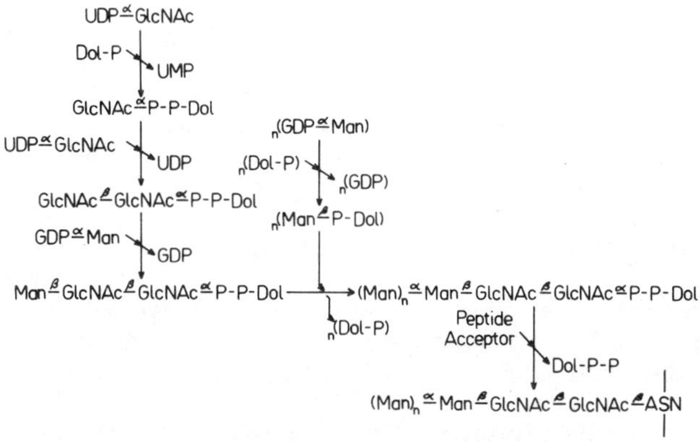

Figure 1. Initiation of Asn-GlcNAc-type oligosaccharides. Oligosaccharides are preassembled as lipid intermediates. Abbreviations: Dol, dolichol; P, phosphate group.

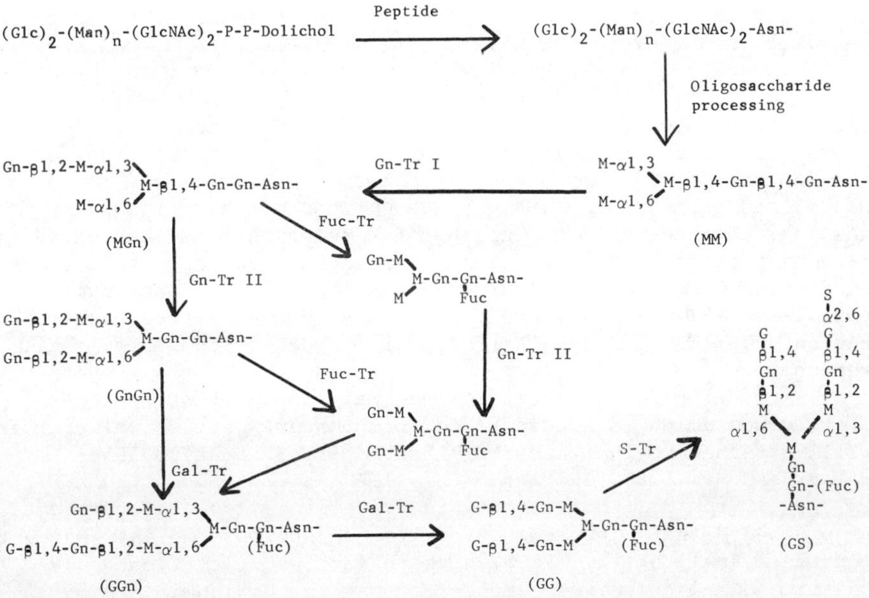

Figure 2. Biosynthesis of Asn-GlcNAc oligosaccharides of the N*-acetyllactosamine type. Initial transfer of oligosaccharide to peptide can involve a glucose-containing intermediate. Processing to the* $Man_3GlcNAc_2Asn$ *core (MM) is then believed to occur, followed by the Golgi-located elongation reactions.*

Abbreviations: M, D*-mannose; Gn,* N*-acetyl-*D*-glucosamine; Asn, asparagine; Fuc,* L*-fucose; G,* D*-galactose; S, sialic acid; Tr, transferase. The glycopeptides are named according to the sugars present at the nonreducing ends; the sugar on the Man-α1,6-branch is named first.*

section). Oligosaccharides destined to become the oligomannoside type (Montreuil, 1975) probably undergo limited processing (loss of glucose residues and possibly some mannose residues). Oligosaccharides destined to become the N-acetyllactosamine type (Montreuil, 1975) undergo extensive processing (removal of all glucose residues and all but 3 mannose residues) to form the core structure $Man_3GlcNAc_2Asn$ (Fig. 2). This core structure then undergoes elongation (the addition of GlcNAc, Fuc, Gal and sialic acid, see section IV below) in the cell's Golgi apparatus. It is not yet understood why some oligosaccharides become the oligomannoside type while others are processed more extensively to the $Man_3GlcNAc_2Asn$ structure. Tabas et al. (1978) have made the interesting suggestion that the oligomannoside structures are more primitive in the evolutionary scale and that higher organisms developed the processing apparatus to enable conversion of oligomannoside structures to N-acetyllactosamine type structures.

III. Subcellular site of Asn-GlcNAc initiation.

Fig. 3 shows the subcellular path of biogenesis of a membrane-bound glycoprotein; a similar scheme has been suggested for secretory glycoproteins (Schachter 1974a,b, 1977) except that these glycoproteins do not remain attached to membrane but are secreted when secretory vesicles fuse with the plasma membrane. There is now considerable evidence that the peptide moieties of glycoproteins are translated on membrane-bound ribosomes (Morrison and Lodish, 1975; Wirth et al., 1977; Toneguzzo and Ghosh, 1975; Schachter 1974a,b, 1978). To explain the transfer of the nascent peptide across the endoplasmic reticulum membrane, Blobel and Dobberstein (1975a,b) have proposed the "signal hypothesis"; this theory proposes that the amino terminal end of the nascent peptide carries a "signal sequence" which somehow causes binding of the ribosome to a receptor site on the endoplasmic reticulum (Fig. 3) and thereby creates a channel for passage of nascent peptide into the intravesicular space of the endoplasmic reticulum.

It has been shown for several tissues that some N-acetylglucosamine is incorporated into nascent peptide while it is still attached to the ribosome (Lawford and Schachter, 1966; Molnar and Sy, 1967; Sherr and Uhr, 1969; Cowan and Robinson, 1970). If the dolichol pathway is operative (Figs. 1,2), it can be concluded that mannose should also be incorporated into nascent peptide. Kiely et al. (1976) have in fact shown that both glucosamine and mannose were present on nascent ovalbumin chains bound to ribosomes. It was not clear from these various experiments whether all incorporation of core takes place at the nascent peptide stage or whether this process also occurs after release of peptide from the ribosome. Conflicting evidence on this point has been obtained; Jamieson (1977) has found that little incorporation of carbohydrate occurs on the nascent peptide of α_1-acid glycoprotein while Rothman and Lodish (1977) suggest that the

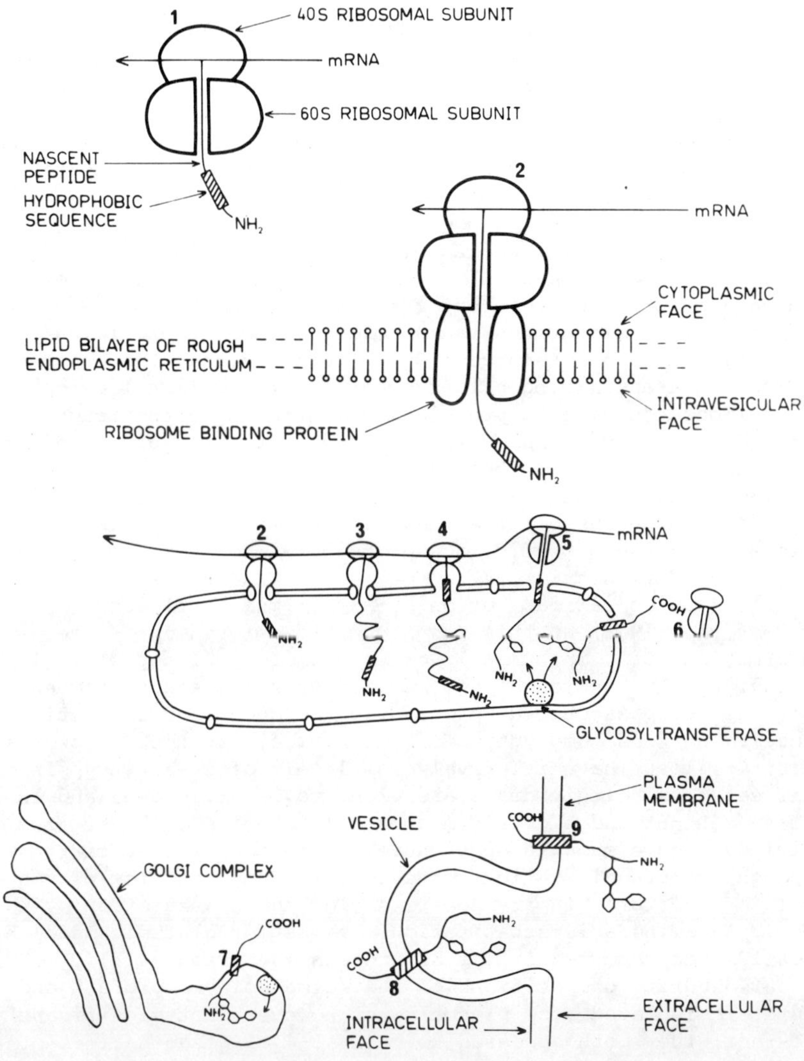

Figure 3. A scheme depicting biosynthesis of a membrane-bound glycoprotein. Translation probably starts on a free ribosome and the nascent peptide carries some sort of "signal" near its amino-terminal end. This signal organizes the appearance of a ribosome-binding protein in the endoplasmic reticulum and thereby creates a channel for passage of nascent peptide through the membrane. Oligosaccharide is incorporated into peptide within the channels of the rough endoplasmic reticulum (stages 5 and 6). The signal sequence is probably cleaved off at some stage in assembly although there can be exceptions. The glycoprotein migrates to the Golgi apparatus for the elongation reactions (stage 7) and then to the plasma membrane (stages 8 and 9). An analogous scheme can be drawn for secretory glycoproteins except that the glycoprotein would not be firmly anchored to the membrane during the assembly process.

entire oligosaccharide core is added at the nascent peptide stage during assembly of the envelope glycoprotein of vesicular stomatitis virus. In fact, the latter workers provide strong evidence that, in their _in vitro_ system, insertion of nascent peptide through the endoplasmic reticulum membrane is coupled to glycosylation; glycosylation does not occur without membrane insertion and both processes are completed well before release of peptide from the ribosome. The reasons for the discrepancy between the data of Jamieson (1977) and Rothman and Lodish (1977) are not known; the former worked with a secretory protein while the latter studied a membrane protein but further studies must be carried out before generalizations can be made. A recent preliminary report by Czichi and Lennarz (1977) indicated that the enzymes involved in transferring oligosaccharide from dolichol pyrophosphate oligosaccharide to protein were enriched in the rough microsome fraction prepared from hen oviduct, providing the first direct evidence for the localization of this process in the rough endoplasmic reticulum. Rat liver microsome polypeptides also served as endogenous oligosaccharide acceptors in a similar _in vitro_ system; these peptide acceptors were, however, not nascent (Vargas and Carminatti, 1977).

Studies carried out on α_1-antitrypsin deficiency by Hercz _et al_. (1978) at this Institute have provided an interesting confirmation that, in human liver, a secretory glycoprotein is glycosylated in the rough endoplasmic reticulum and subsequently undergoes "oligosaccharide processing" as described in section II (above). Persons homozygous for the Z allele at the P_i (protease inhibitor) locus have relatively low levels of α_1-antitrypsin in their sera; such individuals are prone to liver, kidney and lung disease. Light and electron microscopic studies of liver from ZZ individuals have shown a substantial retention of α_1-antitrypsin in rough endoplasmic reticulum vesicles. Thus, the low plasma levels of this protein are due to a blockage in secretion by the liver. An amino acid substitution (lysine for glutamic acid) has recently been reported in the ZZ protein (Jeppsson _et al_., 1978). The carbohydrate compositions of the normal plasma protein and the ZZ protein isolated from plasma were, however, not significantly different; they both appeared to carry 3-4 oligosaccharides of the N-acetyllactosamine type. Hercz _et al_. (1978) isolated the ZZ protein from an autopsy specimen of human liver and showed that it contained no fucose, glucose, sialic acid or galactose but carried decreased levels of N-acetylglucosamine and increased levels of mannose; the data were compatible with 3 oligosaccharides per mole, each oligosaccharide containing on the average 2 residues of GlcNAc and 7 residues of mannose. These results strongly suggest that a mannose-rich oligosaccharide is normally incorporated into polypeptide in the rough endoplasmic reticulum; normal individuals process this oligosaccharide by removal of mannose residues (to 3 mannose residues per oligosaccharide) whereas ZZ livers cannot, for some reason, process the protein to completion. In normal livers, the protein travels to the Golgi apparatus for elongation (addition of GlcNAc, Gal and sialic acid residues, see section IV below; neither normal nor ZZ proteins contain fucose which is usually added in the Golgi apparatus). The defect in the ZZ liver is not known; since other plasma glycoproteins are processed normally, the processing apparatus is not defective and it must be postulated that the amino acid substitution in ZZ α_1-antitrypsin interferes either with normal glycosylation or subsequent oligosaccharide processing, or prevents movement of the protein out of the rough endoplasmic reticulum to a region of the cell where processing occurs. The blockage in this pathway is not complete since some ZZ α_1-antitrypsin is processed, moves to the Golgi apparatus for addition of external sugars and is secreted into the plasma.

IV. Elongation within the Golgi apparatus of Asn-GlcNAc oligosaccharides of the N-acetyllactosamine type.

As pointed out in section II (above), the factors which curtail processing of oligosaccharides destined to become the oligomannoside type are not known; this type of oligosaccharide will not be considered further. In this section, we deal with the fate of oligosaccharides which are processed to the core structure:

```
Man-α1,3
         \
          Man-β1,4-GlcNAc-β1,4-GlcNAc-Asn
         /
Man-α1,6
```

Fig. 2 shows the subsequent elongation of this structure. This process involves addition of four sugars: GlcNAc, Gal, sialic acid and Fuc; the enzymes catalyzing addition of these sugars are located in the Golgi apparatus, as will be discussed briefly at the end of this section.

The addition of the first sugar, a GlcNAc residue, to the core structure (MM, see Fig. 2 for nomenclature) has been clarified primarily by the use of a phytohemagglutinin-resistant mutant of Chinese hamster ovary cells (Stanley et al., 1975; Narasimhan et al., 1977). The mutant cell line was resistant to the toxic action of several lectins because lectin-binding glycoproteins on the cell surface were incompletely glycosylated. It has now been shown that wild type Chinese hamster ovary cells contain at least two N-acetylglucosaminyltransferases, one acting on glycopeptides with the structure MM (see Fig. 2) which we designate GlcNAc-transferase I, and the other acting on glycopeptides with the structure MGn (see Fig. 2) which we designate GlcNAc-transferase II. Evidence indicating the presence of both transferases in wild type cells and the absence of GlcNAc-transferase I in lectin-resistant cells is summarized in Table I (data from Narasimhan et al., 1977).

The glycopeptides used for the assays shown in Table I were derived from a glycopeptide with the structure GS (see Fig. 2). Glycopeptide GS was obtained from a human multiple myeloma immunoglobulin G by pronase digestion and subsequent purification on ion exchange and gel filtration columns. Its structure was established by carbohydrate analysis, sequential glycosidase digestions (Narasimhan _et al._, 1977) and by 220 and 360 MHz high resolution proton magnetic resonance spectra carried out by A. Grey and J.P. Carver, Department of Medical Genetics, University of Toronto. The assignment of the sialic acid residue to N-acetyllactosamine at the Man-α1,3- branch point was especially important and was carried out with the aid of NMR data from Strecker _et al._ (1977**a**).

GlcNAc-transferase I adds a GlcNAc in β1,2 linkage to MM. The β-linkage was determined by susceptibility to β-N-acetylglucosaminidase. The product of the enzyme reaction adhered strongly to Concanavalin A-Sepharose columns indicating the GlcNAc-Man linkage to be 1,2 (Narasimhan _et al._, 1977; Ogata _et al._, 1975; Narasimhan, Wilson and Schachter, in preparation). We have recently studied the susceptibility of the product of the GlcNAc-transferase I reaction to the action of _Cl.perfringens_ endo-β-N-acetylglucosaminidase CI (Ito _et al._, 1975); we have found that the product is completely resistant to the action of this enzyme indicating that all the GlcNAc has been incorporated into the Man-α1,3- terminus rather than into the Man-α1,6- terminus of MM (Wilson, Narasimhan and Schachter, in preparation). We can thus state that the product of GlcNAc-transferase I is MGn (Fig. 2) rather than GnM. It is, however, likely that GlcNAc-transferase I can also add a GlcNAc residue to a Man-α1,6- terminus if the Man-α1,3- terminus is not available (Narasimhan _et al._, 1977). The pathway shown in Fig. 2 is thus a preferred rather than an absolute path.

The addition of a GlcNAc residue by GlcNAc-transferase I controls the entire elongation process. Both the addition of the next GlcNAc residue and of Fuc depend on the action of GlcNAc-transferase I (Fig. 2), as does subsequent addition of Gal and sialic acid residues. It is therefore not surprising that the absence of this enzyme in the lectin-resistant Chinese hamster ovary cell mutant described above leads to incomplete glycoprotein glycosylation and resistance to several lectins.

GlcNAc-transferase II acts only on MGn (Narasimhan _et al._, 1977); however, we have not yet had any GnM available for testing and it is not certain if GlcNAc-transferase II is specific for the Man-α1,6- terminus. The linkage synthesized is β1,2; this was determined as described above for GlcNAc-transferase I (Narasimhan _et al._, 1977). The product of the reaction is therefore GnGn (Fig. 2).

The relationship between the two GlcNAc-transferases described above is presently under study. Mixed substrate (competition) studies have indicated that MM and MGn compete for a common enzyme active site (Narasimhan et al., 1977). Although this finding does not disprove the existence of two different catalytic sites, it nevertheless suggests that GlcNAc-transferases I and II may share a common catalytic subunit; GlcNAc-transferase I would then have to have an additional regulatory subunit which is absent in the lectin-resistant mutant described above.

Hunt et al. (1977) and Tabas et al. (1978) have studied the biosynthesis of vesicular stomatitis virus envelope glycoprotein in wild type and lectin-resistant Chinese hamster ovary cells. In wild type cells, the envelope glycoprotein of the mature virus carries only N-acetyllactosamine-type oligosaccharides (3 mannose residues per oligosaccharide). However, virus grown in the lectin-resistant cell line lacks sialyl-N-acetyllactosamine arms, as would be predicted from the absence of GlcNAc-transferase I (see Fig. 2). Surprisingly, the mature virus from the mutant cells carries more mannose than expected (about 5 mannose residues per oligosaccharide); this suggests that oligosaccharide processing (see section II, above) is incomplete in the GlcNAc-transferase I-deficient cells. The connection between the absence of GlcNAc-transferase I and processing is not known; one possibility is that the postulated regulatory subunit of GlcNAc-transferase I (see above) is somehow involved in removing the last two mannose residues during oligosaccharide processing.

As soon as a GlcNAc has been added to MM (Fig. 2), Fuc can be incorporated into the GlcNAc nearest the Asn residue (Fig. 2). The enzyme catalyzing this reaction is widely distributed (Jabbal and Schachter, 1971; Munro et al., 1975; Munro and Schachter, 1973; Kessel et al., 1977; Chou et al., 1977) and is usually assayed with sialidase-, β-galactosidase-treated α_1-acid glycoprotein as acceptor. Wilson et al. (1976) showed that the enzyme was in fact adding Fuc to the most internal GlcNAc residue by using glycopeptide acceptors and analyzing the product by endo-β-N-acetylglucosaminidase CI; radioactive Fuc migrated with the charged Asn-GlcNAc moiety after endoglycosidase cleavage rather than with uncharged oligosaccharide. More recent work (Wilson and Schachter, in preparation) has shown that the fucosyltransferase will not act on MM but can act on both MGn and GnGn (see Fig. 2). Thus, although this Fuc residue is added near the polypeptide chain, addition is a late Golgi-localized process because it cannot occur until GlcNAc-transferase I has acted. This scheme explains why oligomannoside-type Asn-GlcNAc oligosaccharides never contain Fuc; we still do not know why some N-acetyllactosamine-type oligosaccharides have a Fuc residue while others do not but the presence or absence of the appropriate fucosyltransferase seems a likely possibility.

Perhaps the most thoroughly characterized glycosyltransferase is the galactosyltransferase which adds Gal in β1,4 linkage to the terminal GlcNAc of GnGn. This enzyme is believed to be related, if not identical, to the A protein component of milk lactose synthetase (Brew et al., 1968). The A protein has a very low affinity for glucose; in the presence of the B protein (α-lactalbumin), affinity for glucose is greatly increased and lactose is synthesized. Since α-lactalbumin is present only in mammary gland, lactose synthesis occurs only in this organ. The A protein (or a very similar protein) is widely distributed and can incorporate Gal into β1,4 linkage to free GlcNAc, GlcNAc-terminal oligosaccharides and glycopeptides, and GlcNAc-terminal glycoproteins (sialidase-, β-galactosidase-treated α_1-acid glycoprotein or fetuin, for example). The function of the enzyme in the absence of α-lactalbumin is almost certainly the elongation of Asn-GlcNAc oligosaccharides. The bovine and human milk, bovine colostrum and rat serum enzymes have been purified to homogeneity by classical methods (Fitzgerald et al., 1970) and by affinity chromatography using either α-lactalbumin-Sepharose or UDP-hexanolamine-Sepharose columns (Trayer and Hill, 1971; Andrews, 1970; Khatra et al., 1974; Geren et al., 1976; Powell and Brew, 1974; Fraser and Mookerjea, 1976).

The bovine milk enzyme contains two major catalytically active forms with molecular weights estimated to be 55,000-59,000 and 42,000-44,000, respectively (Magee et al., 1974); the two forms are similar with regard to K_M for substrate, heat inactivation and inhibition by sulfhydryl reagents and it is believed that the smaller form is derived from the larger form by proteolysis due to a plasmin-like protease in bovine milk (Magee et al., 1976). Powell and Brew (1974) found that a single form of the enzyme of molecular weight 51,000 could be isolated from bovine colostrum which contains potent protease inhibitors; this enzyme could be degraded to a smaller molecule with molecular weight 41,000 by the action of trypsin. Galactosyltransferase has also been isolated from the Golgi membranes of lactating sheep mammary glands (Smith and Brew, 1977) and from bovine colostrum fat globule membranes (Powell et al., 1977); the mammary gland enzyme was solubilized with 1% Triton X-100 and, on purification, showed two components, a major component of molecular weight 65,000-69,000 and a minor component of molecular weight 53,000-55,000. The enzymic properties of the mammary gland enzyme resembled the properties of the milk enzyme. It was suggested that the soluble milk enzyme is produced by proteolysis of the Golgi-bound transferase.

The various forms of the galactosyltransferase are glycoproteins and contain about 10-15% carbohydrate (Geren et al., 1977; Lehman et al., 1975; Trayer and Hill, 1971). Detailed kinetics of this enzyme have been published and the order of addition of Mn2+, UDP-galactose, α-lactalbumin, and acceptor have been studied (Ebner, 1973; Tsopanakis and Herries, 1976; Geren et al., 1975a,b; Bell et al., 1976; Powell and Brew, 1975; Khatra et al., 1974; Kitchen and Andrews, 1974a,b).

The scheme in Fig. 2 depicts addition of Gal to the Man-α-1,6- terminus prior to addition to the Man-α-1,3- terminus. There is no direct evidence that the galactosyltransferase shows such preferntial specificity. It is interesting, however, that both bovine immunoglobulin G (Tai et al., 1975) and human multiple myeloma immunoglobulin G (Baenziger and Kornfeld, personal communication) contain the structure GGn (Fig. 2) rather than GnG. Also, Rao et al. (1976) have reported that a galactosyltransferase purified from swine mesentary lymph nodes by p-aminophenyl-β-GlcNAc-Sepharose affinity chromatography transferred Gal more rapidly to GnGn prepared from porcine immunoglobulin G than to other GlcNAc-terminal glycopeptides.

Elongation is completed by the addition of a sialic acid residue to GG (Fig. 2). Sialic acid-galactose linkages in various Asn-GlcNAc type oligosaccharides have been reported to be primarily α2,3 and α2,6 (Montreuil, 1975); however, occasional reports of other linkages have appeared, e.g., Sato et al. (1967) have reported a sialyl-α2,4-galactose structure in α_1-acid glycoprotein and Isemura and Schmid (1971) have suggested a sialyl-α2,2-galactose linkage for the same protein. Various tissues have been shown to contain a Golgi-localized sialyltransferase capable of incorporating sialic acid from CMP-sialic acid into sialidase-treated α_1-acid glycoprotein (Schachter and Rodén, 1973; Schachter, 1974a,b, 1978). It had until recently been assumed that these assays, using crude membrane preparations, were in fact measuring more than one sialyltransferase. Support for this idea came from the finding that crude colostrum (Bartholomew et al., 1973; Paulson et al., 1977b) and rat, pork, bovine and human liver (Hudgin and Schachter, 1972) catalyze the synthesis of both sialyl-α2,3-lactose and sialyl-α2,6-lactose; further, the differential development of these two activities in embryonic rat liver indicated that two separate sialyltransferases were involved (Hudgin and Schachter, 1972). However, analysis of the product of rat liver sialyltransferase action on sialidase-treated α_1-acid glycoprotein (Stoffyn et al., 1977; Van den Eijnden et al., 1977b) showed that the only linkage synthesized was α2,6; there was no indication of α2,3 linkage. Thus, the transferase which makes sialyl-α2,3-lactose is probably not involved in sialic acid incorporation into Asn-GlcNAc oligosaccharides. There is evidence (see section V below) that this 3'-sialyltransferase may be involved in sialic acid incorporation into Ser(Thr)-GalNAc oligosaccharides. The transferases which incorporate sialic acid into Asn-GlcNAc oligosaccharides in linkages other than α2,6 remain to be discovered.

Paulson et al. (1977a) have recently reported the purification to homogeneity of CMP-sialic acid: β-D-galactoside α2-6 sialyltransferase from bovine colostrum. The enzyme was purified 440,000-fold by affinity chromatography on CDP-hexanolamine-Sepharose and CDP-ethanolamine-Sepharose. Like the galactosyltransferase discussed above, this sialyltransferase was found in two forms of equal specific activity, with molecular weights of

56,000 and 43,000 respectively; it was postulated, but not shown, that the smaller enzyme might be a degradation product of the larger enzyme. Both purified enzymes were highly specific for Gal-β1,4-GlcNAc as an acceptor and showed low activity with Gal-β1,3-GlcNAc, Gal-β1,6-GlcNAc, Gal-β1,4-Glc and other β-galactosides; similar substrate specificities had previously been observed with crude enzyme preparations (Schachter and Rodén, 1973). Both enzymes were highly active with sialic acid-free α_1-acid glycoprotein, fetuin, IgG and IgM but not with various sialic acid-free mucins; the former contain Gal-β1,4-GlcNAc termini whereas the latter have Gal-β1,3-GalNAc termini. The purified sialyltransferase synthesizes only sialyl-α2,6-lactose; as mentioned above, crude preparations from colostrum and liver make both the α2,6 and α2,3 isomers of sialyl-lactose.

We have shown that the N-acetylglucosaminyltransferases, fucosyltransferase, galactosyltransferase and sialyltransferase described above are highly enriched in the Golgi apparatus of rat and pork liver (Munro et al., 1975; Schachter et al., 1970) and of other tissues (Letts et al., 1974; Schachter 1974a,b, 1978). Although these enzymes may occur in other subcellular organelles (in particular, on the surfaces of certain cells), the evidence that the major site of elongation is the Golgi apparatus appears highly conclusive. Supportive evidence comes from work in which intact cells are pulsed with radioactive sugars and measurements are made of incorporation of radioactivity into glycoproteins either by biochemical techniques or by autoradiography (Schachter 1974a,b, 1978; Bennett and Leblond, 1977).

Once elongation is completed within the Golgi apparatus, the finished glycoprotein migrates towards the plasma membrane within a transport vesicle (see Fig. 3). The vesicle membrane fuses with the plasma membrane and secretory glycoproteins are extruded from the cell whereas membrane glycoproteins probably become part of the plasma membrane by a lateral diffusion process.

V. Biosynthesis of serine (threonine)-N-acetyl-D-galactosamine type oligosaccharides.

Approaches similar to those described above have been applied to studies on the synthesis of Ser(Thr)-GalNAc-linkage type oligosaccharides. Almost all the reported work has dealt with secretory glycoproteins or "mucins" and can be classified broadly into two over-lapping categories: (i) the synthesis and secretion of mucins by goblet cells and mucous glands of the gastro-intestinal, respiratory and genito-urinary tracts, and (ii) the synthesis of blood group active mucins, in particular mucins with human ABO-Lewis antigenicity. No attempt will be made to discuss the highly specialized field of human blood group glycoprotein and glycolipid biosynthesis since various detailed reviews are available (Watkins, 1974; Schachter and Rodén, 1973; Schachter 1978). A brief discussion of ovine and porcine submaxillary mucin biosynthesis will serve to illustrate the general principles involved in mucin synthesis (Fig. 4).

TABLE I

N-acetylglucosaminyltransferase levels of wild type and lectin-resistant Chinese hamster ovary cells using as acceptors glycopeptides prepared from human IgG; see Figure 2 for glycopeptide nomenclature.

IgG glycopeptide acceptor	N-acetylglucosaminyltransferase activity (nmol/mg/hour)	
	Wild type cells	Lectin-resistant cells
MM	3.7	< 0.2
MGn	21.9	21.3
GnGn	< 0.2	< 0.2

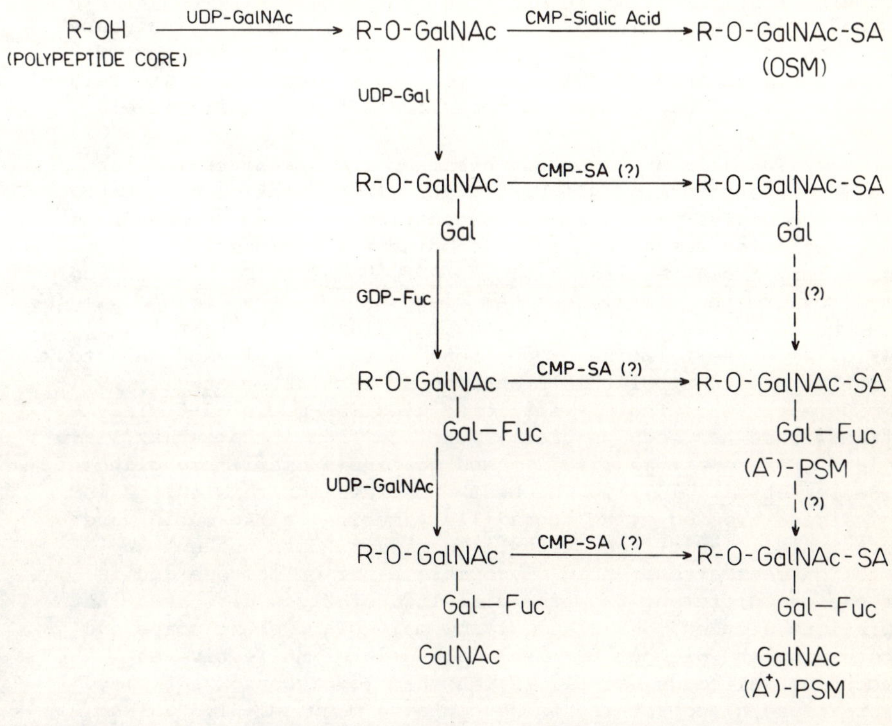

Figure 4. Assembly of ovine (OSM) and porcine (PSM) submaxillary mucins

The first important point is that initiation does not involve pre-assembly of oligosaccharide as a lipid intermediate. It has been reasonably established in several systems that the Ser(Thr)-GalNAc linkage is synthesized by the incorporation of a GalNAc from UDP-GalNAc into a high molecular weight acceptor. The transferase has a high specificity for acceptor; for example, the enzyme from salivary gland will donate GalNAc to mucin treated with glycosidases to remove all carbohydrate but not to a large number of other high and low molecular weight acceptors. An interesting exception is the transfer of GalNAc to a basic protein isolated from bovine myelin (Hagopian *et al.*, 1971); this protein is not normally a glycoprotein and the significance of its acceptor activity is not known. Although pronase treatment of carbohydrate-free mucin acceptor destroys acceptor activity, a recent report has shown that the enzyme will transfer GalNAc to tryptic peptides of carbohydrate-free ovine submaxillary mucin (Hill *et al.*, 1977); further, sequence studies around the serine and threonine residues which carry O-glycosidic oligosaccharides have indicated no homologies to define the substrate specificity of the transferase (Marshall, 1974; Hill *et al.*, 1977). The enzyme requires high molecular weight acceptors but it is not clear why some hydroxyamino acids are glycosylated but others are not. It is interesting that the porcine submaxillary gland UDP-GalNAc:polypeptide N-acetylgalactosaminyltransferase does not adhere to UDP-hexanolamine-Sepharose and has thus far been purified only 30-fold; however, this preparation was relatively free of other mucin glycosyltransferases (Hill *et al.*, 1977).

There is an important branch point immediately after GalNAc incorporation. If sialic acid incorporation occurs to form the sialyl-α2,6-GalNAc disaccharide, further carbohydrate incorporation ceases and the predominant form in ovine submaxillary mucin results (Fig. 4). If, however, Gal is incorporated to form the Gal-β1,3-GalNAc disaccharide, the various oligosaccharides present in porcine submaxillary mucin result (Fig. 4). The relative proportions of the sialyl- and galactosyltransferases control this branch point (Schachter *et al.*, 1971; McGuire, 1970). The CMP-sialic acid:GalNAc-mucin α2-6 sialyltransferase has been partially purified from sheep submaxillary gland (Carlson *et al.*, 1973) and porcine submaxillary gland (Sadler *et al.*, 1977). The best substrate for this enzyme is sialidase-treated ovine submaxillary mucin (GalNAc-mucin) and other sialic acid-free mucins. The enzyme does not act on sialidase-treated α_1-acid glycoprotein nor on lactose and is therefore different from the sialyltransferases discussed in the previous section; a variety of low molecular weight compounds both with and without terminal GalNAc were ineffective as acceptors although Gal-β1,3-GalNAc had low acceptor activity. Antifreeze glycoprotein, which contains many Gal-β1,3-GalNAc units, is an excellent acceptor (Sadler *et al.*, 1977).

The key enzyme in controlling synthesis appears to be the UDP-Gal:GalNAc-mucin β1-3 galactosyltransferase (Schachter et al., 1971). This enzyme is tightly bound to membrane and all attempts at purification have been unsuccessful. The enzyme will not act if GalNAc is substituted with a sialic acid residue (Fig. 4). However, if Gal is incorporated, other sugar residues can be added. Thus, sialic acid can then be added to the internal GalNAc. Further, Fuc can be incorporated in α1,2 linkage to the Gal residue; this enzyme is similar to the human blood group H fucosyltransferase and can transfer Fuc to a variety of high and low molecular weight compounds with a terminal β-galactoside residue (McGuire, 1970). The fucosyltransferase has recently been purified to near homogeneity from porcine submaxillary glands using GDP-hexanolamine-Sepharose (Beyer et al., 1977). It is interesting that although the fucosyltransferase will act on a variety of β-galactosides, it prefers Gal-β1,3-GalNAc and Gal-β1,3-GlcNAc (McGuire, 1970; Beyer et al., 1977).

The final enzyme in the porcine submaxillary mucin synthesis scheme is UDP-GalNAc:Gal-mucin N-acetylgalactosaminyltransferase. This enzyme is analogous to the human blood group A-dependent N-acetylgalactosaminyltransferase and occurs only in those pigs genetically disposed towards making A-positive mucin. The enzyme incorporates GalNAc in α1,3 linkage to the Gal residue of Fuc-α1,2-Gal-terminal acceptors. The porcine submaxillary gland enzyme was purified 38,000-fold by affinity chromatography on UDP-hexanolamine-Sepharose (Schwyzer and Hill, 1977a) and detailed kinetic studies have been carried out (Schwyzer and Hill, 1977b). The pure enzyme has a molecular weight of about 100,000 and may contain 2 subunits.

It is clear from the above that mucins are assembled by step-by-step addition of monosaccharides; no pre-assembly seems to occur. The subcellular path of biosynthesis has not been worked out in as great detail as for Asn-GlcNAc oligosaccharides but passage through the Golgi apparatus probably occurs (Bennett and Leblond, 1977; Schauer et al., 1974).

Glycophorin and other glycoproteins have been reported to carry a sialyl-α2,3-Gal-β1,3-(sialyl-α2,6)-GalNAc tetrasaccharide structure (Marchesi et al., 1976). The sialyltransferase catalyzing this α2,3 linkage to galactose is probably the same enzyme that catalyzes synthesis of sialyl-α2,3-lactose; the synthesis of the α2,3 and α2,6 isomers of sialyl-lactose by colostrum and liver were discussed in the previous section. Sadler et al. (1977) purified two distinct sialyltransferases from porcine submaxillary gland by the use of CDP-hexanolamine-Sepharose affinity chromatography; one enzyme was the CMP-sialic acid:GalNAc-mucin α2-6 sialyltransferase discussed above, the other was an enzyme acting on a variety of acceptors to make a sialyl-α2,3-Gal linkage. Among the acceptors found active with the latter enzyme were lactose, Gal-β1,3-

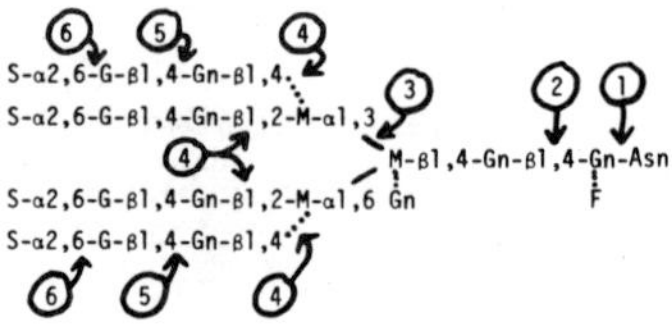

DEFICIENT ENZYME	IN-BORN ERROR	STRUCTURES ACCUMULATING	REFERENCES
4-L-aspartyl-glycosylamine amidohydrolase (enzyme 1)	Aspartyl-glycosyl-aminuria (AGU)	Gn-Asn M-α1,6-M-β1,4-Gn-β1,4-Gn-Asn	Pollitt and Pretty (1974) Pollitt and Jenner (1969) Lundblad et al. (1976) Akasaki et al. (1976)
α-mannosidase (enzyme 3)	Mannosidosis	M-α1,3-M-β1,4-Gn	Nordén et al. (1974)
β-N-acetyl-hexosaminidases A and B (enzyme 4)	G_{M2}-gangliosidosis variant O (Sandhoff-Jatzkewitz disease)	Gn-β1,2-M-α1,3-M-β1,4-Gn Gn-β1,4-M-α1,3-M-β1,4-Gn Gn-β1,2-M-α1,3 \ Gn-β1,4-M-β1,4-Gn / Gn-β1,2-M-α1,6 Gn-β1,4 \ Gn-β1,2-M-α1,3 or 6 \ M-β1,4-Gn / Gn-β1,2-M-α1,6 or 3	Strecker et al. (1977b) Ng Ying Kin and Wolfe (1974)
β-galactosidase (enzyme 5)	G_{M1}-gangliosidosis	G-β1,4-Gn-β1,4 \ G-β1,4-Gn-β1,2-M-α1,3 or 6 \ M-β1,4-Gn / G-β1,4-Gn-β1,2-M-α1,6 or 3	Ng Ying Kin and Wolfe (1975) Wolfe et al. (1974)
Sialidase (enzyme 6)	Mucolipidosis I Mucolipidosis II (I Cell Disease)	S-α2,6-G-β1,4-Gn-β1,2-M-α1,3 \ M-β1,4-Gn / S-α2,6-G-β1,4-Gn-β1,2-M-α1,6	Michalski et al. (1977) Strecker et al. (1976)

Figure 5. Scheme depicting accumulation of glycopeptides and oligosaccharides derived from Asn-GlcNAc oligosaccharides in the urine and organs of patients with in-born errors of metabolism involving the absence of glycosidases. A typical Asn-GlcNAc oligosaccharide of the N-*acetyllactosamine type is shown at the top of the figure. Abbreviations: S, sialic acid; G,* D-*galactose; Gn,* N-*acetyl-*D-*glucosamine; M,* D-*mannose, F,* L-*fucose; Asn, asparagine.*

GalNAc, antifreeze glycoprotein and sialidase-treated porcine submaxillary mucin. Pork liver, previously shown to contain an enzyme capable of making sialyl-α2,3-lactose (Hudgin and Schachter, 1972; see also section IV above) has also been shown to contain a sialyltransferase making sialyl-α2,3-Gal-β1,3-GalNAc-Ser(Thr)- (Van den Eijnden et al., 1977a). Therefore, at least two sialyltransferases act on Ser(Thr)-GalNAc type oligosaccharides. These enzymes act not only on secretory glycoproteins (mucins) but also on membrane-bound glycoproteins such as glycophorin. It is therefore reasonable to predict that these two enzymes will be widely distributed.

VI. Glycoprotein Catabolism.

Space limitations preclude extensive discussion of this topic. Many laboratories have shown the presence in microorganisms, plants and both invertebrate and vertebrate animals of a variety of exo- and endo- glycosidases. The subject is especially relevant to a group of human in-born errors of metabolism involving storage of glycolipids, mucopolysaccharides, glycopeptides and oligosaccharides. We will consider briefly some disorders involving accumulation in organs and urine of structures which appear to be derived from Asn-GlcNAc oligosaccharides.

Fig. 5 summarizes some of this data. It is evident that the exoglycosidases normally remove sugars from the non-reducing end one at a time. The absence of a particular exoglycosidase prevents the action of the enzyme next in line. It is also apparent that endo-β-N-acetylglucosaminidases (enzyme No.2, Fig. 5) and an asparaginase (enzyme No.1, Fig. 5) are required to separate oligosaccharide from polypeptide backbone. The exact sequence of action of the exo- and endo- enzymes is not known. Pathological studies of these diseases indicate that the degradation process is located in the lysosomes.

LITERATURE CITED

Akasaki, M., Sugahara, K., Funakoshi, I., Aula, P. and Yamashina, I. (1976). FEBS Letters 69, 191-194.

Andrews, P. (1970). FEBS Lett. 9, 297-300.

Bartholomew, B.A., Jourdian, G.W., and Roseman, S. (1973). J. Biol. Chem. 248, 5751-5762.

Behrens, N.H., Parodi, A.J. and Leloir, L.F. (1971). Proc. Nat. Acad. Sci. USA 68, 2857-2860.

Bell, J.E., Beyer, T.A. and Hill, R.L. (1976). J. Biol. Chem. 251, 3003-3013.

Bennett, G. and Leblond, C.P. (1977). Histochemical J. 9, 393-417.

Beyer, T.A., Prieels, J.P. and Hill, R.L. (1977). Proc. International Symp. Glycoconjugates, Woods Hole, Mass., in press

Blobel, G. and Dobberstein, B. (1975a). J. Cell Biol. 67, 835-851.

Blobel, G. and Dobberstein, B. (1975b). J. Cell Biol. 67, 852-862.

Brew, K., Vanaman, T.C. and Hill, R.L. (1968). Proc. Nat. Acad. Sci. USA 59, 491-497.

Carlson, D.M., McGuire, E.J., Jourdian, G.W., and Roseman, S. (1973). J. Biol. Chem. 248, 5763-5773.

Chou, T.H., Murphy, C. and Kessel, D. (1977). Biochem. Biophys. Res. Communs. 74, 1001-1006.

Cowan, N.J. and Robinson, G.B. (1970). FEBS Lett. 8, 6-8.

Czichi, U. and Lennarz, W.J. (1977). J. Biol. Chem. 252, 7901-7904.

Ebner, K.E. (1973). In "The Enzymes" (P.D. Boyer, editor), Edit 3, Vol. 9, Part B, pp. 363-377, Academic Press, New York.

Fitzgerald, D.K., Brodbeck, U., Kiyosawa, I., Mawal, R., Colvin, B., and Ebner, K.E. (1970). J. Biol. Chem. 245, 2103-2108.

Fraser, I.H. and Mookerjea, S. (1976). Biochem. J. 156, 347-355.

Geren, C.R., Geren, L.M. and Ebner, K.E. (1975a). Biochem. Biophys. Res. Communs. 66, 139.

Geren, C.R., Magee, S.C. and Ebner, K.E. (1975b). Biochemistry 14, 1461-1463.

Geren, C.R., Magee, S.C. and Ebner, K.E. (1976). Arch. Biochem. Biophys. 172, 149-155.

Geren, C.R., Geren, L.M., Lee, D. and Ebner, K.E. (1977). Biochim. Biophys. Acta 497, 128-132.

Hagopian, A., Westall, F.C., Whitehead, J.S., and Eylar, E.H. (1971). J. Biol. Chem. 246, 2519-2523.

Hercz, A., Katona, E., Cutz, E., Wilson, J.R., and Barton, M. (1978). Submitted for publication.

Herscovics, A., Bugge, B. and Jeanloz, R.W. (1977a). J. Biol. Chem. 252, 2271-2277.

Herscovics, A., Golovtchenko, A.M., Warren, C.D., Bugge, B. and Jeanloz, R.W. (1977b). J. Biol. Chem. 252, 224-234.

Hill, H.D. Jr., Schwyzer, M., Steinman, H.M. and Hill, R.L. (1977). J. Biol. Chem. 252, 3799-3804.

Hudgin, R.L. and Schachter, H. (1972). Can. J. Biochem. 50, 1024-1028.

Hunt, L.M., Etchison, J., Robertson, J., Robertson, M. and Summers, P.F. (1977). Abstracts 174th. National Meeting of the Am. Chem. Soc., Chicago, Ill.

Isemura, M. and Schmid, K. (1971). Biochem. J. 124, 591-604.

Ito, S., Muramatsu, T., and Kobata, A. (1975). Arch. Biochem. Biophys. 171, 78-86.

Jabbal, I. and Schachter, H. (1971). J. Biol. Chem. 246, 5154-5161.

Jamieson, J.C. (1977). Can. J. Biochem. 55, 408-414.

Jeppsson, J-O, Laurell, C-B., and Fagerhol, M. (1978). Eur. J. Biochem. 83, 143-153.

Kessel, D., Sykes, E. and Henderson, M. (1977). J. Natl. Cancer Inst. 59, 29-32.

Khatra, B.S., Herries, D.G. and Brew, K. (1974). Eur. J. Biochem. 44, 537-560.

Kiely, M.L., McKnight, G.S., and Schimke, R.T. (1976). J. Biol. Chem. 251, 5490-5495.

Kitchen, B.J. and Andrews, P. (1974a). Biochem. J. 141, 173-178.

Kitchen, B.J. and Andrews, P. (1974b). Biochem. J. 143, 587-590.

Lawford, G.R. and Schachter, H. (1966). J. Biol. Chem. 241, 5408-5418.

Lehman, E.D., Hudson, B.G. and Ebner, K.E. (1975). FEBS Lett. 54, 65-69.

Letts, P.J., Pinteric, L. and Schachter, H. (1974). Biochim. Biophys. Acta 372, 304-320.

Lundblad, A., Masson, P.K., Norden, N.E., Svensson, S. Ockerman, P.A., and Palo, J. (1976). Eur. J. Biochem. 67, 209-214.

Magee, S.C., Mawal, R. and Ebner, K.E. (1974). Biochemistry 13, 99-102.

Magee, S.C., Geren, C.R. and Ebner, K.E. (1976). Biochim. Biophys. Acta 420, 187-194.

Marchesi, V.T., Furthmayr, H. and Tomita, M. (1976). Ann. Rev. Biochem. 45, 667-698.

Marshall, R.D. (1974). Biochem. Soc. Symp. 40, 17-26.

McGuire, E.J. (1970). In "Blood and Tissue Antigens" (D. Aminoff, editor), pp. 461-478, Academic Press, New York.

Michalski, J.C., Strecker, G., Fournet, B., Cantz, M. and Spranger, J. (1977). FEBS Lett 79, 101-104.

Molnar, J. and Sy, D. (1967). Biochemistry 6, 1941-1947.

Montreuil, J. (1975). Pure and Applied Chemistry 42, 431-477.

Morrison, T.G. and Lodish, H.F. (1975). J. Biol. Chem. 250, 6955-6962.

Munro, J.R. and Schachter, H. (1973). Arch. Biochem. Biophys. 156, 534-542.

Munro, J.R., Narasimhan, S., Wetmore, S., Riordan, J., and Schachter, H. (1975). Arch. Biochem. Biophys. 169, 269-277.

Narasimhan, S., Stanley, P. and Schachter, H. (1977). J. Biol. Chem. 252, 3926-3933.

Ng Ying Kin, N.M.K. and Wolfe, L.S. (1974). Biochem. Biophys. Res. Communs. 59, 837-844.

Ng Ying Kin, N.M.K. and Wolfe, L.S. (1975). Biochem. Biophys. Res. Communs. 66, 123-130.

Nordén, N.E., Lundblad, A., Svensson, S. and Autio, S. (1974). Biochemistry 13, 871-874.

Ogata, S., Muramatsu, T. and Kobata, A. (1975). J. Biochem. (Tokyo) 78, 687-696.

Parodi, A.J., Behrens, N.H., Leloir, L.F. and Dankert, M. (1972) Biochim. Biophys. Acta 270, 529-536.

Paulson, J.C., Beranek, W.E., and Hill, R.L. (1977a). J. Biol. Chem. 252, 2356-2362.

Paulson, J.C., Rearick, J.I. and Hill, R.L. (1977b). J. Biol. Chem. 252, 2362-2371.

Pless, D.D. and Lennarz, W.J. (1977). Proc. Nat. Acad. Sci. USA 74, 134-138.

Pollitt, R.J. and Jenner, F.A. (1969). Clin. Chim. Acta 25, 413-416.

Pollitt, R.J. and Pretty, K.M. (1974). Biochem. J. 141, 141-146.

Powell, J.T. and Brew, K. (1974). Eur. J. Biochem. 48, 217-228.

Powell, J.T. and Brew, K. (1975). J. Biol. Chem. 250, 6337-6343.

Powell, J.T., Jarlfors, U. and Brew, K. (1977). J. Cell Biol. 72, 617-627.

Rao, A.K., Garver, F., and Mendicino, J. (1976). Biochemistry 15, 5001-5009.

Robbins, P.W. (1977). Proc. Fourth International Symp. Glycoconjugates, Woods Hole, Mass., in press.

Rothman, J.E. and Lodish, H.F. (1977). Nature 269, 775-780.

Sadler, J.E., Rearick, J.I., Paulson, J.C. and Hill, R.L. (1977). Proc. International Symp. Glycoconjugates, Woods Hole, Mass., in press.

Sato, T., Yosizawa, Z., Masubuchi, M. and Yamauchi, F. (1967). Carbohyd. Res. 5, 387-398.

Schachter, H. (1974a). Adv. Cytopharmacol. 2, 207-218.

Schachter, H. (1974b). Biochem. Soc. Symp. 40, 57-71.

Schachter, H. (1977). In "Mucus in Health and Disease" (M. Elstein and D.V. Parke, editors), pp. 103-129, Plenum Publish. Co.

Schachter, H. (1978). In "The Glycoconjugates" (M. Horowitz and W. Pigman, editors), Vol. 2, Academic Press, New York, in press.

Schachter, H. and Rodén, L. (1973). In "Metabolic Conjugation and Metablic Hydrolysis" (W.H. Fishman, editor), Vol. 3, pp. 1-149, Academic Press, New York.

Schachter, H., Jabbal, I., Hudgin, R.L., Pinteric, L., McGuire, E.J., and Roseman, S. (1970). J. Biol. Chem. 245, 1090-1100.

Schachter, H., McGuire, E.J. and Roseman, S. (1971). J. Biol. Chem. 246, 5321-5328.

Schauer, R., Buscher, H.-P., and Casals-Stenzel, J. (1974). Biochem. Soc. Symp. 40, 87-116.

Schwyzer, M. and Hill, R.L. (1977a). J. Biol. Chem. 252, 2338-2345.

Schwyzer, M. and Hill, R.L. (1977b). J. Biol. Chem. 252, 2346-2355.

Sherr, C.J. and Uhr, J.W. (1969). Proc. Nat. Acad. Sci. USA 64, 381-387.

Smith, C.A. and Brew, K. (1977). J. Biol. Chem. 252, 7294-7299.

Spiro, M.J., Spiro, R.G. and Bhoyroo, V.D. (1976a). J. Biol. Chem. 251, 6400-6408.

Spiro, M.J., Spiro, R.G. and Bhoyroo, V.D. (1976b). J. Biol. Chem. 251, 6420-6425.

Spiro, R.G., Spiro, M.J. and Bhoyroo, V.D. (1976). J. Biol. Chem. 251, 6409-6419.

Stanley, P., Narasimhan, S., Siminovitch, L., and Schachter, H. (1975). Proc. Nat. Acad. Sci. USA 72, 3323-3327.

Stoffyn, P., Van den Eijnden, D. and Stoffyn, A. (1977). Fed. Proc. 36, 744.

Strecker, G., Hondi-Assah, T., Fournet, B., Spik, G., Montreuil, J., Maroteaux, P., Durand, P., and Farriaux, J-P. (1976). Biochim. Biophys. Acta 444, 349-358.

Strecker, G., Fournet, B., Spik, G., Montreuil, J., Dorland, L., Haverkamp, J., Schut, B.L. and Vliegenthart, J.F.G. (1977a). Proc. Fourth International Symp. Glycoconjugates, Woods Hole, Mass., in press.

Strecker, G., Herlant-Peers, M-C., Fournet, B., Montreuil, J., Dorland, L., Haverkamp, J., Vliegenthart, J.F.G., and Farriaux, J-P. (1977b). Eur. J. Biochem. 81, 165-171.

Struck, D.K. and Lennarz, W.J. (1977). J. Biol. Chem. 252, 1007-1013.

Tabas, I., Schlesinger, S. and Kornfeld, S. (1978). J. Biol. Chem 253, 716-722.

Tai, T., Ito, S., Yamashita, K., Muramatsu, T. and Kobata, A. (1975). Biochem. Biophys. Res. Communs. 65, 968-974.

Tonneguzzo, F. and Ghosh, H.P. (1975). FEBS Lett. 50, 369-373.

Trayer, I.P. and Hill, R.L. (1971). J. Biol. Chem. 246, 6666-6675.

Tsopanakis, A.D. and Herries, D.G. (1976). Biochem. Biophys. Res. Communs. 68, 1102.

Turco, S.J., Stetson, B. and Robbins, P.W. (1977). Proc. Nat. Acad. Sci. USA 74, 4411-4414.

Van den Eijnden, D.H., Dieleman, B. and Schiphorst, W.C.M. (1977a). Proc. Fourth International Symp. Glycoconjugates, Woods Hole, Mass., in press.

Van den Eijnden, D.H., Stoffyn, P., Stoffyn, A. and Schiphorst, W.E.C.M. (1977b). Eur. J. Biochem. 81, 1-7.

Vargas, V.I. and Carminatti, H. (1977). Mol. Cell. Biochem. 16, 171-176.

Waechter, C.J. and Lennarz, W.J. (1976). Ann. Rev. Biochem. 45, 95-112.

Watkins, W.M. (1974). In "The Red Cell" (D.M. Surgenor, editor), 2nd edition, Vol. 1, pp. 293-360, Academic Press, New York; Biochem. Soc. Symp. 40, 125.

Wilson, J.R., Williams, D. and Schachter, H. (1976). Biochem. Biophys. Res. Communs. 72, 909-916.

Wirth, D.F., Katz, F., Small, B. and Lodish, H.F. (1977). Cell 10, 253-263.

Wolfe, L.S., Senior, R.G. and Ng Ying Kin, N.M.K. (1974). J. Biol. Chem. 249, 1828-1838.

RECEIVED April 7, 1978.

3

Structure and Metabolism of Glycolipids

C. C. SWEELEY, Y.-K. FUNG, B. A. MACHER, J. R. MOSKAL, and H. A. NUNEZ

Department of Biochemistry, Michigan State University, East Lansing, MI 48824

Glycolipids have been implicated in the molecular mechanisms of some important biological phenomena that occur at the cell surface, such as contact inhibition, intercellular adhesion, immunochemical tissue specificity, hormone receptor function, and internalization of various external macromolecular and supramolecular materials. Close to a hundred different glycolipids have been isolated and characterized from various terrestrial and marine organisms to date, providing sufficient structural diversity for these proposed functional roles. Interest in the chemistry of these substances was stimulated by the discovery of a family of inherited abnormalities in the metabolism of some of the glycolipids (1-3). Subsequently they became a subject of intense study by cell biologists and immunologists when it was generally recognized that they are membrane constituents of most tissues; current areas of research interest are the immunochemical specificity of glycolipids, their localization and mobility in bilayer membranes, lectin-binding properties, and viral transformation-specific changes in their composition. The aim of this review is to provide an introduction into the chemistry, methods of analysis and metabolism of the major classes of glycolipids that occur in eukaryotic organisms.

Structure

Unlike the complex lipopolysaccharides of bacterial membranes or the glycosides of hydroxylated fatty acids in yeasts, the glycolipids of eukaryotic organisms are of two major types, called glycosphingolipids and glycoglycerolipids. Glycosphingolipids are composed of a family of long-chain aliphatic bases (sphingosines), fatty acids and carbohydrates, whereas the glycoglycerolipids consist of glycerol, fatty acids (or fatty ethers) and carbohydrates. The glycosphingolipids can be subdivided into several classes on the basis of gross molecular features. Thus, there are neutral glycosphingolipids, gangliosides (glycosphingolipids containing sialic acid), sulfoglycosphingolipids (containing sulfate ester groups on the carbohydrate moiety) and phosphoglycosphingolipids. There are also neutral and sulfoglycoglycerolipids. Complete review of the structures and occurrence of the glycosphingolipids and glycoglycerolipids is not possible in this brief presentation. Recent reviews with more extensive

0-8412-0452-7/78/47-080-047$09.75/0

information about specific topics are cited whenever possible to assist the reader with a need for more than a casual introduction to the field. An excellent text (4) on the isolation, structure analysis and metabolism of glycolipids, with an emphasis on details of methods and procedures, should also be consulted.

Ceramide is the hydrophobic portion of a glycosphingolipid that is anchored in a bilayer membrane. It consists of a mixture of molecular species in which the long-chain bases such as sphing-4-enine are attached through amide linkages to several fatty acids. The most common long-chain bases are shown below, but a variety of other chain-length homologs, branched sphingosines and multiply unsaturated types are known to exist (5). They are all derived from sphinganine, previously known as dihydrosphingosine, which is D-erythro-2-aminooctadecane-1,3-diol or 2S,3R-2-aminooctadecane-1,3-diol (6). Methods have been described for their analysis as aldehydes after periodate oxidation (7), N-acetyl-trimethylsilyl derivatives (8), dinitrophenyl derivatives (9), and as the alcohols obtained from intact glycolipids possessing unsaturated bases by ozonolysis and reduction (10).

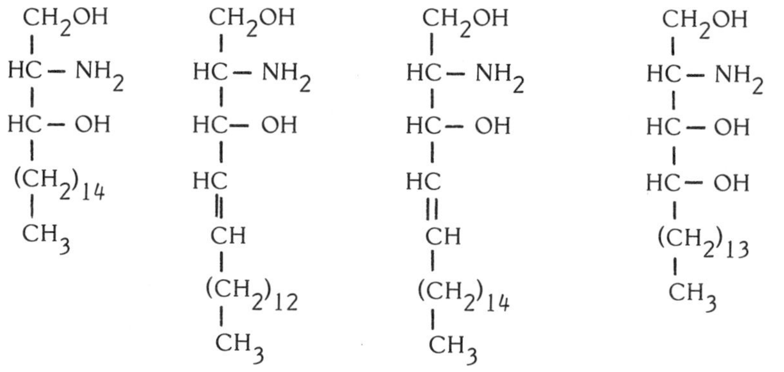

Sphinganine 4-Sphingenine 4-Eicosasphingenine 4-D-Hydroxy-sphinganine

The fatty acids occurring in the ceramide moiety are a variable mixture of saturated and monounsaturated homologs from C_{16} to C_{26} and the α-hydroxy derivatives of these fatty acids. Lignoceric (24:0) and nervonic (24:1, Δ^{15}) acids are generally major components, along with substantial amounts of palmitic (16:0), stearic (18:0), arachidic (20:0), behenic (22:0) and tricosanoic (23:0) acids. In some cases the α-hydroxy fatty acids are the dominant species. Comparisons of the fatty acid composition of a wide variety of glycosphingolipids have been reported (11).

In the glycosphingolipids, ceramide is attached to a mono- or oligosaccharide by a glycosidic linkage with the primary hydroxyl group of the long-chain base. Although it has only been proved conclusively in a few instances by nmr spectroscopy or chemical synthesis, it is assumed that the glycosidic linkage between ceramide and the carbohydrate moiety has the β configuration, and that the sugar residues exist in the pyranose ring form. An example is shown in Figure 1, galactosyl-(β1→4)-glucosyl-β1→1')-ceramide (lactosylceramide). A class of phosphoglycosphingolipids occurs in higher plants and fungi; these structures consist of carbohydrate groups with an inositol residue linked to ceramide through a phosphodiester bridge. A substance of this type from corn was proposed to be GlcNα1→4GlcUAα1→6inositol(2←1αMan)-1-0-phosphorylceramide (169). A related compound from tobacco leaves (170) has been shown to be GlcNAcα1→4GlcUAα1→2*myo*-inositol-1-O-phosphorylceramide (171).

In the glycoglycerolipids studied thus far, the carbohydrate residues are attached to C-3 of *sn*-glycerol by either α- or β-glycosidic linkages. Three important types of glycoglycerolipids are 3-β-galactosyl-1,2-diacyl-*sn*-glycerol (12), 3'-sulfo-3-β-galactosyl-1-alkyl-2-acyl-*sn*-glycerol (seminolipid) (13,14), and a sulfotriglycosyldiglyceride of human gastric secretions, 3-α-(6'-sulfoglucosyl-(α1→6)-glucosyl-(α1→6)-glucosyl)-1-alkyl-2-acyl-*sn*-glycerol (15).

Structural variation of the oligosaccharide moiety of complex glycosphingolipids is so diverse that a special nomenclature has become necessary in referring to these substances. A semisystematic nomenclature has been recommended by the IUPAC-IUB Commission on Biochemical Nomenclature (16) and has been reviewed in some detail (6,17,18). Compounds containing a single monosaccharide unit were historically called cerebrosides. Now they are identified by the sugar they contain; examples are galactosylceramide (GalCer) and glucosylceramide (GlcCer). Similarly, the diglycosylceramide containing a lactose unit is called lactosylceramide (LacCer).

More complex glycosphingolipids are given names that identify a particular portion of the carbohydrate moiety, indicating the family to which it belongs and the number of glycose units in the chain. Prefixes (given below) for the various families imply a specific structure, including the sequential arrangement of the glycose units as well as the positions and anomeric configuration of the glycosidic linkages. Thus, the glycosphingolipid of the globo type that contains all four sugar residues is referred to as globotetraglycosylceramide and is abbreviated GbOse$_4$Cer. Similarly, the glycosphingolipid of the ganglio type with three sugar units is called gangliotriglycosylceramide (GgOse$_3$Cer); this substance is GalNAcβ1→4Galβ1→4GlcCer. Substituents that are not part of the root oligosaccharide are given at the beginning of the name, using a Roman numeral to indicate the monosaccharide unit (counting from ceramide) on which the substituent is located and a superscript Arabic numeral to indicate the position of the glycosidic linkage. According to this system, Forssman hapten (GalNAcα1→3GalNAcβ1→3Galα1→4Galβ1→4GlcCer) should be called IV^3-α-*N*-acetylgalactosaminyl-globotetraglycosylceramide (IV^{3}αGalNAc-GbOse$_4$Cer).

Methods of Analysis of Glycolipids

Determination of the complete chemical structure of a glycolipid usually requires a combination of several techniques. Some of the methods can be used with nanomolar amounts of sample, whereas others

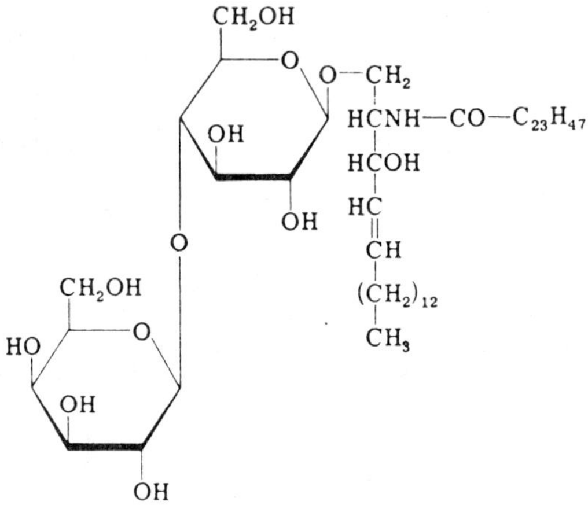

Figure 1. Lactosylceramide (LacCer), galactopyranosyl-(β1 → 4)-glucopyranosyl—(1 → 1′)-ceramide

Prefix	Abbreviation	Structure
lacto	Lc	Gal(β1→3)GlcNAc(β1→3)Gal(β1→4)Glc
lactoneo	Lcn	Gal(β1→4)GlcNAc(β1→3)Galβ1→4)Glc
muco	Mc	Gal(β1→3)Gal(β1→4)Gal(β1→4)Glc
gala	Ga	GalNAc(1→3)Gal(1→4)Gal(α1→4)Gal*
globo	Gb	GalNAc(β1→3)Gal(α1→4)Gal(β1→4)Glc
globoiso	Gbi	GalNAc(β1→3)Gal(α1→3)Gal(β1→4)Glc
ganglio	Gg	Gal(β1→3)GalNAc(β1→4)Gal(β1→4)Glc

*Substances more complex than Galα1→4GalCer in this family have not so far been isolated from biological material.

require about one micromole; carbon-13 nmr spectroscopy is applicable at present only if a large amount (ca. 150 mg) of sample is available. Most of the structural studies reported in the past five years have involved chromatographic separation of the glycolipid (column and thin-layer chromatography), followed by several commonly used analytical procedures: fatty acid, long-chain base and carbohydrate composition by gas-liquid chromatography after acid-catalyzed methanolysis; stepwise degradation of the carbohydrate moiety with specific glycosidases; and analysis of the partially methylated alditol acetates derived from permethylated glycolipid. Frequent use of mass spectrometry has been made in recent studies to validate the structures of the partially methylated alditol acetates and to determine some structural features of the intact permethylated glycolipid. Some of the newest techniques are reviewed briefly in the following section.

Excellent separations of glycolipids according to fatty acid differences have been obtained by high pressure liquid chromatography of perbenzoylated or O-acetyl-N-p-nitrobenzoyl derivatives (19-21), requiring only nanomolar amounts of samples for detection at 254 nm with a suitable flow-through detector. Mixtures of natural glycolipids from GlcCer to $GbOse_4Cer$ can be separated by chromatography on Sorbax Sil (22). Though the technique appears to be ideally suited for the separation of substances with different molecular weights, sufficient studies have not been made to determine whether isomeric glycolipids (different sugar arrangement, different positions and/or configuration of glycosidic linkages, different epimeric components) can be separated effectively.

An alternative chromatographic approach involves the careful analysis by thin-layer chromatography of the effluent material from a column. This technique appears to be most effective when use is made of an ion exchange material such as DEAE-cellulose or DEAE-Sephadex and when a very large number of samples from the column are spotted adjacent to each other on the thin-layer plates. The resulting map looks like a two-dimensional chromatogram with excellent resolution of components that would overlap if analyzed together on the plate. This procedure has been exploited in the study of blood group glycolipids from human erythrocytes (23) and for analyses of the ganglioside composition of brain (24,25). Additional fractionation of gangliosides has been obtained on a preparative scale by DEAE-Sephadex column chromatography and subsequent separation on a new spherical silica, called Iatrobeads (26-29).

Analysis of carbohydrate composition by gas-liquid chromatography is common practice. Trimethylsilyl derivatives of methyl glycosides (obtained by acid-catalyzed methanolysis) or alditol acetates (obtained by acetolysis and hydrolysis) can be used with suitable internal standards to obtain molar ratios of the sugar components with about 100 nanomoles of sample (30-32). These conditions cannot be used to distinguish between N-acetyl and N-glycolyl derivatives of neuraminic acid in the intact glycolipids. Mild acid-catalyzed methanolysis (0.05 N methanolic HCl at 80° for 1 hr) has been used instead, giving high yields of the β-methyl ketosides of N-acetyl and N-glycolylneuraminic acid, the trimethylsilyl derivatives of which separate readily by gas chromatography (33).

The sequential arrangement of the sugar residues in complex glycolipids can be deduced enzymatically or by mild acid hydrolysis or methanolysis. The enzymatic procedure involves the stepwise hydrolysis of the glycolipid with specific glycosidases. This gives conclusive evidence for the nature of the terminal glycose residue at each step as well as the anomeric configuration of the glycosidic linkage (34). It is customary to first remove sialic residues from gangliosides by 1 M formic acid hydrolysis for 1 hr at 100° (35), a procedure which also liberates fucose (36). Negative results in the enzymatic analysis need to be interpreted with some care since they might be caused by improper detergent in the reaction mixture or a glycosidase specificity that gives poor Km or maximal velocity values with a particular structure (34). It is important to realize that the procedure requires a set of highly pure glycosidases. The other method involves partial methanolysis with 0.3 M HCl in chloroform/methanol at 60° for 40 min; it has been used effectively for analyses of a heptaglycosylceramide (37) and various blood group A-active glycolipids from gastric mucosa (36).

Two methods make use of permethylation for the analysis of glycolipids and can be used together to obtain information about the sequential arrangement of carbohydrate residues and the positions of glycosidic linkages. Permethylation is generally carried out in dimethylsulfoxide with a mixture of sodium dimethylsulfinylcarbanion and methyl iodide (38), a mild procedure (room temperature) which gives high yields of completely methylated product. Precautions have to be taken to exclude plasticizers from the starting materials, particularly dimethylsulfoxide. Mass spectra of the permethylated intact glycolipid reveal several kinds of structural detail, including the composition of the ceramide moiety, the nature of the terminal glycose unit, the total number of glycose units and, to some extent, the sequence of sugars (39-45). Spectra of the product obtained by reduction of the N-methyl amide linkage with lithium aluminum hydride are somewhat superior as ions at high mass are stabilized and the volatility of the derivative increases to some extent (39-45). Useful information results from the cleavage reactions (at glycosidic linkages) that result from random charge localization on ring oxygen atoms. Unfortunately these ions are obscured to some extent and often represent weak contributions to the total mass spectrum, due to the presence of the ceramide moiety.

An alternative approach involves conversion of the glycolipid to an oligosaccharide by ozonolysis (46) or treatment with osmium tetroxide (47), both of which attack at the double bond of the sphingosine moiety, yielding a product that is readily hydrolyzed to the free oligosaccharide. The permethylated oligosaccharide has excellent mass spectral characteristics, as illustrated in the case of lacto-N-tetraose and difucohexaose I in Figures 2 and 3. The molecular ion was not observed in either case upon electron impact ionization but may be present in chemical ionization mass spectra. The ion at highest mass in the case of the lacto-N-tetraose derivative is at m/e 858; it results from the loss of a C-6 substituent (·CH_2OMe) from whichever ring bears the charge (on oxygen) in the molecular ion. Three ions (below) of approximately equal intensity, at m/e 219, 464 and 668 in Figure 2, are predicted for the sequence hexose-N-acetylhexosamine-hexose-R by arithmetic calculations; they result from the cleavage of different glycosidic linkages, with charge retention on the ring oxygen and loss of the remainder of the chain as a free radical.

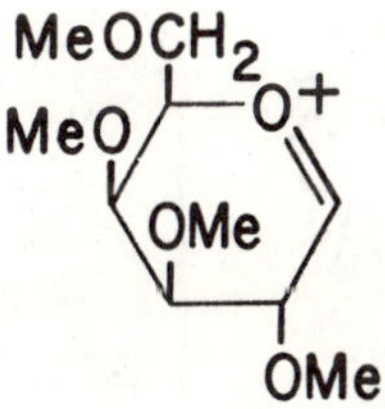

1

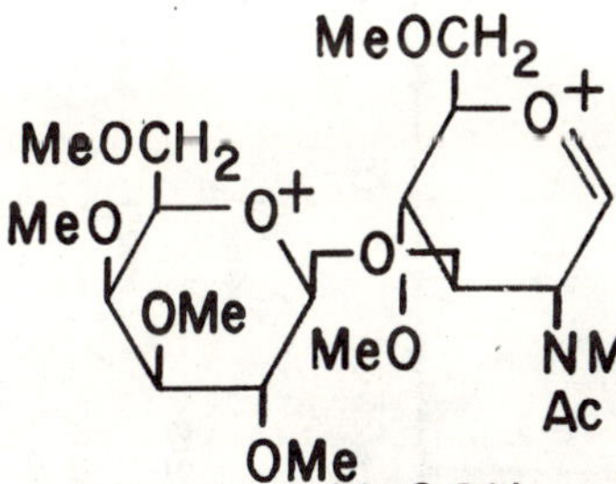

2

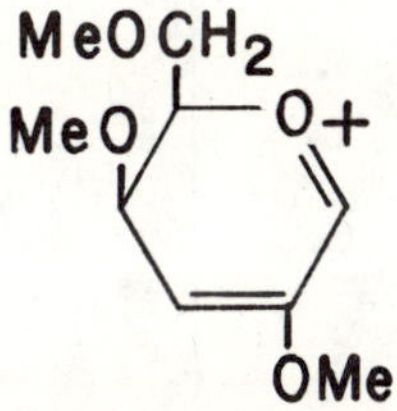

4

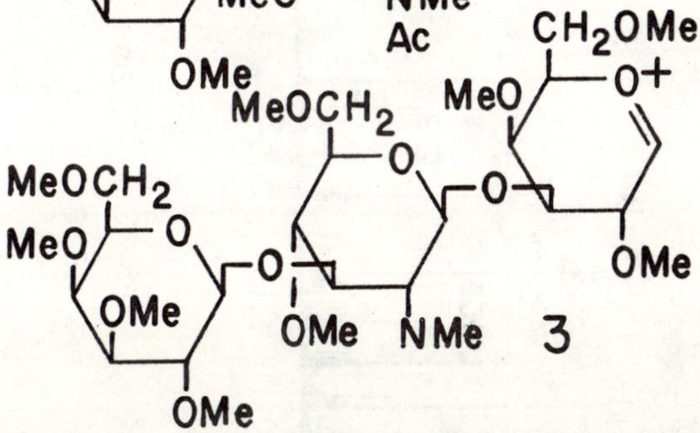

3

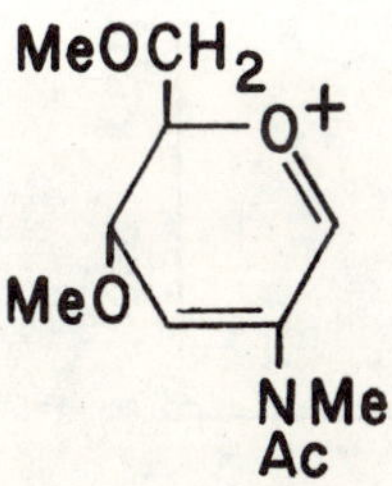

5

Further loss of methanol from 1 and a disaccharide from 3 both give m/e 187 (4), whereas loss of the terminal hexose residue from 2 gives an intense ion at m/e 228 (5). The presence of m/e 464 (a hexosamine-containing disaccharide moiety) and the absence of m/e 260 (a terminal N-acetylhexosamine unit) are strong evidence for the hexose-hexosamine

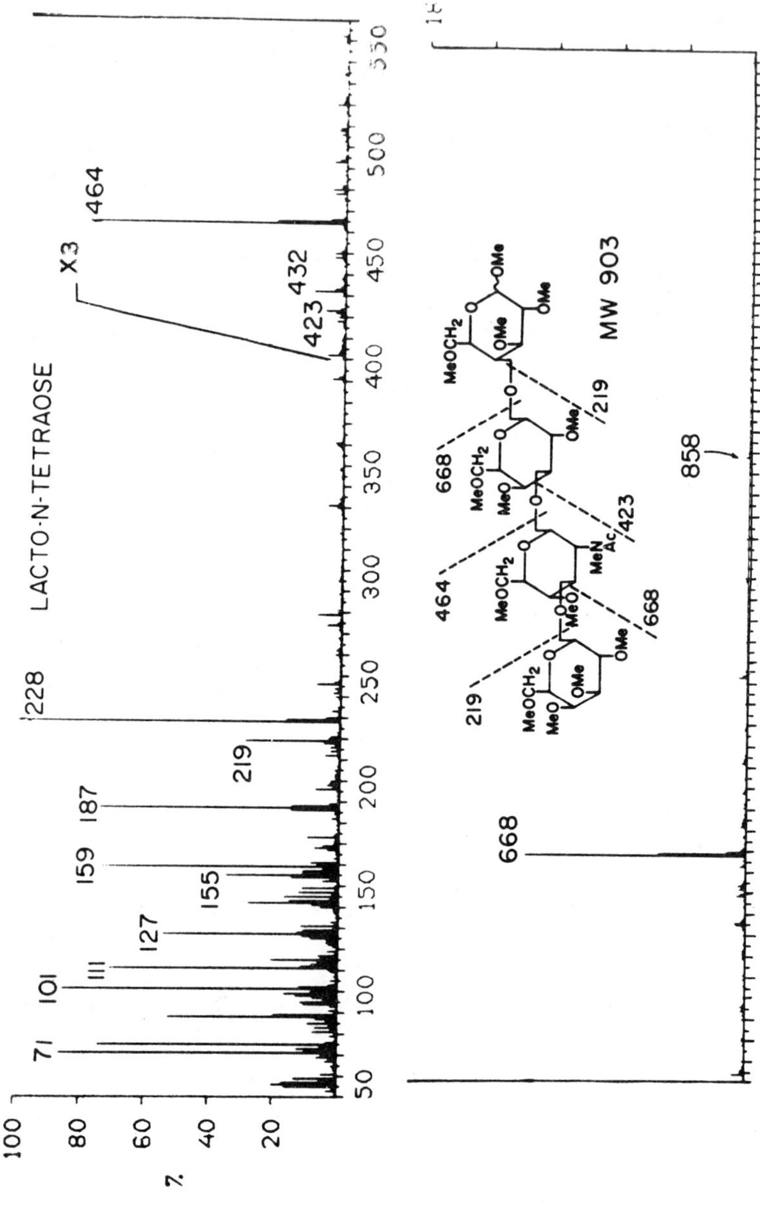

Figure 2. Mass spectrum of permethylated lacto-N-tetraose, obtained by electron impact ionization at 70 eV with a Varian MAT CH-5 double-focusing mass spectrometer (direct probe)

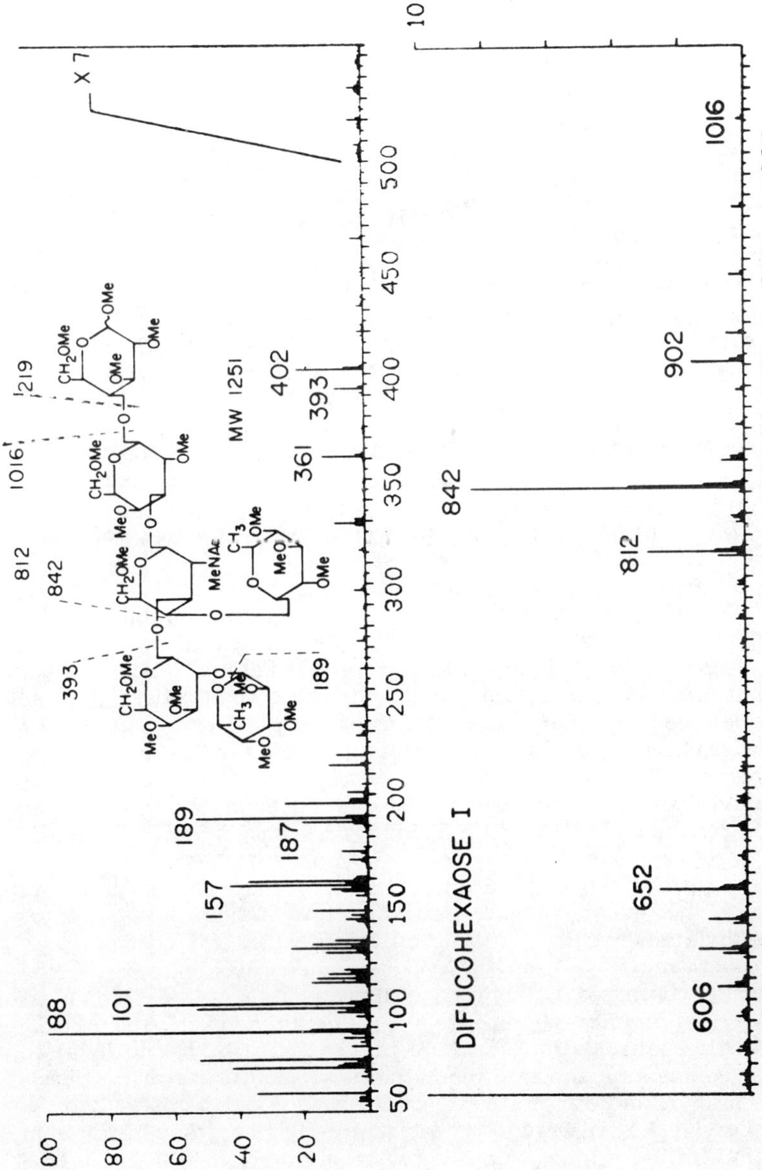

Figure 3. Mass spectrum of permethylated difucohexaose I (conditions given in Figure 2)

arrangement. The presence of a strong fragment ion at m/e 228 is good evidence for a 1,3 linkage of the terminal hexose unit to the hexosamine, since this product is stabilized by the pair of electrons on the nitrogen. The product from loss of terminal hexose in a 1,4 linked disaccharide would not be similarly stabilized and would likely lose methanol to give an intense ion at m/e 196. In fact, this is the case with the doubly substituted hexosamine moiety in the difucohexaose (Fig. 3); further conversion of the ion at m/e 812 to m/e 606 by loss of the C-4 substituent results in a very weak ion, compared with that obtained by loss of the C-3 substituent from m/e 812 to give m/e 402. The ion at m/e 196, for loss of both of C-3 and C-4 substituents from m/e 812, is strong, as expected on the basis of these arguments.

More precise information about the positions of glycosidic linkages can be obtained by the analysis of the monosaccharides liberated from the permethylated glycolipids by acid-catalyzed methanolysis (48) or by hydrolysis and reduction to partially methylated alditols (49-51). In either case free hydroxyl groups in the products are acetylated, and the resulting derivatives are analyzed by combined gas chromatography-mass spectrometry (50,52-54). A large number of model compounds have been analyzed and their retention behavior and mass spectral characteristics reported (55). The mass spectra are similar on casual inspection but generally have at least one ion that is uniquely derived from a particular structure. Epimeric structures are differentiated on the basis of their retention behavior rather than their mass spectra.

Use of this technique to distinguish between terminal, 1,3- and 1,4-substituted N-acetylhexosamines is illustrated in Figures 4-6. Similarities in the mass spectra result from cleavage between C-2 and C-3 in all three cases to give an intense ion at m/e 158, which loses ketene to give another intense ion at m/e 116. As a rule strong ions are formed when cleavage can occur between two carbons substituted with OCH_3 groups, while virtually no ionization occurs by cleavage between two carbons bearing acetoxy groups. The three kinds of hexosamine are most readily distinguished by the presence of m/e 202 and 205 in the case of the product from terminal GalNAc, m/e 233 from 1,4-linked GalNAc and m/e 318, which is strong only in the product of 1,3-linked GalNAc.

Mass chromatography is a powerful tool that can be used to advantage for the analysis of the partially methylated alditol acetates from permethylated glycolipids, using a computerized gas chromatography-mass spectrometry system. Quick and accurate analysis of the products can be obtained without careful inspection of the total mass spectra. For example, Forssman pentaglycosylceramide (GalNAcα1→3GalNAcβ1→3Galα1→4Galβ1→4GlcCer) would be converted by permethylation analysis to equimolar amounts of N-methyl-N-acetyl-1,5-di-O-acetyl-3,4,6-tri-O-methylgalactosaminitol from the terminal GalNAc and N-methyl-N-acetyl-1,3,5-tri-O-acetyl-4,6-di-O-methylgalactosaminitol from the internal GalNAc residue. Mass chromatograms of m/e 205, 233 and 318 (Fig. 7) clearly show that the terminal and 1,3-substituted GalNAc products are present in the mixture and that the 1,4-substituted GalNAc product is not. Previous discussion of this technique by Laine (private communication, ref. 171 and Abstracts, Society for Complex Carbohydrates, New Orleans, October 8, 1976) suggests that it might be used with

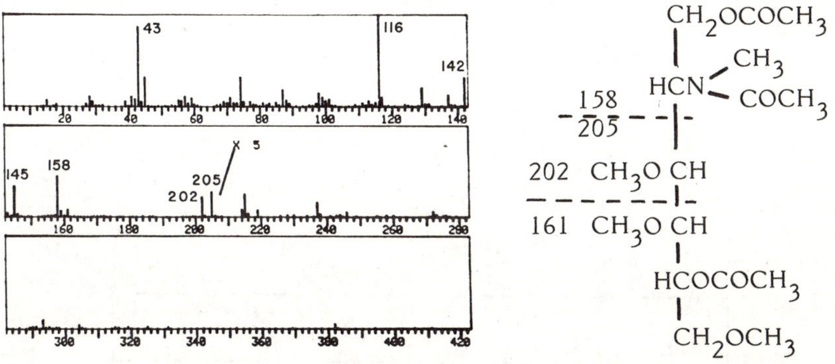

Figure 4. Mass spectrum of N-*methyl*-N-*acetyl-3,4,6-tri*-O-*methyl-1,5-di*-O-*acetyl-galactosamintol*

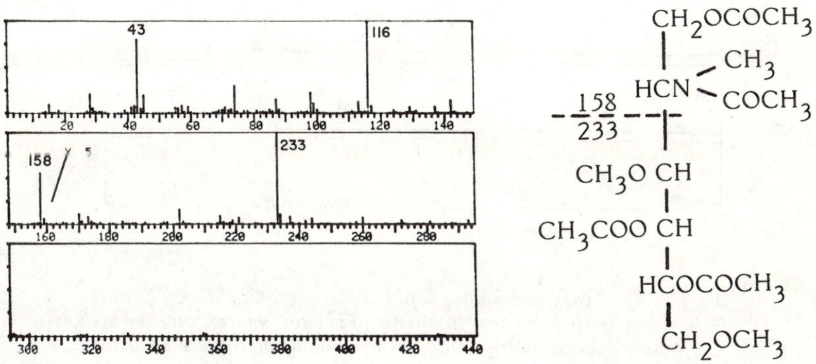

Figure 5. Mass spectrum of N-*methyl*-N-*acetyl-3,6-di*-O-*methyl-1,4,5-tri*-O-*acetylgalactosaminitol*

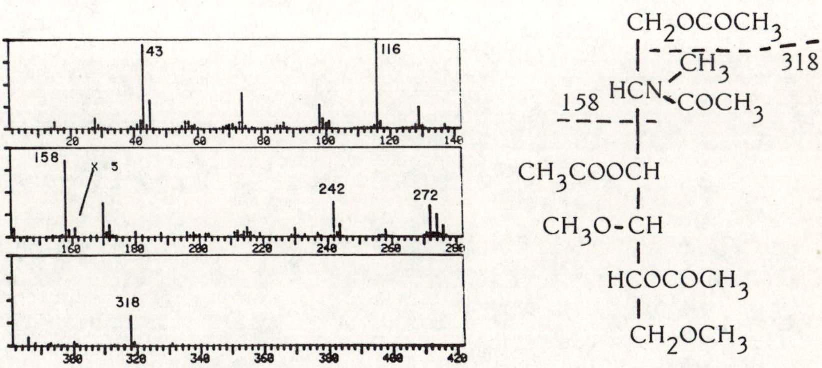

Figure 6. Mass spectrum of N-*methyl*-N-*acetyl-4,6-di*-O-*methyl-1,3,5-tri*-O-*acetyl-galactosaminitol*

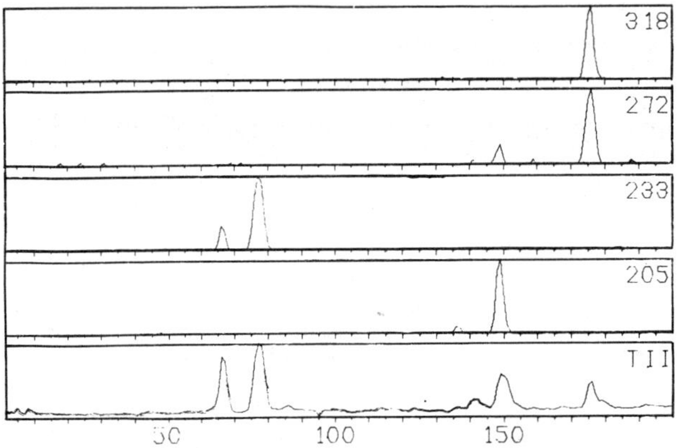

Figure 7. Mass chromatography of m/e *205, 233, 272, and 318 along with total ion intensity (TII) for gas chromatography–mass spectrometry analysis of partially methylated alditol acetates from Forssman pentaglycosylceramide. Ion peaks of* m/e *272 and 318 at scan 176 are specific for 1,3-linked GalNAc* (see *Figure 6), and the peak of* m/e *205 at scan 149 is specific for a terminal GalNAc* (see *Figure 4).*

nanomolar amounts of glycolipid.

Carbon-13 nuclear magnetic resonance spectroscopy offers some unique capabilities in the structural analysis of glycolipids, not the least of which may be the possibility to analyze interactions at the cell surface. Large samples are required to obtain a natural abundance Fourier-transform ^{13}C nmr spectrum, using a high field instrument to resolve the complex of resonances for ring carbons not involved in glycosidic bonds or otherwise derivatized. Assignments have recently been made by Sillerud et al. (56) for the resonances of every carbon nucleus of G_{M1} ganglioside (II^3 α-N-acetylneuraminosyl-gangliotetraglycosylceramide), based on careful comparisons of the spectrum of this ganglioside with those of neuraminosyllactose, ceramide and several related glycolipids. We have used the assignments proposed by Sillerud et al. (56) for G_{M1} ganglioside to analyze the ^{13}C nmr spectrum of a complex neutral glycosphingolipid, globotetraglycosylceramide, and present some of the assignments in Figure 8. A higher resolution spectrum and data for a few model compounds are necessary for definitive assignments.

In the tetrasaccharide moiety the glycosidic carbons resonate in the range typical of α and β glycosidic linkages at 100 to 106 ppm. The hydroxymethyl carbons appear at 61 to 63 ppm, the sugar ring carbons of a glycone linkage appear at 79 to 83 ppm and the ring sugar carbons not involved in inter-sugar linkages fall in the 72 to 77 ppm region, with the exception of the C-2 of N-acetyl-D-galactosamine and C-4 of the galactosyl moieties. The ceramide moiety gives a set of resonances that resemble very closely those observed for ganglioside G_{M1} micelles (56). The alkyl chain shows a large and broad peak for the bulk methylene carbons and sharper peaks for the terminal methyl and neighboring methylene carbons in the region of 14 to 40 ppm. The rest of the carbons show resonances typical of the carbonyl group (175 ppm), of a carbon-carbon double bond (132 ppm), of a hydroxymethyl carbon linked to a glycosidic oxygen (69 ppm), of a secondary alcohol (72 ppm), and of a nitrogen-linked methinyl carbon (55 ppm).

Glycosphingolipid Biosynthesis

The biosynthesis of glycosphingolipids should begin with studies dealing with the synthesis of ceramide (57-59). However, from a biological standpoint, it is the composition and sequence of the glycose units that is the most interesting and most thoroughly investigated.

A number of investigators have studied the biosynthesis of gluco-and galactosylceramide (60-64) and to date it is still unclear whether these compounds are made in vivo by conversion of sphingosine to psychosine and then to cerebroside, or by acylation of sphingosine to yield ceramide, from which cerebroside is derived. However, the synthesis of lactosylceramide from UDP-galactose and glucosylceramide has been established in embryonic chicken brain by Basu et al. (64) and confirmed by Hauser and coworkers (65) in rat spleen. It should be noted here that it was primarily the finding that embryonic chicken brain homogenates are rich in various glycosyltransferases and that detergents (such as Triton X-100) greatly facilitated their measurement (66) that gave the means and impetus to

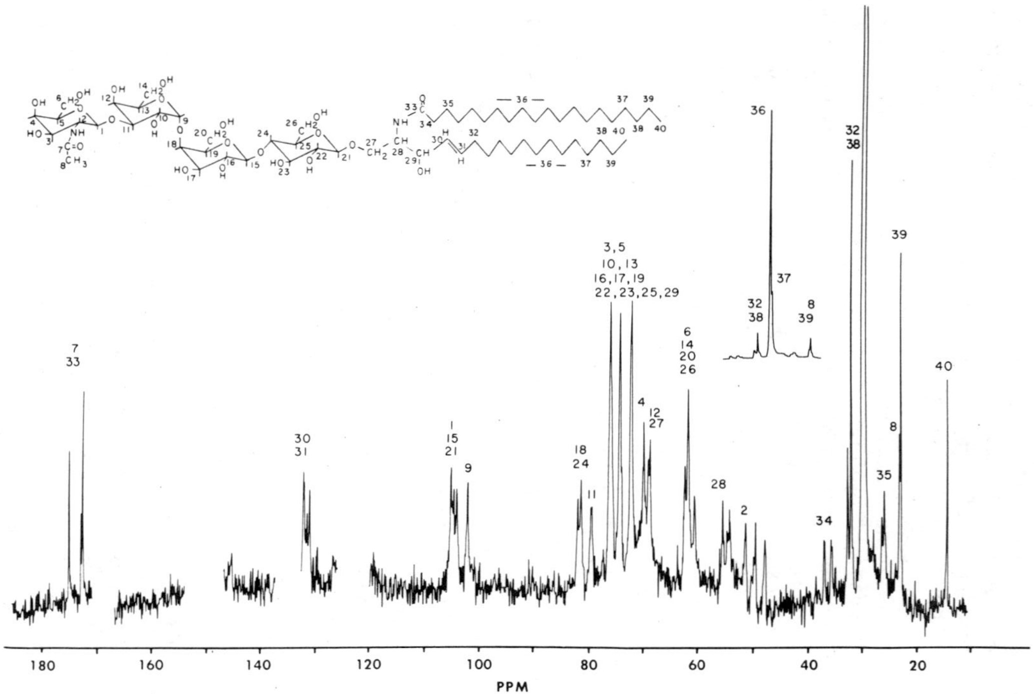

Figure 8. Carbon-13 NMR spectrum of globotetraglycosylceramide from porcine erythrocytes. The 15.08 MHz, proton-decoupled spectrum was obtained in a Bruker WP-60 at 35°C, using 100 mg sample in 1.5 mL pyridine-d$_6$ (TMS internal standard). Transients (25,000) were accumulated at a sweep width of 3000 Hz and a 45°C pulse (8 μsec). (Sweeley, Barker, and Nunez, manuscript in preparation)

further study glycosphingolipid biosynthesis.

Lactosylceramide (LacCer) appears to be the precursor for the various classes of more complex glycosphingolipids (see Figure 9). Glycosylation of this molecule by a galactosyltransferase (65) gives globotriglycosylceramide, the precursor of globotetraglycosylceramide and globopentaglycosylceramide (Forssman hapten). Sialylation of LacCer by sialyltransferase gives II^3-α-N-acetylneuraminosyl lactosylceramide (66-68) and N-acetylgalactosaminyltransferase converts LacCer to gangliotriglycosylceramide (69), a precursor of the gangliosides. Finally, the addition of N-acetylglucosamine to LacCer by the appropriate transferase gives lactoneotriglycosylceramide, and subsequent reactions lead to the blood group-active glycosphingolipids (70).

As mentioned above, studies by Basu, Steigerwald, Kaufman and Roseman (71-75) historically were the first to be reported on the *in vitro* biosynthesis of glycosphingolipids, in particular the gangliosides. From their work the pathway in Figure 9 was proposed. Later this pathway was corroborated in frog brain (76), rat brain (77-79) and cultured murine neuroblastoma clones (80). Even further support for the ganglioside pathway in Figure 9 was obtained using an *in vivo* strategy consisting of incubating cultured murine neuroblastoma cells with radiolabeled precursors, followed by isolation of the glycolipid products (81).

The synthesis of globotriglycosylceramide, the precursor of globotetraglycosylceramide, has been established in rat spleen (65). Heat inactivation patterns and inhibition by different glycosphingolipids showed that this galactosyltransferase was different from that which catalyzes the transfer of UDP-galactose to glucosylceramide. Chien *et al.* (82) demonstrated a β-N-acetylgalactosaminyltransferase that would catalyze the synthesis of globotetraglycosylceramide from globotriglycosylceramide and UDP-GalNAc. More recently Kijimoto *et al.* (70) and Ishibashi *et al.* (83) have reported an α-N-acetylgalactosaminyltransferase activity in guinea pig tissues that can convert globotetraglycosylceramide to globopentaglycosylceramide (Forssman hapten) by addition of N-acetylgalactosamine via UDP-GalNAc. Yeung *et al* (84) demonstrated the synthesis *in vitro* of both globoside and Forssman hapten in a clone of cultured adrenal tumor (Y-1) cells. Dawson and Sweeley (85) concluded that the synthesis of globotetraglycosylceramide of porcine erythrocytes occurrs in bone marrow cells.

Studies on the biosynthesis of the blood group-active glycosphingolipids have also been made (see Figure 10). Basu and coworkers have reported the synthesis of lactoneotetraglycosylceramide from lactotriglycosylceramide (the core blood group structure) and a blood group B-active pentaglycosylceramide from lactoneotetraglycosylceramide by β-galactosyltransferase (86) and an α-galactosyltransferase (87) isolated from rabbit bone marrow cells, respectively. They also reported the presence of galactosyltransferase activities involved in the synthesis of this blood group B-active glycolipid in cultured mouse neuroblastoma cells (80).

GDP-fucose:glycolipid fucosyltransferase activities related to blood group structures have also recently been demonstrated. The biosynthesis *in vitro* of blood group H_1 Gal(2←1αFuc)β1→4GlcNAcβ1→3Galβ1→4GlcCer (88,89) and human B-type Galα1→3Gal(2←1αFuc)β1→4GlcNAcβ1→3Galβ1→-4GlcCer (90) glycosphingolipids have been described in a Golgi-enriched

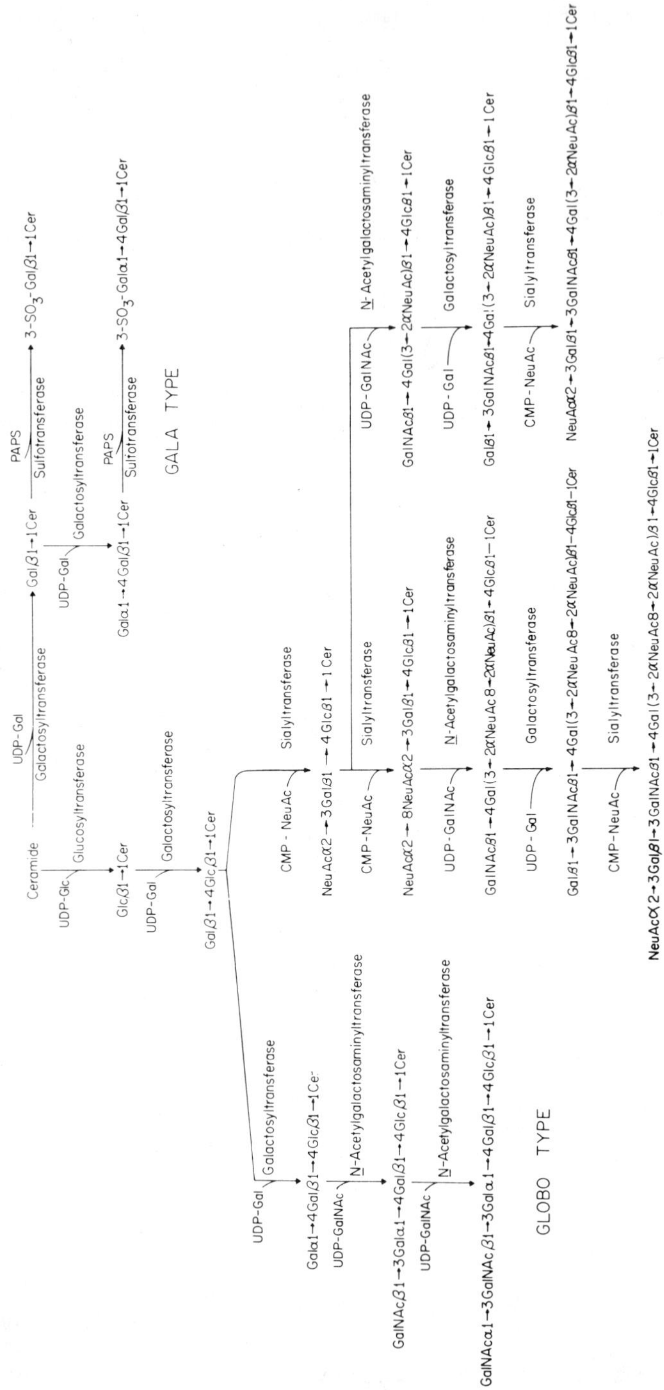

Figure 9. Pathways for the biosynthesis of sulfo-, neutral (Forssman hapten), and acidic (gangliosides) glycosphingolipids

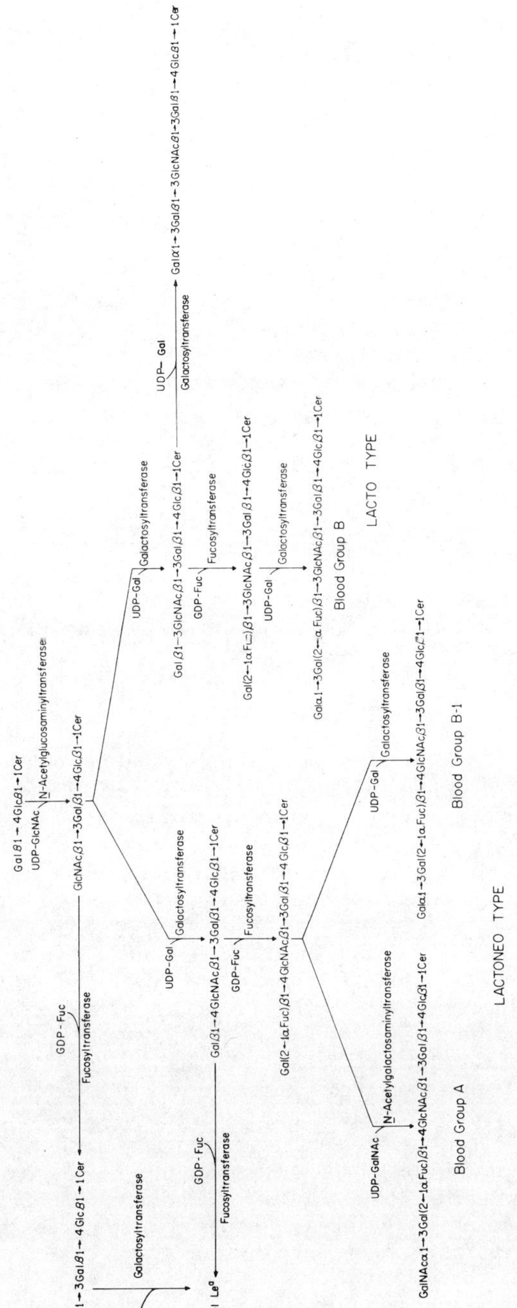

Figure 10. Pathways for the biosynthesis of blood group-related glycosphingolipids

fraction isolated from bovine spleen. Pacuzka and Koscielak (91) also demonstrated the synthesis of these glycolipids with enzymes from human serum. Moreover, Presper et al (92) found a fucosyltransferase in human (IMR-32) neuroblastoma cells that is involved in the synthesis of types H-1 and B-1 glycolipids as well as a tetraglycosylceramide precursor of a novel Le^a blood group glycosphingolipid.

Many other glycosphingolipids have been purified and structurally characterized but their biosynthesis in tissues or cultured cells have not yet been determined. Two particularly interesting groups are the branched blood group-active glycolipids such as those reported by Watanabe et al. (93) and Slomiany and Slomiany (94) and the "macro-glycolipids" such as those characterized by Gardas and Koscielak (95).

The above data establish that glycosphingolipids are synthesized by the stepwise addition of activated sugars directly to the appropriate glycolipid acceptor. It will be interesting to see if this holds true for the "macro-glycolipids" as well, since they appear to contain a number of repeating units. Perhaps intermediate carriers will be involved similar to those described in glycoprotein biosynthesis.

In conclusion, increasing numbers of reports are appearing in which glycolipid biosynthetic pathways are demonstrated in various cell culture systems. Using such model systems in concert with the characterization of endogenous and cell surface glycoconjugates should be very helpful in furthering our understanding of the role of glycosphingolipids in cellular processes and tumorigenesis.

Acceptor specificity of glycosyltransferases

The synthesis of the carbohydrate moiety of glycosphingolipids apparently takes place in a sequential manner and is catalyzed by glycosyltransferases with specificity for both nucleotide sugar donor and glycolipid acceptor. Until recently (97), purified glycosyltransferases have not been available to test for acceptor specificity with glycolipid substrates. Therefore, judgments have been made on the basis of indirect evidence (Table I)

Hildebrand and Hauser (65) compared the heat lability and metal ion requirements of galactosyltransferases from rat spleen. A galactosyl transferase utilizing glucosylceramide as an acceptor molecule was inactivated by heat treatment (50°C for 5 minutes), while lactosylceramide galactosyltransferase activity was only slightly affected (20% reduction). Although both transferase activities were stimulated by manganese, only glucosylceramide galactosyltransferase was enhanced by magnesium. A third difference between the two transferase activities was their response to the addition of various sphingolipids. Lactosylceramide synthesis was inhibited by sphingosine, galactosylsphingosine, lactosylsphingosine and ceramide, whereas globotriglycosylceramide synthesis was inhibited by lactosylsphingosine, ceramide and galactosylceramide.

Substrate competition experiments have shown that rabbit bone marrow contains two galactosyltransferase activities, β-galactosyltransferase synthesizing lactotetraglycosylceramide and an α-galactosyltransferase synthesizing lactopentaglycosylceramide (87). When comparisons

TABLE I

Acceptor Specificity of Glycosyltransferases

Enzymes	Acceptors	Evidence for Acceptor Specificity	Reference
Gal-transferase	GlcCer,LacCer	1. Metal ion 2. Heat lability 3. Inhibition by sphingolipids	Hildebrand and Hauser (1969)
Gal-transferase	$LcOse_3Cer$,$LcOse_4Cer$	Substrate competition	Basu and Basu (1973)
Fuc-transferase	$LcnOse_4Cer$ asialoganglioside	Gene expression	Pacuszka and Koscielak (1976)
NeuAc-transferase	LacCer,$II^3\alpha$NeuAc-$GgOse_4Cer$	Heat lability	Kaufman et al. (1963)
NeuAc-transferase	LacCer,$II^3\alpha$NeuAc-$GgOse_4Cer$	Viral transformation	Fishman et al. (1972)
GalNAc-transferase	Blood group glycolipids, A, Le^b,H-mega and H	Pure enzyme	Schwyzer and Hill (1977)

were made of ^{14}C-galactose incorporation (from UDP-galactose) in the presence of each acceptor individually and when both acceptors were present together, an additive relationship was found. The authors therefore concluded that two galactosyltransferases were present.

Fucosyltransferase activities have been measured in human serum, with lactoneotetraglycosylceramide and "asialoganglioside" as acceptors (91). Serum of ABO and Lewis blood group contained fucosyltransferase activity with both acceptors, while serum preparations from two Bombay blood donors (O_h) lacked asialogangliofucosyltransferase activity. Thus, the two fucosyltransferase activities appear to be products of different genes.

Reports by Kaufman et al. (73) and Fishman et al. (96) demonstrated that glycosyltransferases with glycolipid products of the same anomerity may be different proteins. They differentiated similar sialyltransferase activities (with lactosylceramide and N-acetylneuraminosylgangliotetraglycosylceramide as acceptors) on the basis of in vitro (heat lability) and in vivo (viral transformation) results. Kaufman et al. (73) found that lactosylceramide:sialyltransferase was more stable to heating at 50° than II^3-α-N-acetylneuraminosylgangliotetraglycosylceramide:sialyltransferase. Fishman et al. (96) demonstrated that Swiss 3T3 cells contain these same sialyltransferase activities. Following transformation by SV-40 virus, N-acetylneuraminylgangliotetraglycosylceramide sialyltransferase was reduced, whereas lactosylceramide:sialyltransferase activity did not change.

In all of the cases cited above, enzyme activities were measured in homogenates or membrane preparations. Recently Schwyzer and Hill (97) reported on the acceptor specificity of a homogeneous preparation of N-acetylgalactosaminyltransferase from porcine submaxillary glands. The enzyme specifically transferred N-acetylgalactosamine from UDP-GalNAc to substrates (both glycoprotein and glycolipid) with a blood group H structure. Fucα1→2Galβ1→4GlcNAcβ1→3Galβ1→4Glc Cer and H-mega glycolipid were both acceptors for the enzyme, while A and Le^b glycolipids did not serve as substrates.

The studies discussed above certainly suggest that glycosyltransferases display a specificity for glycolipid acceptors; however, some references can be cited which argue for nonspecificity. Chien et al. (82) found a single N-acetylgalactosaminyltransferase from embryonic chicken brain that could utilize lactosylceramide, globotriglycosylceramide and II^3-N-acetylneuraminosyl-lactosylceramide as glycolipid acceptors, based on the results of substrate competition experiments. The same workers (88) have used a similar experimental approach to demonstrate multiple acceptor recognition with a fucosyltransferase from bovine spleen.

Although several studies have dealt with the question of acceptor specificity, little is known concerning the possible role of protein modifiers (e.g., "A protein" in lactose synthase) and chemical alteration (i.e., phosphorylation, adenylation) in glycosyltransferase reactions. In the reports discussed above, the acceptor specificity (or lack of) may be a consequence of the presence or absence of such modifications or alterations. These ideas will be discussed further in the section dealing with the regulation of glycosyltransferases.

Glycosyltransferase localization

Roseman has proposed that glycosyltransferases on the surface of mammalian plasma membranes may be involved in intercellular glycosylation, i.e., glycosylation of surface glycoproteins and glycolipids of one cell by glycosyltransferases on the surface of an adjacent cell, and that enzyme-substrate complexes of these enzymes could be the key mechanism governing intercellular adhesion and thus, possibly, tumorigenesis (75). Since that time a large body of data has emerged, both in support of this idea (see review by Shur and Roth (110)) as well as against it (see review by Keenan and Morre (111)).

In order to establish the presence of cell surface glycosyltransferases, a number of criteria must be fulfilled. The purity of isolated membranes must be very carefully established by both membrane marker enzymes and electron microscopy. When using whole cells, one must control for possible hydrolysis of sugar nucleotides and transport of free sugars into the cell, and enzyme activity due to lysis or secretion must be carefully monitored. Further, glycosyltransferase assays must be optimized under all conditions examined. A number of investigators arguing for the presence of cell surface glycosyltransferases have controlled for sugar nucleotide breakdown (112-114). Others have controlled for intracellular glycosyltransferase "leakage" (112,115-117). Upon closer inspection of the experimental design of these various studies, it can be seen that none has included all of the appropriate controls, as discussed above.

Keenan and Morre and their coworkers have presented the strongest evidence against cell surface glycosyltransferases. Figure 11 depicts the membrane flow hypothesis put forth by Morre (118,119). In essence, membrane biosynthesis includes the actual transfer of membranous material from one subcellular site to another, finally reaching (via fusion) the cell surface. Endoplasmic reticulum membranes are transferred to the Golgi-apparatus where they undergo modification to become plasma membrane-like. Completed membrane material is then incorporated into secretory vesicles which move to and fuse with the plasma membrane, discharging their intravesicular contents into the extracellular milieu. Morre goes on to postulate such a route for cell surface glycolipids and glycoproteins.

Initial data to support the ideas above came from electron microscopic autoradiography (120) where it was shown that glycosylation of glycoproteins takes place in the Golgi-apparatus. Studies with highly purified and well characterized membrane fractions shortly followed. High specific activities for several glycoprotein (121) and glycolipid (122) glycosyltransferases in Golgi-apparatus preparations were established. More recently, lactose synthase (123) and glycosyltransferases for ganglioside biosynthesis (124) have been demonstrated in mammary gland and liver Golgi, respectively. Glycosyltransferase activities concentrated in the Golgi-apparatus of small intestines (125), pancreas (126), thyroid (127), and many other tissues have also been reported. The strength of this work relies on (1) the careful demonstration of purified membranes, (2) the significantly higher specific activity of glycosyltransferases in

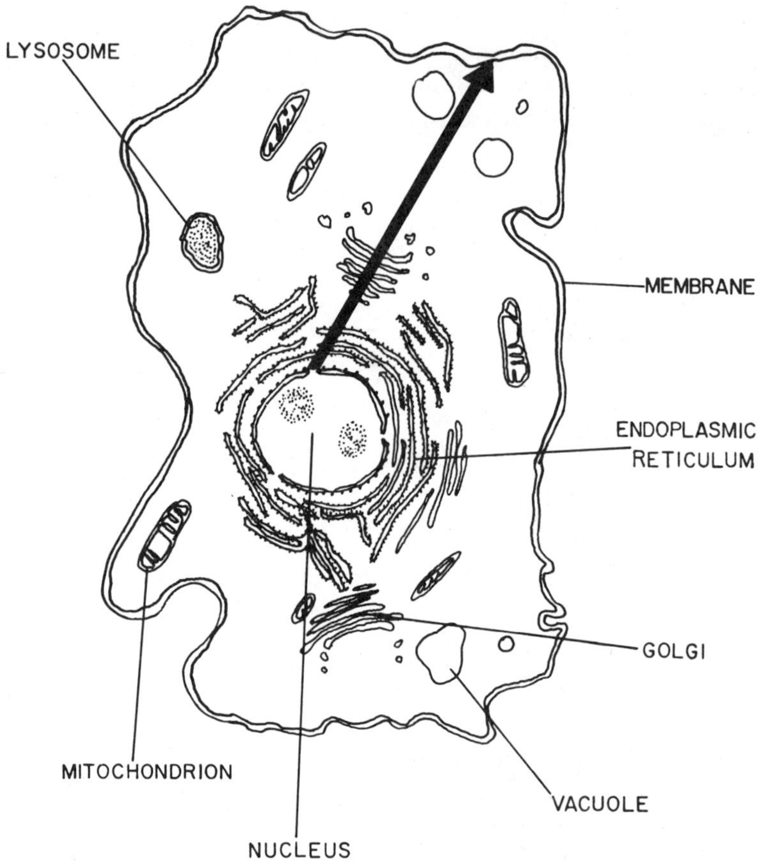

Figure 11. Synthesis of cell surface glycoconjugates according to the "membrane-flow" hypothesis

Golgi-membranes compared to plasma membranes or homogenatess and (3) autoradiographic localization of radiolabeled glycose moieties in Golgi and endoplasmic reticulum.

To date no conclusive evidence has been given for intercellular glycosylation of lipids or proteins by surface glycosyltransferases. However, many reports have appeared which implicate cell surface glycosyltransferases in a wide varity of recognition phenomena (116,128-133).

In conclusion, it seems quite reasonable to assume that there are glycosyltransferases on the cell surface, based on the number and diversity of reports, even if all of the appropriate controls have not been adequately considered. However, conclusions as to their function are tentative at best. The fact that glycosylation enzymes are found in the Golgi-apparatus seems to be well established. Moreover, the idea that some glycosyltransferases appear on the cell surface via "membrane flow" and function only as membrane repair systems cannot be ruled out. Clearly, this area of research needs much more investigation. Hopefully, further studies will provide critical information about the function of complex glycoconjugates, irrespective of whether the functionally most important glycosyltransferases are on the cell surface or in the Golgi-apparatus.

Regulation of glycosyltransferases

Several studies (98-101) with glycoprotein glycosyltransferases have demonstrated an *in vitro* effect of nucleotide phosphates. Both inhibitory and stimulatory effects have been observed. Enzymes utilizing glycolipid acceptors are also affected by adenine- and uridine-containing compounds (Table II). In a study by Ishibashi *et al.* (102) both positive and negative modulating effects on N-acetylgalactosaminyltransferases from guinea pig kidney were noted with uridine nucleotides. Interestingly, the α-transferase which catalyzes the synthesis of globopentaglycosylceramide was markedly inhibited by UDP while the β-transferase (globotetraglycosylceramide synthesis) was activated. UDP-glucose was similar to UDP in its effects on the transferases, whereas several other uridine compounds suppressed both activities. Other information has been presented on the possible regulatory effects of substrates and products of glycosyltransferases and is referred to in Table II.

Lactose synthase provides an interesting example of glycosyltransferase regulation (103). *In vitro*, α-lactalbumin causes galactosyltransferase to become a lactose synthase. In the absence of α-lactalbumin, glucose is a very poor acceptor (Km > 1M) and the enzyme serves as a glycoprotein galactosyltransferase. The possibility that protein modifiers exist for the regulation of glycolipid synthesis can now be examined. The availability of techniques to obtain purified glycosyltransferases, along with cell systems, in which viral transformation (see section on Acceptor Specificity) appears to result in a specific loss of a glycosyltransferase, should provide the means to test such a possibility.

The regulation of glycosyltransferase activity has been demonstrated in intact animal and cell culture systems (Table II). The level of

TABLE II

REGULATION OF GLYCOSYLTRANSFERASES

GLYCOSYLTRANSFERASE(S)	EFFECTOR	EFFECT (+ or -)	REFERENCE
	In Vivo		
UDP-Gal:Ceramide Galactosyltransferase	Testosterone	(+)	Gray (1971)[a]
UDP-Gal:GM1 Galactosyltransferase, UDP-Gal:Glucosylceramide Galactosyltransferase, UDP-Gal:Lactotriosylceramide Galactosyltransferase	dBcAMP	(+)	Basu *et al*. (1976)[b]
CMP-NAN:Lactosylceramide Sialyltransferase	Butyrate	(+)	Simmons *et al*. (1975)[c] Macher, *et al*. (1977)[d]
GDP-Fucose:neolactotetraosylceramide Fucosyltransferase	6-mercaptoguanosine	(+)	Presper *et al*. (1978)[e]
	In Vitro		
Lactose synthase	alpha-lactalbumin (low concentrations) (high concentrations)	(+) (+) (-)	Bell *et al*. (1976)[f]
	substrate inhibition		Morrison, J.F. and Ebner K.E. (1971)[g]
Porcine blood group A:N-acetylgalactosaminyltransferase	UDP Fucosyllactose	(-) (-)	Schwyzer and Hill (1977)[h]
UDP-Gal:Glucosylceramide and Lactosylceramide Galactosyltransferases	ATP	(-)	Chandrabose, K.A., and MacPherson, I.A. (1976)[i]
UDP-GalNAc:Triglycosylceramide N-Acetylgalactosaminyltransferase	Uridine UDP UDP-Glucose UDP-Galactose	(-) (+) (+) (-)	Ishibashi *et al*. (1976)[j]
UDP-GalNAc:Globoside N-Acetylgalactosaminyltransferase	Uridine UDP UDP-Glucose UDP-Galactose	(-) (-) (-) (-)	

[a] G.M. Gray (1971) *Biochem. Biophys. Acta*, 239, 494-500.
[b] Basu, S., Moskal, J.R. and Gardner, D.A., (1976) "Ganglioside Function:Biochemical and Pharmacological Implications", (Porcellati, G., ed.) Plenum Publishing Corp., N.Y. p. 45-63.
[c] Simmons, J.L., Fishman, P.H., Freese, E., and Brady, R.O., (1975) *J. Cell Biol.*, 66, 414-424.
[d] Macher, B.A., Lockney, M., Moskal, J.R., Fung, Y-K. and Sweeley, C.C., (1978) *Exp. Cell Res.*, unpublished.
[e] Presper, K.A., Basu, M., and Basu, S., (1978) *Proc. Nat. Acad. Sci. U.S.A.*, 75, 289-293.
[f] Bell, J.E., Beyer, T.A. and Hill, R.L., (1976) *J. Biol. Chem.*, 251, 3003-3013.
[g] Morrison, J.F. and Ebner, K.E., (1971) *J. Biol. Chem.*, 246, 3992-3998.
[h] Chandrabose, K.A. and MacPherson, I.A., (1976) *Biochim. Biophys. Acta*, 429, 96-111.
[i] Schwyzer, M.and Hill, R.L., (1977) *J. Biol. Chem.*, 252, 2346-2355.
[j] Ishibashi, T., Atsuta, T. and Makita, A., (1976) *Biochim. Biophys. Acta*, 429, 759-767.

digalactosylceramide in the kidneys of female C3H/He and C57/BL mice was considerably lower than in male mice of the same strains (104). Administration of testosterone to the female mice resulted in a remarkable increase in digalactosylceramide levels. Enzymatic studies demonstrated a regulatory effect by testosterone on UDP-galactose:ceramide galactosyltransferase in these mice. Further information concerning steroid regulation of glycosyltransferases has not been published.

Basu et al. (105) assayed a number of glycosyltransferase activities in clones (NIE-115, adrenergic and NS-20, cholinergic both derived from C-1300 tumors of A/J mice) of mouse neuroblastoma cells and found that three enzymes involved in glycolipid synthesis were stimulated by $N^6,O^{2'}$-dibutyryladenosine cyclic 3',5'-monophosphate (dBcAMP) (Table II). Lactoneotetraglycosylceramide synthesis was increased 5-fold in NIE-115 cells and 2-fold in NS-20 cells. UDP-galactose:glucosylceramide galactosyltransferase activity was increased 3-fold in NS-20 cells and remained unchanged in NIE-115 cells, while the synthesis of II^3α-N-acetylneuraminosylgangliotetraglycosylceramide was slightly increased in dBcAMP-treated NS-20 cells. It is possible that dBcAMP may exert its stimulatory effect on glycosyltransferases via phosphorylation. Alternatively, these changes may reflect alterations in cell cycle distributions of the cell populations. This suggestion would be in line with findings presented by Wolf and Robbins (106) and Chatterjee et al. (107) that glycolipid synthesis is greatest during certain portions of the cell cycle (M, G_1).

Our own studies (Ref. 108 and Fig. 12) and those of Simmons et al. (109) on the effect of butyrate in cultured human cells demonstrated that glycosyltransferase levels can be adjusted within hours of drug treatment. Simmons et al. (109) originally showed that butyrate treatment of HeLa cells for 18-24 hr resulted in a 15-20 fold increase in CMP-sialic acid:lactosyltransferase activity and that this enzyme activity returned to control levels when the butyrate was removed. Macher et al. (108) have studied the same enzyme in KB cells and found a similar induction. They have also demonstrated that KB cells apparently are synchronized by butyrate treatment (Fig. 13). Furthermore, release from butyrate treatment and addition of fresh medium resulted in a remarkable induction of UDP-Gal:lactosylceramide galactosyltransferase (Fig. 14). Therefore, it seems that butyrate treatment and release results in differential activation of sialyltransferase and galactosyltransferase activities in human cells.

Glycosphingolipid Catabolism

Virtually all the enzymes necessary for the stepwise catabolism of glycolipids have been studied *in vitro* and in many cases they have been partially or completely purified. These enzymes, which are exoglycosidases, have received much attention because of their involvement in the hereditary lipidoses. Recently, the potential of using enzyme replacement therapy for the treatment of these diseases has been examined. Another effort in the field has been directed toward an understanding of the kinetic properties and specificity of these enzymes. A brief discussion of the exoglycosidases involved in glycolipid metabolism is given below.

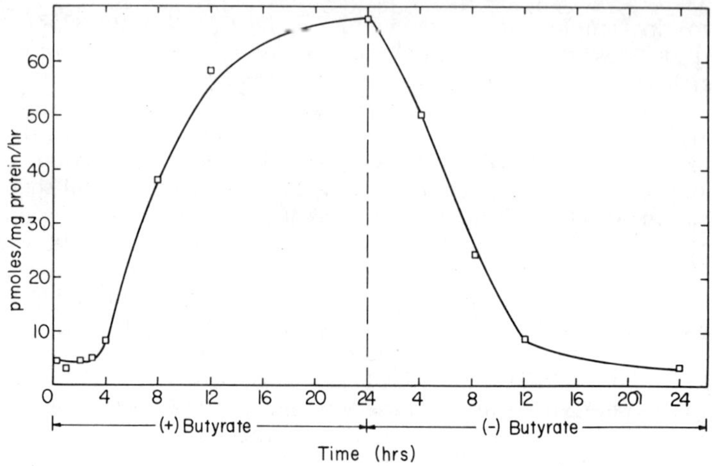

Figure 12. ^{14}*C–thymidine uptake after release from butyrate and double thymidine block in KB cells. Redrawn from Simmons et al.,* J. Cell Biol. *(1975)* **66**, *414.*

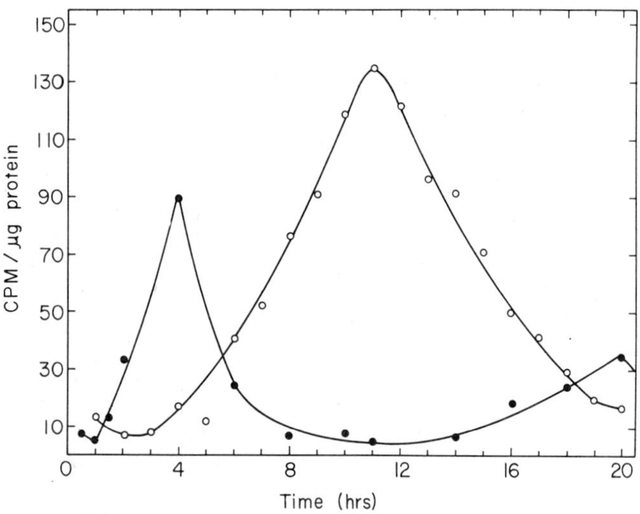

Figure 13. The effect of butyrate on CMP-NAN:lactosylceramide sialyltransferase activity in HeLa cells. (○—○) Release from butyrate; (●—●) release from thymidine block.

Occurrence and inducibility

Glycosidases are widely distributed in microorganisms, animals and plants. Neuraminidase, for example, was first discovered in a virus (134) and later found to occur in a variety of microorganisms, avian and mammalian tissues. The induction of specific bacterial glycosidases in the presence of the corresponding substrate is well documented. For example, α-galactosidase and β-galactosidase are not constitutive but are inducible in *E. coli* (135). With regard to tissue and cellular distribution, the glycosidases occur widely in organs (brain, kidney, liver and spleen), bone marrow, plasma, spermatozoa, platelets and erythrocytes. Although some glycosidases have been found in mitochondria (136), plasma membranes (137) and chloroplasts (138), the lysosome is the only organelle that contains the full complement of hydrolase activities necessary for the complete degradation of glycolipids and glycoproteins.

Purification and physical properties

Virtually all glycosidases purified to date are glycoproteins; several kinds of purification procedures are available that are based on affinity methods. In recent years, immobilized substrate analogs have been used to facilitate the purification of several glycosidases (139). Their purification can also be accomplished by the use of immobilized plant lectins specific for certain carbohydrate units on the enzymes. Concanavalan A-Sepharose, for example, has been used for the isolation of glycosidases such as α-N-acetylgalactosaminidase (140) and the β-N-acetylhexosaminidases. Variation in the composition of the carbohydrate moiety leads to microheterogeneity, for example molecular weight and charge, in an otherwise homogeneous glycosidase preparation. Variation in sialic acid content has often been cited as a cause of multiple isoelectric forms of these enzymes. Multiple forms of L-fucosidase , for example, have been shown (141) to be convertible to the neutral forms by neuraminidase treatment. An excellent discussion on the issue of homogeneity of glycoproteins has appeared elsewhere (142). Another problem of the multiple forms, which we wish to emphasize, has to do with substrate specificity. Often, the presence of a contaminating glycosidase cannot be readily detected unless the incubation time is sufficiently long. This is especially important to realize when glycosidases are used for structural eludication of glycoconjugates. Li and Li, for example, have shown (34) in studies of the core glycopeptide of ovalbumin that the liberation of one of the mannosyl units was due to the presence of 0.06% β-mannosidase in an apparently homogeneous α-mannosidase preparation.

Substrate specificity

The substrate specificities of glycosidases are much more complicated than their names imply. Figure 15 shows the structures of several glycosphingolipids (globopentaglycosylceramide, II^3,IV^3-α,α-di-N-acetylneuraminosyl-gangliotetraglycosylceramide, and fucolipid A^a) and the

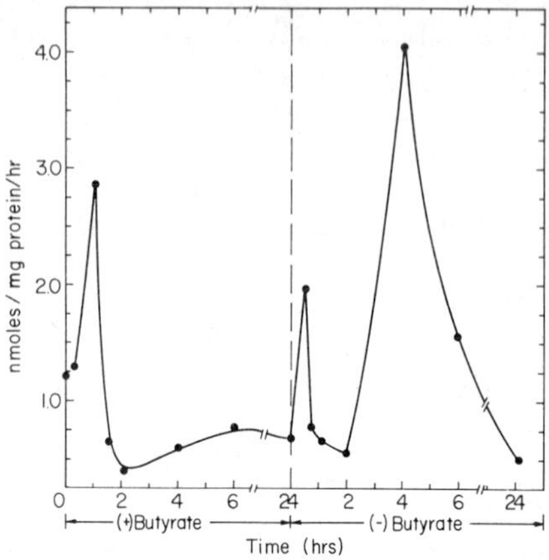

Figure 14. The effect of butyrate on UDP-gal:lactosylceramide α-galactosyltransferase activity in KB cells

Globopentaglycosylceramide

A B C D E
GalNAc(α1→3)GalNAc(β1→3)Gal(α1→4)Gal(β1→4)Glc(β1→1')Cer

II^3,IV^3-α,α-Di-N-acetylneuraminosyl-gangliotetraglycosylceramide

G H I K
Gal(β1→3)GalNAc(β1→4)Gal(β1→4)Glc(β1→1')Cer
|(α2→3) F |(α2→3) J
NeuAc NeuAc

Fucolipid A[a]

L N O Q R
GalNAc(α1→3)Gal(β1→4)GlcNAc(β1→3)Gal(β1→4)Glc(β1→1')Cer
|(α1→2) M |(α1→3) P
Fuc Fuc

LINKAGE	ENZYME
A, L	α-N-Acetylgalactosaminidases
B, H, O	β-Hexosaminidases
C	α-Galactosidases
D, G, I, N, Q	β-Galactosidases
E = K = R	β-Glucosidases
F, J	Neuraminidases
M, P	α-Fucosidases

Figure 15. Structures of three glycosphingolipids and the exoglycosidases which catalyze the sequential hydrolysis of glycolipid glycosidic linkages (A to R)

corresponding exoglycosidases involved in their metabolism. Several points regarding the mode of action of these enzymes follow.

Glycosidases from different sources often have different specificities. Several kinds of neuraminidase have been isolated, each of which has unique properties. Degradation of G_{D1a} begins with the removal of the sialic acid linked to the galactose at the non-reducing end (Figure 15, linkage F), using neuraminidase from *Vibrio cholerae*. The other sialic acid of G_{D1a} is resistent to this enzyme (as is G_{M1}, the product) (143). Removal of the vicinal N-acetylgalactosamine unit from G_{M1} is necessary before this neuraminidase can hydrolyse the ketosidic linkage shown in (Figure 15, linkage J). It has been shown, however, that neuraminidase from *Clostridium perfringens* (144) can remove this sialic acid (linkage J) from G_{M1} directly; thus the specificity of a given enzyme often varies depending upon its source.

While each glycosidase has an absolute specificity toward the anomeric linkage, absolute epimeric specificity is not always observed. This is shown by comparison of α-N-acetylgalactosaminidase and β-N-acetylhexosaminidase. α-N-acetylgalactosaminidase lacks specificity for the C-2 substituent as well as its configuration on the carbohydrate position at the C-2 (145,146). As shown in Figure 16, the enzyme can hydrolyze all three substrates (at different rates) which differ in the substituent and/or absolute configuration (shaded portions) at C-2. The enzyme, however, does show absolute specificity at the C-4 position. In contrast, β-N-acetylhexosaminidase has an absolute requirement for an equatorial acetamido moiety at C-2 but the configuration at C-4 is not important, as shown in Figure 17. Moreover, α-galactosidase A has been shown (147) to have an absolute requirement at both C-2 and C-4 positions and is thus distinguished from the B form of α-galactosidase (15,16). Dean *et al.* (147) and Schram *et al.* (148) have shown that α-galactosidase B is an α-N-acetylgalactosaminidase.

Multiple forms of a given enzyme may exist which differ in their ability to hydrolyze glycolipids but not artificial water soluble substrates. Two forms of human β-galactosidase have been shown to hydrolyze the artificial substrate, 4-methylumbelliferyl-β-D-galactopyranoside (149,-150). One of these forms (G_{M1}-ganglioside:β-galactosidase) was more specific for G_{M1}- and asialo G_{M1}-ganglioside, while the other form (galactosylceramidase) was more specific for galactosyl- and lactosylceramide. Similarly, two forms of β-N-acetylhexosaminidase have been found which catalyze the hydrolysis of globotetraglycosylceramide and 4-methylumbelliferyl-β-N-acetylhexosaminide. Only one (β-N-acetylhexosaminidase A) of these forms was able to hydrolyze II^3-N-acetylneuraminosyl-gangliotriglycosylceramide. Two forms of human β-glucosidase have been identified. Both hydrolyze 4-methylumbelliferyl-β-D-glucopyranoside but only glucocerebroside:β-glucosidase is capable of hydrolyzing glucocerebroside (151,152). This form of the enzyme appears to be lysosomal in origin.

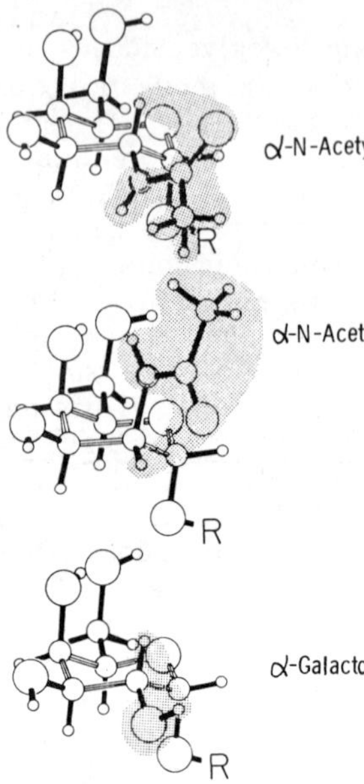

Journal of Cell Biology

Figure 16. Three-dimensional projections of terminal carbohydrate residues that bind to α*-*N*-acetylgalactosaminidase.* R *represents the remaining portion of the sphingolipid. Substituents that the enzyme cannot distinguish are shaded* (109).

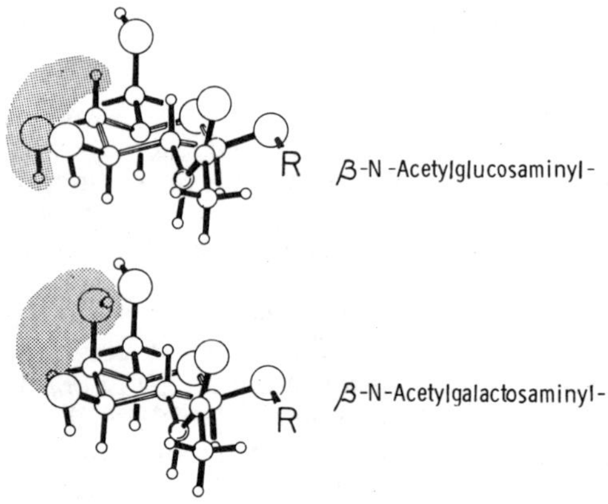

Journal of Cell Biology

Figure 17. Three-dimensional projections of terminal carbohydrate residues which bind to β*-*N*-acetylhexosaminidases.* R *represents the remaining portion of the sphingolipid. Substituents that the enzyme cannot distinguish are shaded* (109).

Lipid-protein interactions

Detergents are essential for the dispersion of most glycolipid substrates in glycosidase assay mixtures. In addition, it has been shown that detergents can stimulate or inhibit glycosidases directly. Presumably, detergents mimic the effect of naturally occurring membrane amphiphilic lipids.

β-Glucocerebrosidase:β-glucosidase, for example, has been shown to be activated by bile salts (151,153,154) such as deoxycholate, glycocholate and taurocholate; acidic phospholipids such as phosphatidylserine, phosphatidylinositol and phosphatidic acid are also active (152,155). Interestingly, the water-soluble β-glucosidase, which does not have activity toward lipid substrates, is inhibited by these amphiphiles. On the other hand, the β-glucocerebrosidase:β-glucosidase, which hydrolyzes lipid substrates was inhibited by choline-containing lipids. Ho (156,157) has proposed a glucocerebrosidase enzyme system composed of glucocerebrosidase (factor C), a glycoprotein activator (factor P), and an acidic phospholipid. The acidic phospholipid appears to be necessary for the association of the activator and the enzyme. The activator-enzyme complex hydrolyzes artificial water-soluble substrates and GlcCer. Sodium taurocholate can substitute for the activator.

Recently, Hechtman (158) has reported the isolation of an activator required for the hydrolysis of II^3-α-N-acetylneuraminosyl-gangliotriglycosylceramide by β-hexosaminidase A. This activator does not affect the rate of hydrolysis of synthetic substrates or gangliotriglycosylceramide.

Modulation of other glycosidase activities has also been reported. Li and Li (159) have isolated an activator which stimulates the hydrolysis of a number of glycolipids. However, this activator may be interacting with substrate rather than enzyme, judging from the fact that a 1:1 molar ratio of activator and substrate is required.

Synthetic lipoidal activators have also been reported (160). Figure 18 shows the structures of some synthetic fatty acyl amides which resemble fatty acyl sphingosine. The hydrolysis of galactosylceramide by β-galactosidase was stimulated by fatty acyl derivatives of 2-amino-2-methyl-1-propanol, with the N-decanoyl derivative being the most active. In general, the V_{max} but not the K_m was changed. Omission of the branched methyl group resulted in inhibition instead of stimulation.

The reports cited above suggest that protein activators stimulate the activity of glycosidases toward glycolipid substrates but do not modify the enzymatic activity toward water-soluble substrates. Our recent studies on the interaction of sodium taurocholate with α-N-acetylgalactosaminidase from canine liver (146,161) indicate that amphiphiles may stimulate the hydrolytic activities of glycosidases toward both artificial water-soluble substrates and glycolipid substrates. Low concentrations of sodium taurocholate stimulate the activity of the enzyme toward the artificial substrate PNP-α-GalNAc. However, as the concentration of taurocholate is increased, inhibition results. A Lineweaver-Burk plot indicates that sodium taurocholate, at all concentrations, lowers the V_{max} and the K_m for the hydrolysis of the artificial substrate. The percent maximal inhibition levels off at the critical micelle concentration of

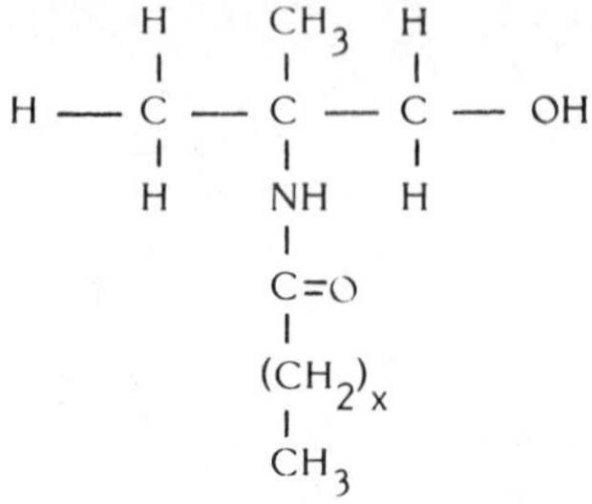

X 6, 8, 10, 12, 14, 16

Figure 18. Fatty acid amides which affect hydrolytic activity of rat brain β-galactosidase toward galactosylceramide. The N-decanoyl derivative was the most effective (160).

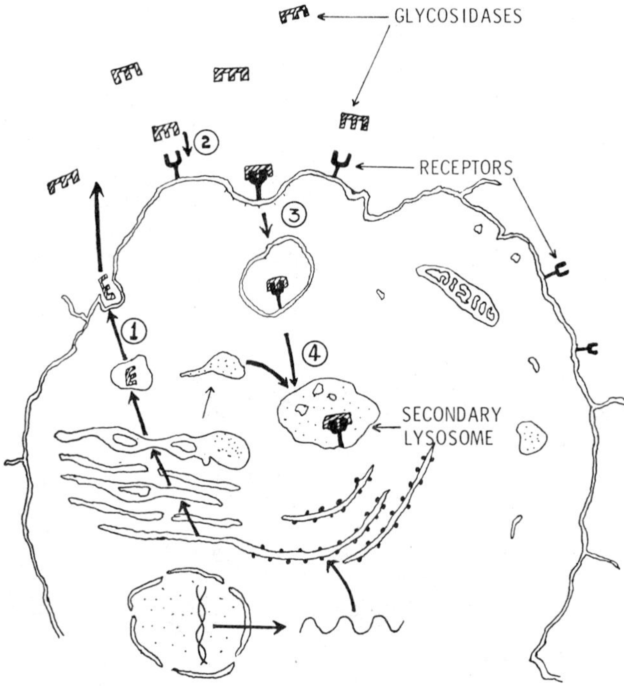

① RELEASE OF GLYCOSIDASES
② RECOGNITION OF ENZYME BY RECEPTOR
③ UPTAKE OF ENZYME BY PINOCYTOSIS
④ FUSION OF PINOCYTIC VESICLE WITH PRIMARY LYSOSOME

Figure 19. Pictorial model of exocytosis and endocytosis of lysosomal glycosidases

taurocholate (7 mM). Moreover, sodium taurocholate at very low concentrations significantly quenches the intrinsic fluorescence of the enzyme. Our data (161) indicate that the enzyme activity is modulated by the interaction with the monomeric but not the micellar form of taurocholate. These studies may shed light on the mode of action of glycosidases in their natural membrane environment.

Physiological significance of glycosidases

Besides the well publicized lipidoses resulting from glycosidase deficiencies, attention has shifted toward the release and uptake of these enzymes from cells. An abundance of evidence indicates that tumors, studied *in vivo*, release large quantities of lysosomal enzymes into their environment (162-164). Presumably this release of degradative enzymes is associated with tumor invasiveness and maintainance of the neoplastic state. Other studies (dealing with I-cell disease) have shown that specific recognition molecules on cell surfaces are responsible for the uptake of lysosomal glycosidases. A diagram of these processes is depicted in Figure 19. The uptake depends upon both the carbohydrate moiety of the glycosidases (165,166) and the presence of specific receptors on the plasma membrane (167,168). Clearly, more detailed studies are needed, results of which should be significant in the design of enzyme replacement therapy.

Acknowledgements

The authors would like to especially acknowledge the secretarial work of Ms. Dorothy Byrne. Mr. Kenneth J. Dean and Dr. S.-S.J. Sung are also gratefully acknowledged for providing data dealing with α-N-acetylgalactosaminidase and α-galactosidase. This work was supported by NIH research grants (AM 12434 and HL 17462).

Literature Cited

1. Lysosomes and Storage Diseases, ed. H.G. Hers and F. van Hoof, Academic Press, New York, 1973.
2. The Metabolic Basis of Inherited Disease, 4th Ed., ed. J.B. Stanbury, J.B. Wyngaarden and D.S. Fredrickson, McGraw-Hill, New York, 1978.
3. Practical Enzymology of the Sphingolipidoses, ed. R.H. Glew and S.P. Peters, Alan R. Liss, Inc., New York, 1977.
4. Glycolipid Methodology, ed. L.A. Witting, Amer. Oil Chemists' Society, Champaigne, 1976.
5. Karlsson, K.A., in Chemistry and Metabolism of Sphingolipids, ed. C.C. Sweeley, North-Holland Publishing Co., Amsterdam, 1970, pp. 6-43.
6. Sweeley, C.C. and Siddiqui, B., in The Glycoconjugates. Vol. I. Mammalian Glycoproteins and Glycolipids, ed. M.I. Horowitz and W. Pigman, Academic Press, New York, 1977, pp. 459-540.
7. Sweeley, C.C. and Moscatelli, E.A., J. Lipid Res., 1, 40 (1959).

8. Gaver, R.C. and Sweeley, C.C., J. Am. Oil Chem. Soc., 42, 294 (1965).
9. Karlsson, K.A., Nature, 188, 312 (1960).
10. Kawamura, N. and Taketomi, T., J. Biochem. (Tokyo), 81, 1217 (1977).
11. Martensson, E., Biochim. Biophys. Acta, 116, 296 (1966).
12. Norton, W.T. and Brotz, M., Biochem. Biophys. Res. Commun., 12, 198 (1963).
13. Kornblatt, M.J., Schachter, H. and Murray, R.K., Biochem. Biophys. Res. Commun., 48, 1489 (1972).
14. Ishizuka, I., Suzuki, M. and Yamakawa, T., J. Biochem. (Tokyo), 73, 77 (1973).
15. Slomiany, B.L., Slomiany, A. and Glass, G.B.J., Eur. J. Biochem., 78, 33 (1977).
16. IUPAC-IUB Comission on Biochemical Nomenclature, Lipids, 12, 455 (1977).
17. Burton, R.M., in Glycolipid Methodology, ed. L.A. Witting, Amer. Oil Chemists' Society, Champaign, 1976, pp. 1-11.
18. Macher, B.A. and Sweeley, C.C., in Methods in Enzymology, ed. V. Ginsburg, Academic Press, New York, in press.
19. Iwamori, M., Moser, H.W., McCluer, R.H. and Kishimoto, Y., Biochim. Biophys. Acta, 380, 308 (1975).
20. McCluer, R.H. and Evans, J.E., J. Lipid Res., 17, 412 (1976).
21. Suzuki, A., Handa, S. and Yamakawa, T., J. Biochem. (Tokyo), 80, 1181 (1976).
22. Suzuki, A., Handa, S. and Yamakawa, T., Abstr. Japan-U.S. Seminar on Structure and Functional Significance of Membrane Glycolipids, Honolulu, October 5, 1977.
23. Hakomori, S. and Watanabe, K., in Glycolipid Methodology, ed. L.A. Witting, Amer. Oil Chemists' Soc., Champaign, 1976, pp. 13-47.
24. Iwamori, M. and Nagai, Y., J. Neurochem., in press.
25. Iwamori, M. and Nagai, Y., Proc. Japanese J. Biochem. Lipids, 19, 257 (1977).
26. Ando, S. and Nagai, Y., Biochim. Biophys. Acta, 424, 98 (1976).
27. Momoi, T., Ando, S. and Nagai, Y., Biochim. Biophys. Acta, 441, 488 (1976).
28. Ueno, K., Ando, S. and Yu, R.K., J. Lipid Res., in press.
29. Ohashi, M. and Yamakawa, T., J. Biochem. (Tokyo), 81, 1675 (1977).
30. Vance, D.E. and Sweeley, C.C., J. Lipid Res., 8, 621 (1967).
31. Chambers, R.E. and Clamp, J.R., Biochem. J., 125, 1009 (1971).
32. Wong, C.G., Sung, S.-S.J. and Sweeley, C.C., in Methods in Carbohydrate Chem., ed. R. Whistler, Academic Press, in press.
33. Yu, R. and Ledeen, R.W., J. Lipid Res., 11, 506 (1970).
34. Li, Y.T. and Li, S.C., in The Glycoconjugates. Vol. I. Mammalian Glycoproteins and Glycolipids, ed. M.I. Horowitz and W. Pigman, Academic Press, 1977, pp. 52-68.
35. Svennerholm, L., Mansson, J.E. and Li, Y.T., J. Biol. Chem., 248, 740 (1973).
36. Slomiany, A. and Slomiany, B.L., Eur. J. Biochem., 76, 491 (1977).

37. Slomiany, A., Slomiany, B.L. and Horowitz, M.I., in Glycolipid Methodology, ed. L.A. Witting, Amer. Oil Chemists' Soc., Champaign, 1976, pp. 49-75.
38. Hakomori, S., J. Biochem. (Tokyo), 55, 205 (1964).
39. Karlsson, K.A., Biochemistry, 13, 3643 (1973).
40. Karlsson, K.A., Pascher, I. and Samuelsson, B.E., Chem. Phys. Lipids, 12, 271 (1974).
41. Smith, E.L., McKibbin, J.M., Karlsson, K.A., Pascher, I. and Samuelsson, B.E., Biochim. Biophys. Acta, 388, 171 (1975).
42. Holm. M., Pascher, I. and Samuelsson, B.E., Biomed. Mass Spect., 4, 77 (1977).
43. Ledeen, R.W., Kundu, S.K., Price, H.C. and Fong, J.W., Chem. Phys. Lipids, 13, 429 (1974).
44. Hanfland, P. and Egge, H., Chem. Phys. Lipids, 16, 201 (1976).
45. Karlsson, K.A., in Glycolipid Methodology, ed. L.A. Witting, Amer. Oil Chemists' Soc., Champaign, 1976, pp. 97-122.
46. Wiegandt, H. and Baschang, G., Z. Naturforsch., Teil B20, 164 (1965).
47. Hakomori, S., J. Lipid Res., 7, 789 (1966).
48. Kundo, S.K., Ledeen, R.W. and Gorin, P.A.J., Carbohyd. Res. 39, 179 (1975).
49. Bjorndal, H., Lindberg, B. and Svensson, S., Acta Chem. Scand., 21, 1801 (1967).
50. Bjorndal, H., Lindberg, B. and Svensson, S., Carbohyd. Res., 5, 433 (1967).
51. Bjorndal, H., Hellerqvist, C.G., Lindberg, B. and Svensson, S., Angen. Chem. Int. Ed., 9, 610 (1970).
52. Bjorndal, H., Lindberg, B., Pilotti, A. and Svensson, S., Carbohyd. Res., 15, 339 (1970).
53. Sung, S.-J.J., Esselman, W.J. and Sweeley, C.C., J. Biol. Chem., 248, 6528 (1973).
54. Stellner, K., Saito, H. and Hakomori, S., Arch. Biochem. Biophys., 155, 464 (1973).
55. Jansson, P.E., Kenne, L., Liedgren, H., Lindberg, B. and Lonngren, J., Chem. Commun. (Univ. of Stockholm), 1976 (No. 8), pp. 1-75.
56. Sillerud, L.O., Prestegard, J.H., Yu, R.K., Schafer, D.E. and Konigsberg, W.H., Biochemistry, in press.
57. Braun, P.E. and Snell, E.E., J. Biol. Chem., 243, 3775 (1968).
58. Stoffel, W., LeKim, D. and Sticht, G., Z. Physiol. Chem., 349, 1637 (1968).
59. Scribney, M., Biochim. Biophys. Acta, 125, 542 (1966).
60. Basu, S., Kaufman, B. and Roseman, S., J. Biol. Chem., 243, 5802 (1968).
61. Basu, S., Kaufman, B. and Roseman, S., J. Biol. Chem., 248, 1388 (1973).
62. Cleland, W.W. and Kennedy, E.P., J. Biol. Chem., 235, 45 (1960).
63. Morell, P. and Radin, N.S., Biochemistry, 8, 506 (1969).
64. Basu, S., Schultz, A., Basu, M. and Roseman, S., J. Biol. Chem., 246, 4272 (1971).
65. Hildebrand, J. and Hauser, G., J. Biol. Chem., 244, 5170 (1969).

66. Basu, S., Ph.D. Thesis, Univ. of Michigan, Ann Arbor, Michigan, 1966.
67. Basu, S. and Kaufman, B., Fed. Proc., 24, 479 (1965).
68. Kaufman, B., Basu, S. and Roseman, S., Methods Enzymol., 8, 365 (1966).
69. Basu, M., Chien, J.L. and Basu, S., Biochem. Biophys. Res. Commun., 60, 1097 (1974).
70. Kijimoto, S., Ishibashi, T. and Makita, A., Biochem. Biophys. Res. Commun., 56, 177 (1974).
71. Basu, S., Kaufman, B. and Roseman, S., J. Biol. Chem., 240, PC4115 (1965).
72. Steigerwald, J.C., Basu, S., Kaufman, B. and Roseman, S., J. Biol. Chem., 250, 6727 (1975).
73. Kaufman, B., Basu, S. and Roseman, S., Proc. of the Third Int. Symposium on the Cerebral Sphingolipidoses, ed. by S.M. Aronson and B.W. Volk, Pergamon Press, New York, 1967, p. 193.
74. Kaufman, B., Basu, S. and Roseman, S., J. Biol. Chem., 243, 5804 (1968).
75. Roseman, S., Chem. Phys. Lipids, 5, 270 (1970).
76. Yip, M.C.M. and Dain, J.A., Biochem. J., 118, 247 (1970).
77. Yip, M.C.M. and Dain, J.A., Biochim. Biophys. Acta, 206, 252 (1970).
78. Hildebrand, J., Stoffyn, P. and Hauser, G., J. Neurochem., 17, 403 (1970).
79. Arce, A., Maccioni, H.J. and Caputto, R., Fed. Proc., 29, 925 (1970).
80. Moskal, J.R., Gardner, D.A. and Basu, S., Biochem. Biophys. Res. Commun., 61, 751 (1974).
81. Kempf, S.F. and Stoolmiller, A.C., J. Biol. Chem., 251, 7626 (1976).
82. Chien, J.L., Williams, T. and Basu, S., J. Biol. Chem., 248, 1778 (1973).
83. Ishibashi, T., Kijimoto, S., and Makita, A., Biochim. Biophys. Acta, 337, 92 (1974).
84. Yeung, K.K., Moskal, J.R., Chien, J.L., Gardner, D.A. and Basu, S., Biochem. Biophys. Res. Commun., 59, 252 (1974).
85. Dawson, G. and Sweeley, CC., J. Biol. Chem., 247, 410 (1972).
86. Basu, M. and Basu, S., J. Biol. Chem., 247, 1489 (1972).
87. Basu, M. and Basu, S., J. Biol. Chem., 248, 1700 (1973).
88. Basu, S., Basu, M. and Chien, J.L., J. Biol. Chem., 250, 2956 (1975).
89. Basu, S., Basu, M., Moskal, J.R., Chien, J.L. and Gardner, D.A., in Glycolipid Methodology, ed. L.A. Whittig, 1976, p. 123-139.
90. Presper, K.A., Basu, M. and Basu, S., Fed. Proc., 35, 1441 (1976).
91. Pacuszka, T. and Koscielak, J., Eur. J. Biochem., 64, 499 (1976).
92. Presper, K.A., Basu, M. and Basu, S., Proc. Natl. Acad. Sci., U.S.A., 75, 789 (1978).
93. Watanabe, K., Laine, R.A. and Hakomori, S., Biochemistry, 14, 2725 (1975).
94. Slomiany, B.L. and Slomiany, A., Biochim. Biophys. Acta, 486, 531 (1977).
95. Gardas, A. and Koscielak, J., FEBS Lett., 42, 101 (1974).

96. Fishman, P.H. and Brady, R.O., in Biological Roles of Sialic Acid ed. A. Rosenberg and C.L. Schengrund, Plenum Publishing Corp., New York, 1976.
97. Schwyzer, M. and Hill, R.L., J. Biol. Chem., 252, 2346 (1977).
98. Letoublon, R., Richard, M., Louisot, R. and Got, R., Eur. J. Biochem., 18, 194 (1971).
99. Ko, G.K.W. and Raghupathy, E., Biochim. Biophys. Acta, 313, 277 (1973).
100. Eylar, E.H. and Cook, G.M.W., Proc. Natl. Acad. Sci. U.S., 54, 1678 (1965).
101. Testas, M., Chao, H. and Molnar, J., Arch. Biochem. Biophys., 138, 135 (1970).
102. Ishibashi, T., Atsuta, T. and Makita, A., Biochem. Biophys. Acta, 429, 759 (1976).
103. Bell, J.E., Beyer, T.A. and Hill, R.L., J. Biol. Chem., 251, 3003 (1976).
104. Gray, G.M., Biochim. Biophys. Acta, 239, 494 (1971).
105. Basu, S., Moskal, J.R. and Gardner, D.A., in Ganglioside Function: Biochemical and Pharmacological Implications, ed. Porcellati, G., Plenum Publishing Corp., New York, 1976, pp. 45-63.
106. Wolf, A. and Robbins, P.W., J. Cell Biol., 61, 676 (1974).
107. Chatterjee, S., Sweeley, C.C. and Velicer, L.F., Biochem. Biophys. Res. Commun., 54, 585 (1973).
108. Macher, B.A., Lockney, M., Moskal, J.R., Fung, Y-K. and Sweeley, C.C., Expt. Cell Res., in preparation.
109. Simmons, J.L., Fishman, P.H., Freese, E. and Brady, R.O., J. Cell Biol., 66, 414 (1975).
110. Shur, B.D. and Roth, S., Biochim. Biophys. Acta, 415, 473 (1975).
111. Keenan, T.W. and Morre, D.J., FEBS Lett., 55, 7 (1975).
112. Roth, S., McGuire, E.J. and Roseman, S., J. Cell. Biol., 51, 536 (1971).
113. Datta, P., Biochemistry, 13, 3987 (1974).
114. Patt, L.M. and Grimes, W.J., J. Biol. Chem., 249, 4157 (1974).
115. Weiser, M.M., J. Biol. Chem., 248, 2542 (1973).
116. Podolsky, D.K., Weiser, M.M., LaMont, J.T. and Isselbacher, K.J., Proc. Natl. Acad. Sci., U.S.A., 71, 904 (1974).
117. Roth, S. and White, D., Proc. Natl. Acad. Sci., U.S.A., 69, 485 (1972).
118. Franke, W.W., Morre, D.J., Deumling, B., Cheetham, R.D., Kartenbeck, J., Jarasch, E. and Zentgraf, H., Z. Naturforsch., 26b, 1031 (1971).
119. Morre, D.J., Keenan, T.W. and Huang, C.M., in Advances in Cytopharmacology ed. B. Cecarelli, F. Clementi and J. Meldolesi, Vol. 2, Raven Press, New York, 1974, pp. 107-126.
120. Bennett, G., LeBlond, C.P. and Haddad, A., J. Cell Biol., 60, 258 (1974).
121. Schacter, H., Jabbal, I., Hudgin, R.L., Pinteric, L., McGuire, E.J. and Roseman, S., J. Biol. Chem., 245, 1090 (1970).
122. Morre, D.J., Merlin, L.M. and Keenan, T.W., Biochem. Biophys. Res. Commun., 37, 813 (1969).
123. Brodbeck, U. and Ebner, K., J. Biol. Chem., 241, 762 (1966).

124. Keenan, T.W., Morre, D.J. and Basu, S., J. Biol. Chem., 249, 310 (1974).
125. Mahley, R.W., Bennett, B.D., Morre, D.J., Gray, M.E., Thistlewaithe, W. and LeQuire, V.S., Lab Invest., 25, 435 (1971).
126. Ronzio, R.A., Arch. Biochem. Biophys., 159, 777 (1973).
127. Chabaud, O., Bouchilloux, S., Ronin, C. and Ferrand, M., Biochimie, 56, 119 (1974).
128. LaMont, J.T., Perrotto, J.L., Weiser, M.M. and Isselbacher, K.J., Proc. Natl. Acad. Sci., U.S.A., 71, 3726 (1974).
129. Podolsky, D.K. and Weiser, M.M., Biochem. J., 146, 213 (1975).
130. Bosmann, H.B., Biochem. Biophys. Res. Commun., 48, 523 (1972).
131. Bosmann, H.B., Beiber, G.F., Brown, A.E., Case, K.R., Gersten, D.M., Klimmerer, T.W. and Lione, A., Nature, 246, 487 (1973).
132. McClean, R.J. and Bosmann, H.B., Proc. Natl. Acad. Sci., U.S.A., 72, 310 (1975).
133. Shur, B. and Roth, S.,Am. Zool., 13, 1129 (1973).
134. Burnet, F.M., McCrea, J.F. and Stone, J.D., Brit. J. Exptl. Pathol., 27, 228(1946).
135. Koppel, J.L., Porter, C.J. and Crocker, B.F., J. Gen. Physiol., 36, 703 (1953).
136. Subba, Rao, K. and Pieringer, R.A., J. Neurochem., 17, 483 (1970).
137. Fleischer, B. and Fleischer, S., Biochim. Biophys. Acta, 183, 265 (1969).
138. Gatt, S. and Baker, E.A., Biochim. Biophys. Acta, 185, 402 (1969).
139. Opheim, D.J. and Touster, O., J. Biol. Chem., 252, 739 (1977).
140. Sung, S.-S.J. and Sweeley, C.C., in Current Trends in Sphingolipidoses and Allied Disorders, Vol. 68, ed. B.W. Volk and L. Schneck, 1976, pp. 323-337.
141. Alhadeff, J.A. and O'Brien, J.S. in Practical Enzymology of the Sphingolipidoses, ed. R.H. Glew, S.P. Peters, A.R. Liss Inc., New York, 1977, pp. 247-281.
142. Martin, I. and Horowitz, M.I., The Glycoconjugates, Vol. 1, 1977, Academic Press ed. M.I. Horowitz and W. Pigman, New York, p. 15-34.
143. Kuhn R. and Wiegandt, H., Chem. Ber., 89, 2013 (1963).
144. Rauvala, H., FEBS Letts., 65, 229 (1976).
145. Dean, K.J., Sung, S.-S.J. and Sweeley, C.C., in Enzymes of Lipid Metabolism, eds. P. Mandel, L. Freyst and S. Gatt, Plenum Publishing Corp., New York, in press.
146. Fung, Y.-K. and Sweeley, C.C., Fed. Proc., in press.
147. Dean, K.J., Sung, S.-S.J. and Sweeley, C.C., Biochem. Biophys. Res. Commun., 77, 1411 (1977).
148. Scham, A., Hamers, M.N. and Tager, J.M., Biochim. Biophys. Acta, 482, 138 (1977).
149. Suzuki, K., in Practical Enzymology of the Sphingolipidoses, eds. R.H. Glew and S.P. Peters, Alan R. Liss, Inc., New York, 1977, pp. 102-132.
150. Tanaka, H., and Suzuki, K., Brain Res., 122, 325 (1977).
151. Ho, M.W., Biochem. J., 136, 721 (1973).
152. Peters, S.P., Coyle, P. and Glew, R.H., Arch. Biochem. Biophys. 175, 589 (1976).
153. Gatt, S., Biochem., J., 101, 687 (1966).

154. Gatt, S. and Rapport, M.M., Biochim. Biophys. Acta, 113, 567 (1966).
155. Blonder, E., Klibansky, C. and DeFries, A., Biochim. Biophys. Acta 431, 45 (1976).
156. Ho, M.W. and Light, D., Biochem. J., 136, 821 (1973).
157. Ho, M.W., in Enzyme Therapy in Lysosomal Storage Disease eds. J.M. Tager, G.J.M. Hooghwinkel and W. Th Daems, Amsterdam, North Holland, 1974, pp. 239-246.
158. Hechtman, P., Can. J. Biochem., 55, 315 (1977).
159. Li, S. and Li, Y.-T., J. Biol. Chem., 251, 1159 (1976).
160. Arora, R.C. and Radin, N.S., Lipids, 7, 56 (1971).
161. Fung, Y.-K. and Sweeley, C.C., in preparation.
162. Bosmann, H.B. and Hall, T.C., Proc. Natl. Acad. Sci., U.S.A., , 71, 1833 (1974).
163. Sylven, B., Eur. J. Cancer, 4, 463 (1968).
164. Koono, M., Ushijima, K. and Hayashi, H., Int. J. Cancer, 13, 105 (1974).
165. Stahl, P., Schlesinger, P.H., Rodman, J.S. and Doebber, T., Nature 264, 86 (1976).
166. Kaplan, A., Fischer, D. and Sly, W.S., J. Biol. Chem., 253, 647 (1978).
167. Hickman, S., Shapiro, L.J. and Neufeld, E.F., Biochem. Biophys. Res. Commun., 57, 55 (1974).
168. Kawasaki, T. and Ashwell, G., J. Biol. Chem., 252, 6536 (1977).
169. Carter, H.E., Strobach, D.R. and Hawthorne, J.N., Biochemistry, 8, 383 (1969).
170. Kaul, K. and Lester, R.L., Plant Physiol., 55, 120 (1975).
171. Hsieh, T.C.Y., Kaul, K., Laine, R.A. and R.L. Lester (private communication).

RECEIVED March 15, 1978.

4

Endo-β-acetylglucosaminidases—Their Metabolic Role in Disease Processes and Their Use in the Study of Glycoprotein Structure and Function

FRANK MALEY, ANTHONY L. TARENTINO, and ROBERT B. TRIMBLE

New York State Department of Health, Division of Laboratories and Research, Albany, NY 12201

In the course of studies on the core structure of glycopeptides from RNase B and ovalbumin, we discovered an enzyme in chitinase extracts (1, 2) capable of hydrolyzing the di-N-acetyl chitobiosyl linkage of oligosaccharides to yield the following products: R-GlcNAc + GlcNAc$(Man)_x(GlcNAc)_y$ where R = Asn; Asn-peptide, Asn-protein, and x = 5 or greater, while y = 0 to 4. This enzyme was called endo-β-N-acetylglucosaminidase-H (endo-H) because its specificity appeared to be directed towards substrates composed of at least five mannose residues.

Further investigation of these initial crude preparations from *Streptomyces plicatus* revealed them to contain two enzymes, one with a specificity for long chain mannosyl oligosaccharides, and the other with a primary specificity for short chain oligosaccharides such as $(Man)_1(GlcNAc)_2$Asn (2). The latter enzyme, which can be separated from endo-H by gel filtration (Figure 1), has been designated endo-L and will be discussed later. We realize now that it was the presence of this enzyme in our early endo-H preparations that enabled us to affect the isolation of Manβ1→4-GlcNAc and to clearly demonstrate for the first time that the core sequence of most Asn-associated glycopeptides is composed of this disaccharide (1, 3, 4). With the former enzyme, designated endo-H, it was shown (1, 5) that the ovalbumin oligosaccharide chain is linked to the distal GlcNAc of the di-N-acetyl chitobiosyl unit and not to the primary GlcNAc as originally believed (6). These high mannosyl or simple oligosaccharides, such as those associated with ovalbumin, ribonuclease B, deoxyribonuclease, carboxypeptidase Y, mungbean nuclease, invertase, immunoglobulin M, and thyroglobulin, were shown by us (6, 7) to be substrates for endo-H and stand in contrast to the resistant "complex" oligosaccharides associated with most immunoglobulins (8, 9, 10) and viral glycoproteins (11) which are composed of a core sequence of

```
Manα1
     ↘6
       Manβ1→4GlcNAcβ1→4GlcNAc-Asn
     ↗3
Manα1
```

0-8412-0452-7/78/47-080-086$05.00/0

where fucose is usually linked to the first GlcNAc.

The manner in which this unit is synthesized remains to be determined, but it appears to result from the processing of larger molecular weight oligosaccharide chains that are rich in mannose (12, 13). It has become apparent that enzymes such as endo-H and a related endoglycosidase from Diplococcus pneumoniae, endo-D (14), will provide valuable assistance in assessing both the structure and function of glycoproteins, their mechanism of biosynthesis and their potential role in disease processes. But, because of the increasing indiscriminate use of these enzymes without regard to their specificity of action and their attendant contaminants, erroneous conclusions may result. It is for this reason that I would first like to review some of the properties of these enzymes which are often overlooked, and to include some of our recent findings which would appear to extend the range of activity of endo-H.

As shown in Figure 2, which compares the specificity of endo-H relative to endo-D, it is clear that endo-H hydrolyzes longer mannosyl oligosaccharides than endo-D, an enzyme which has a much greater specificity for the branched complex oligosaccharide core $(Man)_3(GlcNAc)_2Asn$ (15). In contrast to endo-D, endo-H can effectively hydrolyze oligosaccharide chains with as many as 50 or more mannosyl residues (6) and even the complex core region, but at a rate that is several orders of magnitude lower than that affected by endo-D, providing fucose is not present on the proximal N-acetylglucosamine of the core region (Figure 3). As indicated, the presence of fucose completely impairs endo-H's already low activity against the complex core oligosaccharide, but has no effect on endo-D activity.

Because of endo-H's capacity to relieve specific glycoproteins of their oligosaccharide units, it can be a valuable asset in estimating molecular weights of glycoproteins by SDS-acrylamide gel electrophoresis, a technique that often yields inaccurate results with glycoproteins. Thus, by carefully utilizing endo-H in the removal of oligosaccharides from invertase (16), a protein with a mass that is 50% carbohydrate, it could be shown that this enzyme consists of two identical subunits, each with a molecular weight of 61,000 (Figure 4) after subtraction of the remaining N-acetylglucosamines. Incomplete removal of the carbohydrate, however, gives the appearance of two nonequivalent chains, one with a molecular weight of 65,000 and the other of 68,000 (Figure 5). But since endo-H is stable in 0.5% SDS, the oligosaccharide chains of the invertase molecule can be made completely accessible to endo-H and almost all of its 18 oligosaccharide chains removed (Table I). In the case of CPase Y, four oligosaccharide chains were found to be associated with a 51,000 dalton monomer (Table II).

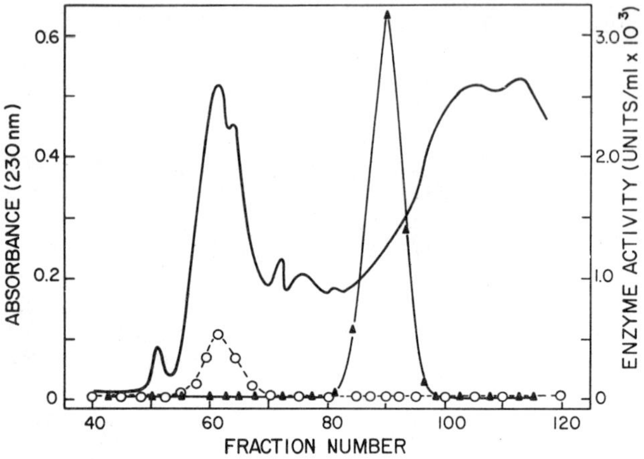

Journal of Biological Chemistry

Figure 1. Separation of S. plicatus *endo-L (○) and endo-H (▲) on a Sephadex G-100 column (1.5 × 220 cm). For further details* see *Ref.* 2.

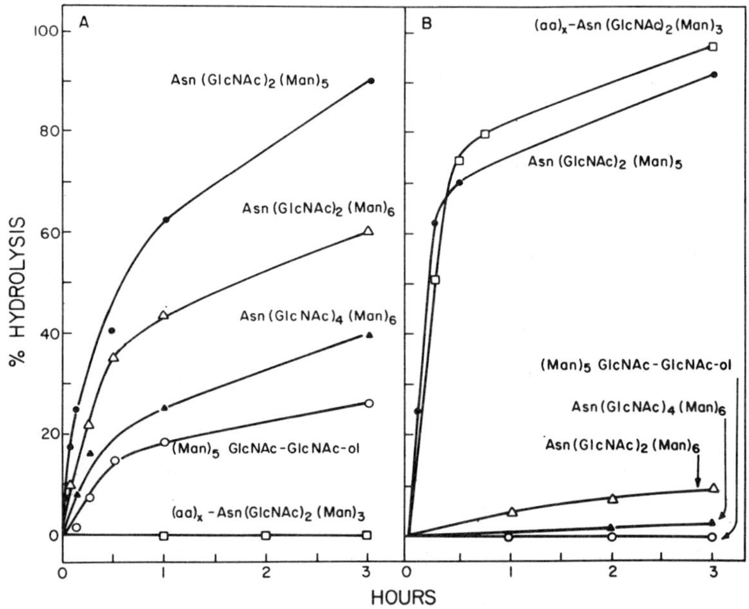

Biochemical and Biophysical Research Communications

Figure 2. A comparison of the rates of hydrolysis of various oligosaccharides by endo-H and endo-D. The described substrates were incubated for the indicated times with (A) 2.5 ng of pure endo-H or (B) 0.5 μg of protein containing endo-D. Adapted from Tarentino and Maley (15).

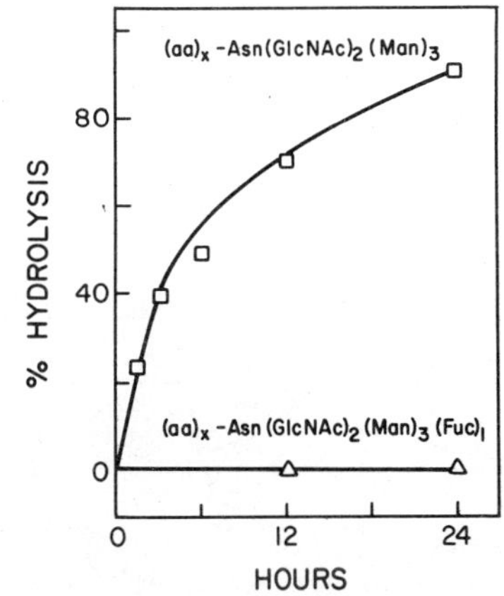

Biochemical and Biophysical Research Communications

Figure 3. The rate of hydrolysis of $(Fuc)_1(Man)_3(GlcNAc)_2Asn\text{-}(aa)_x$ and its defucosylated derivative by endo-H. The [^{14}C]-N-acetylated derivatives were incubated with 10 times more endo-H than those in Figure 1. Adapted from Tarentino and Maley (15).

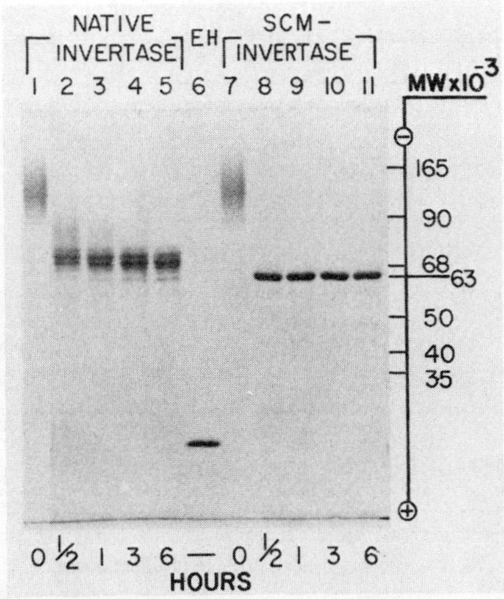

Journal of Biological Chemistry

Figure 4. Gel electrophoresis of native (lanes 1–5) and S-carboxymethylated (SCM) invertase (lanes 7–11) after incubation with endo-H. Endo-H (0.5 μg, lane 6) was used to treat 2 mg of native or S-carboxymethylated invertase in a 1 mL reaction mixture. Aliquots were electrophoresed in a flat-plate discontinuous SDS-polyacrylamide gel system. Adapted from Trimble and Maley (16), which should be referred to for further details.

TABLE I

Carbohydrate content of invertase preparations

Preparation	Enzyme treatment	GlcN	Man	Chains/ holoenzyme
		mol/120,000 g protein		
Native invertase	None	35.8	570	18
	Endoglycosidase	22.3	89	4
	Endoglycosidase + α-mannosidase	21.9	18	4
(CM)-invertase	None	35.9	565	18
	Endoglycosidase	18.9	16	1

All values are based on an $A_{280}^{1\ cm}$ of 2.25 and a holoenzyme molecular weight of 120,000.

Journal of Biological Chemistry

TABLE II

Carbohydrate content of carboxypeptidase Y fractions before and after endoglycosidase treatment

Enzyme preparation	CPase Y molar ratio[a]		
	GlcN	Man	Mannosyl chains
Native	7.90	55	4
Endoglycosidase-treated	4.95	15	1
Denatured and endoglycosidase-treated	3.72	0	0

[a]Based on a peptide molecular weight of 51,000.

Biochemical and Biophysical Research Communications

Alternatively, molecular weights of the released oligosaccharides can be estimated by Biogel P-4 chromatography providing appropriate standards are used (17).

The significance of the role of carbohydrate attached to protein has been of great concern for some time and while removal of the carbohydrate has not greatly influenced the enzymic and physical properties of most of the glycoproteins we and others have studied, an alteration in such physical parameters as solubility has been noted, particularly in the case of invertase which possesses such a high mass of this material. Recently it has been shown by Wang and Hirs (18) that carbohydrate influences the spectral properties of ovine RNase B. Even more recently, we have found the stability of carbohydrate depleted invertase to be markedly altered at pH 4.0 relative to native invertase (Figure 6). This parameter is also reflected in the decreased capacity of the carbohydrate depleted enzyme to renature when denatured by guanidine hydrochloride, suggesting that the carbohydrate in this case influences the capacity of the polypeptide chain to refold. Because of the rather high content of carbohydrate in this enzyme, it is difficult to extrapolate these studies to other glycoproteins with considerably lower amounts of carbohydrate.

Caution must be exercised in interpreting these findings however, and the potential for contaminants associated with even highly purified endoglycosidases makes this even riskier. Thus, even the presence of small quantities of exoglycosidases or proteases can greatly influence the results obtained considering the lengthy incubations often used by investigators. Witness the case of endo-D which was initially believed to release complex oligosaccharides (19) until it was realized on further purification that peripheral sugars as sialic acid, galactose, and N-acetylglucosamine must first be released by contaminating exoglycosidases before the $(Man)_3GlcNAc$ chain could be released (14).

Even when purified about 4,000 fold to apparent homogeneity by the procedure shown in Table III, endo-H contains traces of proteolytic activity which are not readily apparent unless more sensitive techniques than ordinarily used are applied, such as measuring the release of radioactivity from hemoglobin labeled with ^{14}C-labeled glycine ethylester. Thus, it is shown in Figure 7, which presents a G-75 elution profile, the last step in the purification of endo-H, that small traces of proteolytic activity are still present in the enzyme peak. Denatured CPase Y is particularly sensitive to this proteolytic activity and it is shown in Figure 8 that this protein is hydrolyzed to varying degrees by different fractions from the endo-H peak, even though the enzyme appears to be homogenous at this stage. However, treatment of those fractions containing proteolytic activity with phenylmethane sulfonyl fluoride almost completely eliminates this contaminant. One must be cautious in assessing the molecular weight of labeled oligosaccharides, that glycopeptides are not in reality being considered.

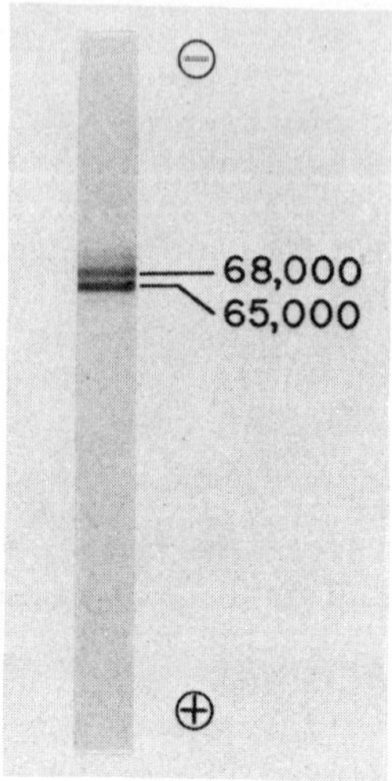

Figure 5. Electrophoresis of native invertase treated with endo-H. The systems used for digestion and electrophoresis were similar to those in Figure 4. The two forms of the enzyme represent incompletely removed oligosaccharide chains associated with the enzyme subunits.

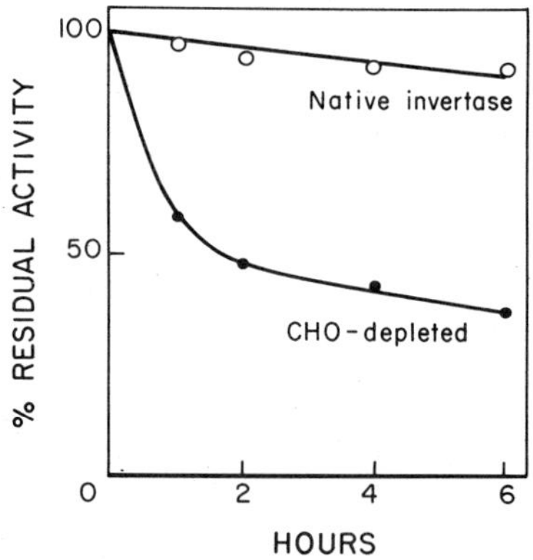

Figure 6. Stability of native and carbohydrate (CHO)-depleted invertase maintained at pH = 4.0. Each enzyme form (1 μg) was incubated in 1 mL of 0.01M sodium citrate pH = 4.0 at 4°C for the indicated times and then assayed (F. K. Chu and F. Maley, in preparation).

TABLE III

Purification of endo-β-N-acetylglucosaminidase H from *Streptomyces plicatus*

Purification step[a]	Volume (ml)	Protein (mg)	Activity (units)	Specific activity (units/mg)	Recovery (%)
1. Cultural filtrate and zinc precipitation	60,300	34,430	189	0.0053	100
2. Ammonium sulfate, 0.4-0.9	1,300	6,750	136	0.02	72
3. DE-52 Cellulose, pH 8.45	680	996	126	0.13	67
4. SP-Sephadex C-25, pH 4.6	404	17	94	5.5	50
5. Sephadex G-75, pH 8.45	29	3.5	69	19.7	37

[a]Step 1 material was prepared in 10-liter batches. Steps 2 through 4 were performed on the zinc paste from 30 liters of cultural filtrate. Step 5 was performed by combining two preparations from the preceeding step.

This procedure represents a modification from that previously described (2), and will appear in a future volume of *Methods in Enzymology*.

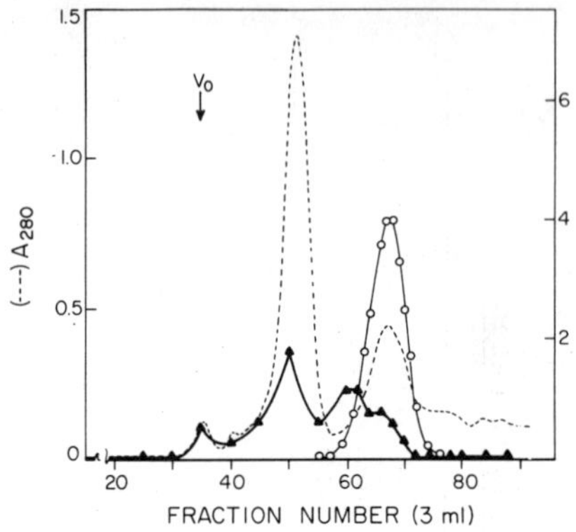

Figure 7. Assessment of residual proteolytic activity present in the Sephadex G-75 elution profile from Step 5 of Table III. The substrate used to measure proteolysis was [^{14}C]-glycine ethylester hemoglobin (13,000 cpm/mg protein). The peak of enzyme activity was associated with a single band of protein on polyacrylamide gel electroprhoresis. (○—○) Endo-H activity (cpm × 10^4) or (▲—▲) protease activity (cpm × 10^2).

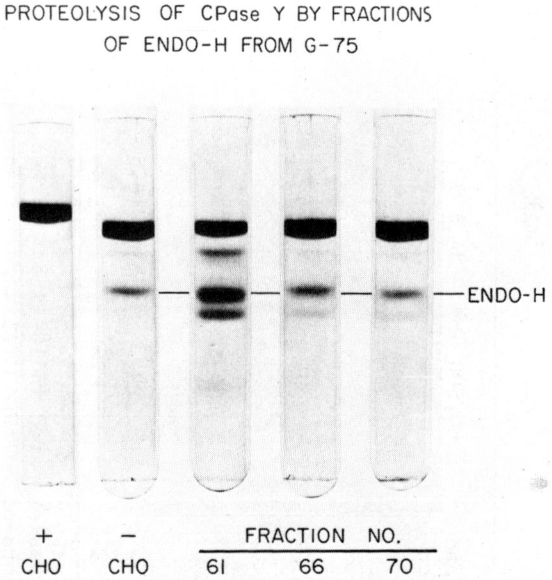

Figure 8. Assay for proteolytic activity in various fractions from the elution profile of Figure 7. Carboxypeptidase Y (1 mg) was incubated with 0.2 unit of endo-H from the indicated fractions in a 1 mL of reaction mixture for 4 hr at 37°C. Equal aliquots from these reactions were denatured with sodium dodecyl sulfate and electrophoresed on SDS-polyacrylamide gel as described in Trimble and Maley (7). The migration distances of untreated (+CHO) and depleted (−CHO) carboxypeptidase Y and endo-H are indicated in the first two lanes at the left.

From what was indicated in Figure 2, endo-H appears most active with substrates such as $(Man)_5(GlcNAc)_2Asn$, but because of the unavailability until now of compounds such as $(Man)_4(GlcNAc)_2$-Asn and $(Man)_2(GlcNAc)_2Asn$, it was not possible to determine the lower limits of the enzyme's specificity. Recently, however, Tai et al. (20) isolated $(Man)_4(GlcNAc)_2Asn$ from an ovalbumin digest by use of the Montgomery column (6) and found that this compound like $(Man)_5(GlcNAc)_2Asn$ was an excellent substrate for endo-H. By treating $(Man)_5(GlcNAc)_2Asn$ with α-mannosidase for limited time periods, we have been able to isolate $(Man)_4(GlcNAc)_2Asn$ and $(Man)_3(GlcNAc)_2Asn$ through the use of Biogel P-4 chromatography. It was initially believed, from the Biogel P-4 elution profile, that the latter compound was $(Man)_2(GlcNAc)_2Asn$, but by comparing its elution time with other glycosyl-Asn derivatives on the amino acid analyzer (Figure 9), it was established that this oligosaccharide was composed of $(Man)_3(GlcNAc)_2Asn$ (Figure 9A, b). The insert to Figure 9A establishes the utility of this chromatographic procedure for characterizing glycosyl-Asn derivatives with up to six mannosyl residues. Methylation of $(Man)_4(GlcNAc)_2Asn$ revealed this compound to differ from that isolated by Tai et al. (19) in that the peripheral mannosyl linkage was α1→6 (Figure 10 and Table IV) instead of α1→3 as shown by these investigators.

In contrast to the branched structure of $(Man)_3(GlcNAc)_2Asn$, associated with the core of the complex oligosaccharides, that isolated from the α-mannosidase digestion of $(Man)_5(GlcNAc)_2Asn$, was determined by methylation (Table IV) and periodate oxidation to be linear and to possess the peripheral structure shown in Figure 10. Both $(Man)_4(GlcNAc)_2Asn$ and linear $(Man)_3(GlcNAc)_2Asn$ appear to be hydrolyzed as rapidly by endo-H as its most active substrate, which to date is $(Man)_5(GlcNAc)_2Asn$, and contrasts sharply with the rather slow hydrolysis by this enzyme of branched $(Man)_3(GlcNAc)_2Asn$ (Figures 2 and 3). By utilizing a modified Smith degradation on $(Man)_6(GlcNAc)_2Asn$ (21, 22), $(Man)_2(GlcNAc)_2Asn$ could be isolated and verified by compositional analysis, methylation (Table IV), and chromatography (Figure 9B). This compound was hydrolyzed at only half the rate of branched $(Man)_3(GlcNAc)_2$-Asn, while $(Man)_1(GlcNAc)_2Asn$ is probably not hydrolyzed at all by endo-H. Thus, the statement in our abstract that $(Man)_2(GlcNAc)_2$-Asn is as good a substrate as $(Man)_5(GlcNAc)_2Asn$ is in error and should be corrected by replacing the former compound with linear $(Man)_3(GlcNAc)_2Asn$. $(Man)_1(GlcNAc)_2Asn$, however, could be very effectively hydrolyzed to Manβ1→4GlcNAc + GlcNAc-Asn by endo-L, the enzyme referred to earlier as enabling us to isolate and to characterize this disaccharide (1). That this enzyme in not chitobiase is indicated by its inability to hydrolyze $(GlcNAc)_2Asn$. Endo-L has now been purified about 300 fold and possesses properties distinctly different from endo-H, not the least of which is their substrate specificities. Other differences are associated with the higher molecular weight of endo-L and its sharper pH optimum (23). A comparison of these enzymes and their capacity to

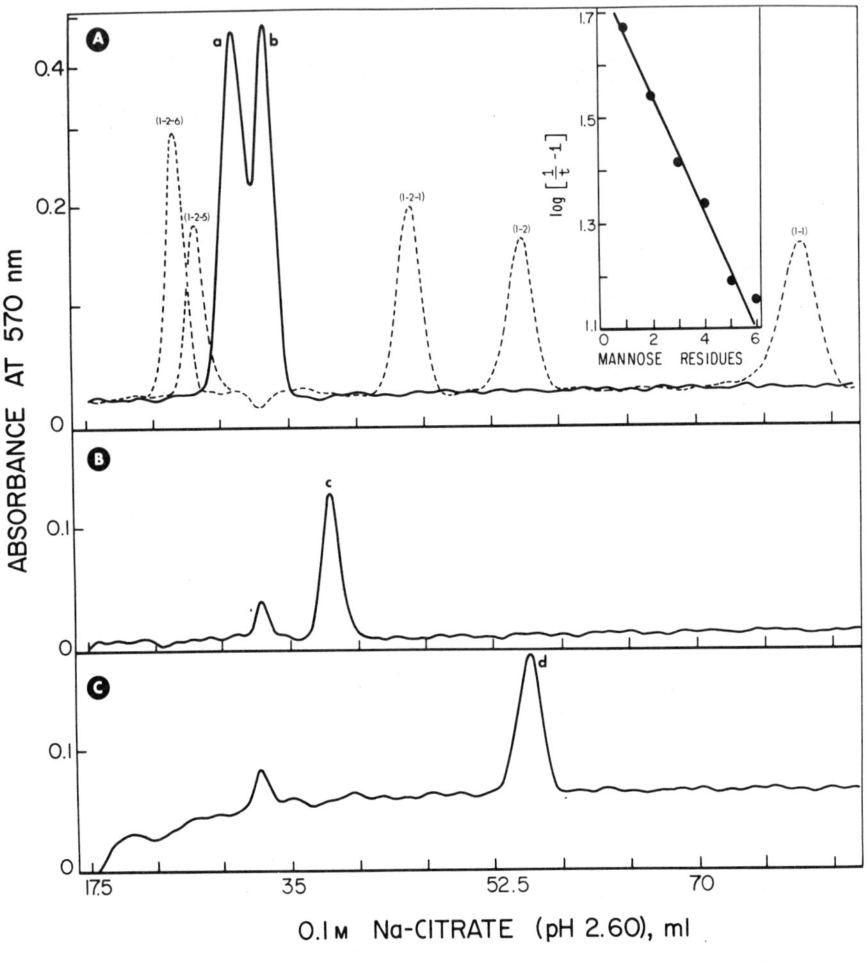

Figure 9. Chromatographic separation of glycosyl asparagine derivatives on the amino acid analyzer using AG-50 X-4 resin.

The various stands used (prepared from ovalbumin) are indicated by the dashed line and represent $(Man)_6(GlcNAc)_2Asn$, 1-2-6; $(Man)_5(GlcNAc)_2Asn$, 1-2-5; $(Man)_1(GlcNAc)_2Asn$, 1-2-1; $(GlcNAc)_2Asn$, 1-2; GlcNAc-Asn, 1-1. Peaks a *and* b *resulted from a limited α-mannosidase digestion of 1-2-5 as shown in Figure 10; peak* c *in B was prepared by the periodate oxidation of 1-2-5; peak* d *in C was obtained by treating 1-2-2 with periodate. The elution times* (t), *measured at peaks of various glycosyl-Asn derivatives, were plotted as shown in the insert vs. the number of mannose residues in the compounds. As indicated, compounds* a, b, *and* c *fall on the line at 4, 3, and 2 mannose residues respectively, which is consistent with their mannosyl content (Table IV). The proposed mannosyl linkages shown in Figure 10 were obtained from the methylation data in Table IV. More specific details on the methodology used are in preparation.*

TABLE IV

Composition of compounds tested as substrates for endo-H

Isolated compound	Analysis			Methylated mannitol products			
	Asn	GlcNAc	Man	(2,3,4,6)	(2,3,4)	(2,4,6)	(2,4)
$(Man)_4(GlcNAc)_2Asn$	1.00	1.85	4.06	2.2	1.3		1.0
$(Man)_3(GlcNAc)_2Asn$	1.00	1.82	2.84	0.84	1.1	1.0	
$(Man)_3(GlcNAc)_2$-R[a]	1.55	2.00	3.00				
$(Man)_2(GlcNAc)_2Asn$[b]	1.00	1.80	1.96	1.0	1.4		

[a]Prepared from IgM (9); R = peptide.

[b]Periodate oxidation product from $(Man)_6(GlcNAc)_2Asn$.

hydrolyze specific dansylated substrates prepared from ovalbumin (Figure 11) reveals that endo-L is not very effective against substrates containing more than two residues of mannose. In a similar vein, endo-H is not effective against compounds containing two mannosyl residues or less. We cannot be certain at present whether the small amount of activity seen with $(Man)_3(GlcNAc)_2Asn$ as substrate is native to endo-L and is not due to traces of endo-H. There is, however, a greater degree of confidence that homogenous endo-H decreases in activity by 10,000 fold in going from linear $(Man)_3(GlcNAc)_2Asn$ to $(Man)_2(GlcNAc)_2Asn$. These recent specificity studies suggest the need for caution in estimating oligosaccharide size based on the sensitivity or resistance of glycopeptides to endo-H.

Following our characterization of the properties of endo-H (2, 6, 15) and that of endo-D by Koide and Muramatsu (14) a number of other useful endoglycosidases have been identified. Thus, an endo-β-N-acetyl-galactosaminidase has been described from cultural filtrates of *Diplococcus pneumoniae* (24), an endo-galactosidase from *Escherichia freundii* (25), two endo-β-N-acetylglucosaminidases from *Clostridium perfringens* (26), one with slightly different specificity than endo-H, and more recently an amidase from almond emulsin (27). The amidase, although not purified to a significant extent, possesses the intriguing property of releasing oligosaccharides from a stem bromelein glycopeptide by hydrolyzing the glucosaminyl asparagine bond. If this enzyme is more generally applicable, it should be extremely useful in the study of glycopeptides and possibly glyocproteins, but not enough in known about its specificity at the present time.

As far as the existence of an animal endoglycosidase with properties similar to those of endo-H or endo-D, its presence can be inferred from the structure of oligosaccharides excreted in the urine and present in the tissues of animals with certain lysosomal storage diseases. Thus, mannosyl oligosaccharides with a single N-acetylglucosamine on the reducing end were found in the urine and tissue of black angus cattle (28) and humans (29) afflicted with α-mannosidosis. Oligosaccharides consisting of 7 to 10 sugar residues with the complex tetrasaccharide core of $(Man)_3GlcNAc$ have also been found in patients with GM_1 or GM_2 gangliosidosis (30, 31, 32), and fibroblasts from patients with α-fucosidosis (32) apparently accumulate the disaccharide Fucα1→6GlcNAc, a disaccharide found at the reducing end of many "complex" type oligosaccharides. Whether these products accumulate due to the above specific exoglycosidase deficiencies combined with cleavages affected by a single endoglycosidase is not known, but we have described an enzyme in animal tissues partially capable of explaining these findings (33).

The enzyme was purified over 1,500 fold from hen-oviduct and as shown in Figure 12 hydrolyzed $[^{14}C]$-N-acetylated glycosyl-Asn derivatives derived from ovalbumin in a manner similar to that described for endo-H. As an added bonus, this enzyme preparation

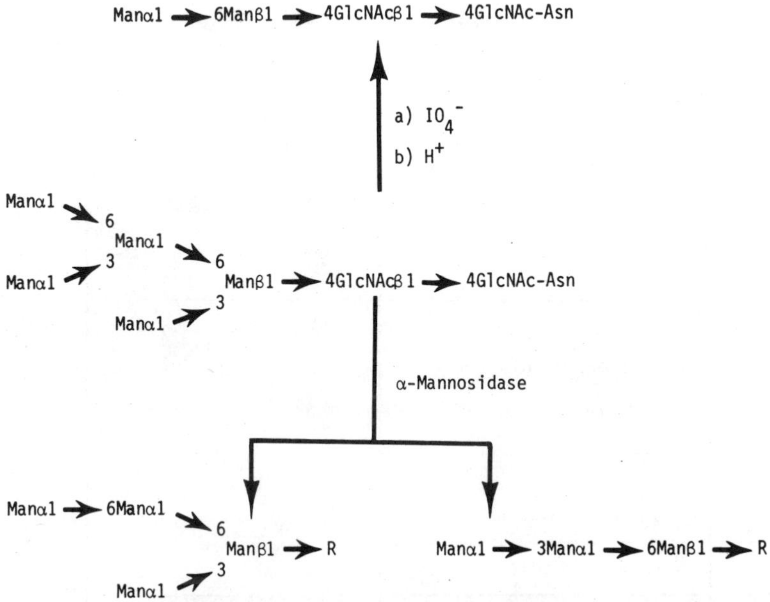

Figure 10. Proposed structures of compounds prepared from $(Man)_5(GlcNAc)_2Asn$ by periodate oxidation and limited α-mannosidase digestion. The structures proposed are based on compositional analyses, in addition to the methylation and periodate oxidation data presented in Figure 9 and Table IV. $R = (GlcNAc)_2$-Asn.

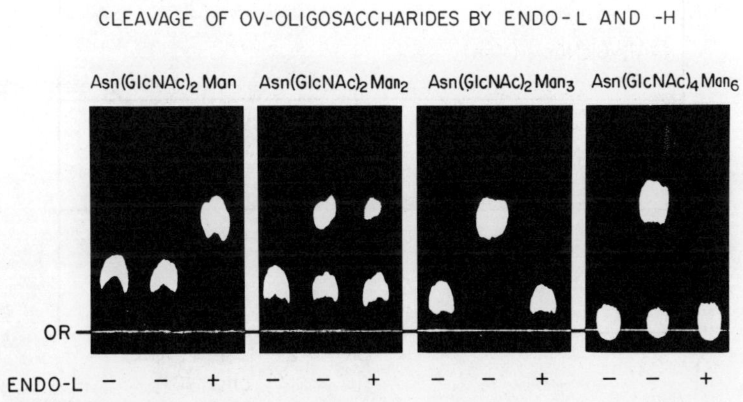

Figure 11. Studies on the substrate specificity of endo-H and endo-L. The corresponding dansyl derivatives of the compounds indicated were treated with endo-L and endo-H as shown, and aliquots were chromatographed ascendingly on Whatman 3-MM with butanol–ethanol–water (2:1:1).

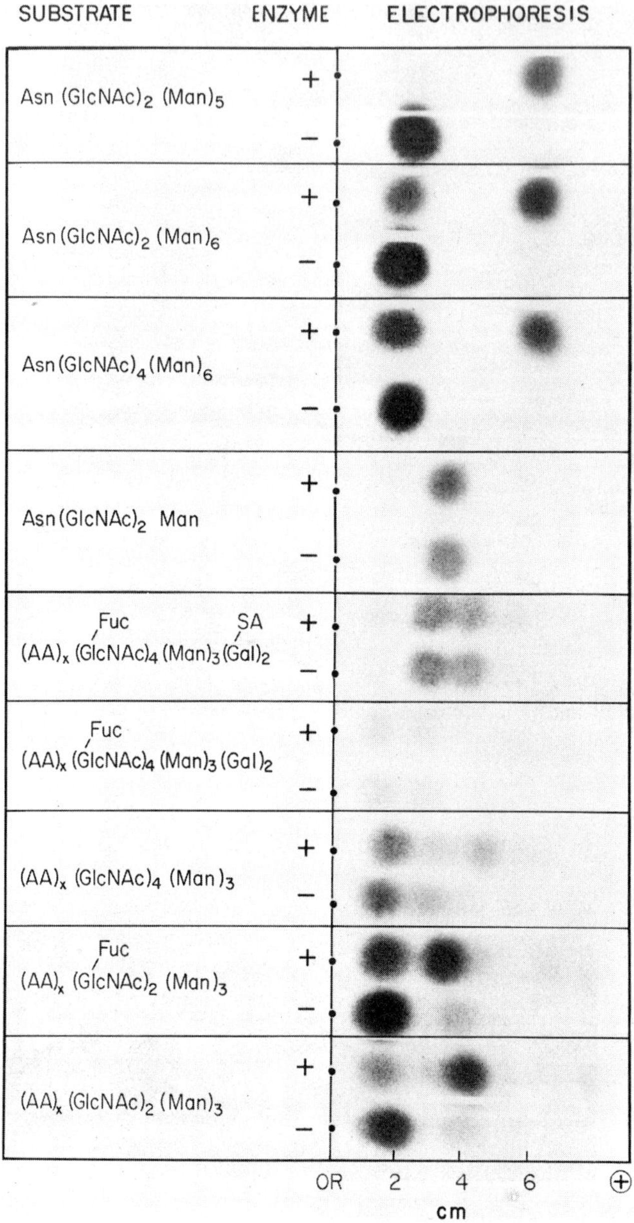

Journal of Biological Chemistry

Figure 12. Substrate specificity of hen oviduct endoglycosidase. The indicated [^{14}C]-N-acetylated derivatives were electrophoresed on paper before and after treatment with this enzyme preparation, followed by the preparation of a radioautogram. Adapted from Trimble et al. (23) which should be consulted for further details.

also hydrolyzed the complex core glycopeptide containing fucose, a property of endo-D but not of endo-H. Also characteristic of endo-D is the finding that the complex branched glycopeptide, $(Man)_3(GlcNAc)_2Asn$, is hydrolyzed at a faster rate than $(Man)_6$-$(GlcNAc)_4Asn$ (33), the reverse of that obtained with endo-H (Figure 2). If an enzyme similar to this one is present in animal tissues from other species, the type of oligosaccharides associated with the above mentioned genetic disorders can easily be rationalized. Through the use of a new assay we have developed where the mannosyl non-reducing end of glycosyl-Asn derivatives, such as $(Man)_6(GlcNAc)_2Asn$-dansyl, are lightly oxidized with periodate and then reduced with sodium borotritide, endoglycosidase activity has been found in several rat tissues, as well as in *Tetrahymena pyriformis* (34). An advantage of this substrate is the resistance of the labeled diol to α-mannosidase, an enzyme which could yield a false positive endoglycosidase activity in crude tissue extracts.

Although an animal genetic disorder implicating endo-β-N-acetylglucosaminidase has not been reported yet, the question arises as to what might be expected if an organism is deficient in this enzyme. In all probability this defect would be difficult if not impossible to assess since the numerous types of exoglycosidases present in animal tissues would limit the accumulation of long chain oligosaccharides associated with glycopeptides involving glucosaminyl-Asn linkage groups. That is unless α-mannosidase was also deficient, in which case oligosaccharides with GlcNAc on the reducing end would not be found and glycopeptides with mannosyl-rich oligosaccharides should accumulate.

At present, the role of endo-H and other similarly related enzymes appears to be restricted to that of degrading glycoproteins and glycopeptides, but their potential contribution to the biosynthesis and processing of glycoproetins should not be ignored. In any event, the discovery of the endoglycosidases has added a new dimension to the study of glycoproteins, which hopefully will continue to provide new and important information on their structure and function.

Acknowledgements

We would like to express our appreciation to Dr. Thomas H. Plummer, Jr. and Dr. Frederick K. Chu for their aid with some aspects of this study.

Literature Cited

1. Tarentino, A. L., Plummer, T. H., Jr., and Maley, F., J. Biol. Chem. (1972) 247, 2629.
2. Tarentino, A. L., and Maley, F., J. Biol. Chem. (1974) 249, 811.
3. Sukeno, T., Tarentino, A. L., Plummer, T. H., Jr., and Maley, F., Biochem. Biophys. Res. Commun. (1971) 45, 219.
4. Tarentino, A. L., Plummer, T. H., Jr., and Maley, F., J. Biol. Chem. (1973) 248, 5547.
5. Tarentino, A. L., Plummer, T. H., Jr., and Maley, F., J. Biol. Chem. (1974) 249, 818.
6. Huang, C.-C., Mayer, H.-E., Jr., and Montgomery, R., Carbohydrate Res. (1970) 13, 127.
7. Trimble, R. B., and Maley, F., Biochem. Biophys. Res. Commun. (1977) 78, 935.
8. Baeziger, J., and Kornfeld, S. J., J. Biol. Chem. (1974) 249, 7260.
9. Tarentino, A. L., Plummer, T. H., Jr., and Maley, F., Biochemistry (1975) 14, 5516.
10. Kornfeld, R., and Kornfeld, S., Annu. Rev. Biochem. (1976) 45, 217.
11. Burke, D. J., and Keegstra, K., J. Virol. (1976) 20, 676.
12. Robbins, P. W., Hubbard, S. C., Turco, S. J., and Wirth, D. F., Cell (1977) 12, 893.
13. Tabas, I., Schlesinger, S., and Kornfeld, S., J. Biol. Chem. (1978) 253, 716.
14. Koide, N., and Muramatsu, T., J. Biol. Chem. (1974) 249, 4897.
15. Tarentino, A. L., and Maley, F., Biochem. Biophys. Res. Commun. (1975) 67, 455.
16. Trimble, R. B., and Maley, F., J. Biol. Chem. (1977) 252, 4409.
17. Etchinson, J. R., Robertson, J. S., and Summers, D. F., Virology (1977) 78, 375.
18. Wang, F.-F. C., Hirs, C. H. W., J. Biol. Chem. (1977) 252, 8358.
19. Muramatsu, T., Atkinson, P. H., Nathenson, S. G., and Ciccarini, C., J. Mol. Biol. (1973) 80, 781.
20. Tai, T., Yamashita, K., and Kobata, A., Biochem. Biophys. Res. Commun. (1977) 78, 434.
21. Montgomery, R., Wu, Y.-C., and Lee, Y. C., Biochemistry (1965) 4, 578.
22. Garner, E. F., Goldstein, I. J., Montgomery, R., and Smith, F., J. Am. Chem. Soc. (1958) 80, 207.
23. Trimble, R. B., Tarentino, A. L., and Maley, F., in preparation.
24. Umenoto, J. Bhavanandan, V. P., and Davidson, E. A., J. Biol. Chem. (1977) 252, 8609.
25. Fukuda, M. N., and Matsumura, G., J. Biol. Chem. (1976) 251, 6218.

26. Tai, T., Yamashita, K., Ito, S., and Kobata, A., J. Biol. Chem. (1977) 252, 6687.
27. Takashi, N., Biochem. Biophys. Res. Commun. (1977) 76, 1194.
28. Lundblad, A., Nilsson, B., Nordén, N. E., Svensson, S., Öckerman, P.-A., and Jolly, R. D., Eur. J. Biochem. (1975) 59, 601.
29. Nordén, N., Lundblad, A., Svensson, S., and Autio, S., Biochemistry (1974) 15, 871.
30. Wolfe, L. S., Senior, R. G., and Kin, N., J. Biol. Chem. (1974) 249, 1828.
31. Kin. N., and Wolfe, L. S., Biochem. Biophys. Res. Commun. (1974) 59, 837.
32. Tsay, G. C., and Dawson, G., Biochem. Biophys. Res. Commun. (1975) 63, 807.
33. Tarentino, A. L., and Maley, F., J. Biol. Chem. (1976) 251, 6537.
34. Tarentino, A. L., and Maley, F., Anal. Biochem. (1977) 77, 185.

RECEIVED March 15, 1978.

Genetic Disorders of Complex Carbohydrate Metabolism

Introduction

VICTOR GINSBURG

National Institutes of Health, Bethesda, MD 20014

The research area of genetic disorders of complex carbohydrate metabolism was firmly established in 1952 with the demonstration by the Coris that the abnormal accumulation of glycogen in von Gierke's disease results from a lack of glucose 6-phosphatase in the liver of these patients. In 1964, Hers and his coworkers showed the lysosomal location of glycogen which accumulates in Pompe's disease resulting from a deficiency in acid maltase. This important finding led to the realization that the lack of various lysosomal enzymes is the basis of many genetic disorders of glycolipid, glycoprotein and mucopolysaccharide metabolism. In 1968, another important advance occurred with the report by Neufeld and coworkers that fibroblasts from patients with Hurler and Hunter syndromes when grown together in culture corrected each others genetic deficiencies in that they did not accumulate mucopolysaccharides in contrast to the results obtained when each cell line was grown separately. This work led to the finding that lysosomal enzymes are normally excreted and selectively taken up by cells, provided a simple test for the diagnosis of the various mucopolysaccharidoses and suggested the possibility of therapy by enzyme replacement.

At present, over 30 specific enzyme deficiencies have been related to various genetic diseases of complex carbohydrate metabolism. Most of the deficient enzymes are hydrolases, generally lysosomal. Known deficiencies in synthetic enzymes which cause disease are rare, perhaps because they are harder to detect experimentally: the missing product of a defective synthetic enzyme is more difficult to recognize than the abnormal accumulation of the substrate of a defective degradative enzyme.

RECEIVED February 21, 1978.

5

Mucous Glycoproteins in Cystic Fibrosis

THOMAS F. BOAT and PIWAN CHENG

Department of Pediatrics, Rainbow Babies and Childrens Hospital, Case Western Reserve University, School of Medicine, Cleveland, OH 44106

Cystic fibrosis (CF), the most common lethal genetic trait of Caucasians, causes much of the chronic progressive lung disease in children and young adults. It also is responsible for intestinal obstruction at birth (meconium ileus), pancreatic insufficiency with maldigestion and failure to grow, and obstructive liver disease in this age group. Furthermore, obliteration of the vas deferens causes aspermia and infertility in most males with CF. Individuals with CF secrete sweat which contains excessive amounts of chloride and sodium. Diagnostic criteria include detection of elevated chloride levels in sweat of individuals with typical pulmonary, gastrointestinal, or genital manifestations of the disease (1).

The pathophysiology of CF has been the subject of extensive investigation (2). However, the basic defect responsible for the disease has not been identified. Claims have been made by several investigators that circulating factors which can be detected by inhibition of oyster gill (3) or rabbit tracheal (4) ciliary motility, or secreted factors which inhibit sodium reabsorption by ductal epithelium in sweat and salivary glands (5), may be related closely to the gene defect. Alteration of short circuit current across rat intestinal epithelium, inhibition of sugar and amino acid transport by intestinal epithelium, augmentation of leukocyte degranulation, inhibition of phagocytosis by alveolar macrophages, and inhibition of debranching enzyme activity have also been ascribed to circulating or secreted "factors" associated with CF. Polyamines, their condensation products (6) and small polycationic peptides (7) have been implicated as "factors" but a substance has not been isolated and characterized which is clearly responsible for any of these effects. Cell membrane properties with the exception of increased resistance of fibroblasts to oubain toxicity (8) appear unaltered. Recent evidence suggests a deficiency of proteolytic activity in CF serum (9) and abnormal interaction of circulating α_2-macroglobulin with proteolytic enzymes (10). Pursuit of initial observations in many of these studies has led to contradictory results, making most studies difficult to inter-

0-8412-0452-7/78/47-080-108$05.00/0

pret or fit into an overall pathophysiologic scheme. One possible explanation for inconsistencies in studies carried out to date is that CF may be due to genetic abberations at more than one locus.

On the other hand the pathogenesis of most clinical manifestations in CF is relatively well understood and appears to be related to two anomalies: 1) abnormal behavior of the mucus-secretions, giving rise to obstruction, and in some cases infection within mucus-secreting organs (lung, intestine, pancreas, biliary tract, salivary glands, and genitourinary tract), and 2) abnormality of the electrolyte composition of eccrine sweat glands, occasionally resulting in heat prostration due to massive salt loss (2). This review will examine the abnormality of mucous secretions in detail and present current thinking about its biochemical basis. Emphasis will be given to lung mucus because airflow obstruction and lung infection are responsible for most of the morbidity and 95% of the mortality related to CF.

Mucus from individuals with CF generally is described as abundant and excessively sticky. Rheologic studies of lung mucus have been carried out largely on sputum (11) with a high content of DNA (12) which masks the viscoelastic properties of the primary secretions. While careful rheologic studies of uninfected mucous secretions with a low DNA content from CF subjects have not been reported, it is reasonable to assume that the obstructive events leading to the clinical syndrome are the result of abnormal properties of mucus, and perhaps of the mucous glycoproteins themselves which seem to be the primary determinant of uninfected dog tracheal mucus (13). Three potential causes for these abnormal properties can be proposed: 1) hypersecretion of normal mucus which overloads clearance mechanisms, resulting in stasis and obstruction; 2) inadequately hydrated mucus or mucus with an altered ion content such that glycoprotein complexes are formed which have abnormal rheologic properties; or 3) alteration of chemical properties of mucous glycoproteins resulting in formation of mucous gels with abnormal rheologic characteristics.

Mucus Hypersecretion: Morphologic examination of several organs from individuals with CF shows marked enlargement of the mucus-secreting glands (hypertrophy) and a marked increase in the number of mucus-secreting goblet cells within the surface epithelium (hyperplasia) (14). This is true not only of lung, but also of intestine, salivary glands, and uterine cervix. In fact, goblet cells appear early in the course of the disease in the surface epithelium of very small airways (bronchioles) where they normally do not reside in substantial numbers. This metaplastic change introduces mucus into airways which are not equipped to handle gel-like secretions. Mucus then impacts in these small airways, producing the earliest recognizable lung lesion (15). Morphologic evidence for a hypersecretory state in CF lungs at birth is contradictory (16,17). However, it is clear that duodenal mucus-secreting glands are enlarged and apparently hyperactive at birth (17). This finding suggests that increased secretion of mucus may be more than a secondary reaction to chronic infection or inflammation.

A search for pathophysiologic determinants of mucous hypersecretion until recently turned up few leads. In the last several years, studies of the effect of serum on ciliary motility led to the observation that large amounts of "mucus" were released from ciliated epithelium in the presence of CF serum (18,19). Recently Czegledy-Nagy and Sturgess (20) studied this response by scanning electron microscopy and observed that rabbit tracheal epithelium released more secretory product in the presence of CF than control serum. These observations are qualitative and in addition, products which were assumed to contain mucous glycoproteins were not identified. We have developed an assay to assess the effect of CF serum and other biological fluids on the release of mucous glycoproteins by rabbit tracheal epithelium. Glycoproteins are labeled *in vitro* (21) with $^{35}SO_4$ and ^{3}H-6-D-glucosamine, and their release into culture medium in the presence of 50% (v/v) CF serum, 50% (v/v) control serum, and no serum is monitored. Preliminary results show that the presence of serum from any source in the medium increases the release of labeled high molecular weight macromolecules which have a carbohydrate composition typical of mucous glycoproteins. Some CF sera transiently induce a higher secretory rate for these labeled macromolecules than is achieved in the presence of age and sex-matched control sera, but overall we have not yet been able to demonstrate a statistically significant effect for CF sera (Figure 1). At longer exposure periods (1 hour or more) no suggestion of a differential CF serum effect on mucous glycoprotein release was found. The nature of the stimulatory effect in some CF sera is not known. Polyamines augment the release of glycoproteins from canine tracheal explants (22) but not human tracheal epithelial explants (23). Polycationic peptides also increase the release of glycoproteins *in vitro* from dog tracheal epithelium (24). Polycationic substances may influence mucous glycoprotein synthesis by stimulating the activity of glycosyltransferases at low Mn^{+2} concentrations (25). Whether this occurs with intact tissues under physiologic conditions is not known.

Explants of post-mortem tracheobronchial epithelium from CF subjects secrete mucous glycoproteins at a higher baseline rate than do explants from children and adults who have died of non-respiratory conditions (26). This observation is consistent with an increased mass of mucus-secreting cells in tissues of subjects with severe bronchitis as it occurs in CF, and does not reflect an increased intrinsic level of function for CF secreting units. Further support for this concept comes from autoradiographic studies of Neutra (27) which show no difference in the rate of movement of labeled glycoprotein precursors through goblet cells of CF, compared with control, rectal epithelial explants.

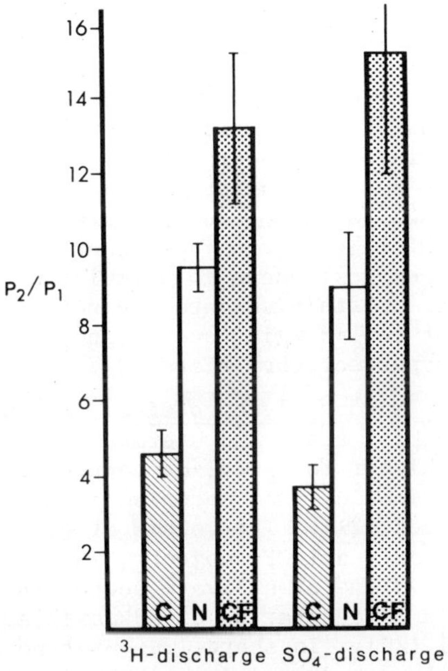

Figure 1. Discharge of $^{35}SO_4$ and ^{3}H-labeled glycoproteins by rabbit tracheal explants in the presence of no serum (C), normal serum (N), and CF serum (CF). Explants were cultured in the presence of 2 × 10^7 dpm of ^{3}H-6-D-glucosamine and 1 × 10^8 dpm of $^{35}SO_4$ in 2 mL of medium for 24 hr (P_1), rinsed thoroughly with unlabeled medium and then cultured for 10 min (P_2) in 2 mL of unlabeled medium containing no serum or 50% (v/v) serum. Media harvested from P_1 and P_2 culture periods, including washings, were treated with 5% trichloroacetic acid–1% phosphotungstic acid and the washed precipitates were solubilized with hydroxide of hyamine–H_2O_2 for scintillation counting (24). Data are expressed as a ratio of the rate of acid-precipitable label released in P_2 to that in P_1 (56) for each experimental condition. Approximately 80% of labeled macromolecules released by rabbit tracheal explants were high molecular weight substances as assessed by BioGel A-5m chromatography and had a carbohydrate composition typical for mucous glycoproteins (21). Constant specific activity of discharged glycoproteins was achieved within 24 hr of exposure to medium containing labeled precursors. C vs. N; $p < 0.01$. C vs. CF; $p < 0.01$. N vs. CF; $0.05 < p < 0.10$. (Student's T test for matched pairs)

Sturgess and Reid observed with autoradiographic techniques that the response of a secretory index (proportion of cells showing incorporation of ^{3}H-glucose and movement of label to the lumenal surface) to cholinomimetic stimulation was higher for hypertrophic glands of CF tracheal explants than for glands from non-bronchitic tracheas (28). However, direct measurement of the secretory rate for mucous glycoproteins from human tracheal explants in our hands has demonstrated a similar secretory response to cholinomimetic agents for CF and non-bronchitic tracheal explants (26). Therefore, a strong case cannot be made for increased intrinsic responsiveness to stimulation of CF mucus secreting elements. Definitive data are not available to support the hypothesis that cholinergic tone is increased in CF, resulting in a chronic hypersecretory state (2).

Abnormal Milieu for Mucous Glycoproteins: Mucous secretions of the CF tracheobronchial tract and uterine cervix have a lower water content and higher total solids content than control secretions (29). Meconium from meconium ileus patients is also relatively dehydrated (29). The volume of pancreatic secretions is diminished in nearly all CF subjects, indicating a decreased water output. The volume of salivary secretions is not diminished, but the protein concentration of submaxillary saliva is higher than normal (29). In short, mucous-type exocrine secretions are relatively dehydrated in cystic fibrosis. A paucity of water in these secretions may explain the inspissation and ductal or small airways obstruction seen prominently in CF. In the lungs, a lack of water could reduce the volume of periciliary fluid and interfere with ciliary motility (1). In addition, relatively "dry" secretions may have sub-optimal viscoelastic properties which would interfere with normal coupling of mucus to ciliary movement and consequently with mucus clearance from the lungs (30). The non-mucous secretions (parotid saliva, sweat) of subjects with CF, however, are not reduced in volume and are not concentrated with respect to organic constituents (29). Therefore, a generalized defect in water secretion is not characteristic of this disease.

Increased concentrations of calcium are found in several secretions of individuals with CF, including submaxillary and parotid saliva, tears, and seminal plasma (29). Formation of complexes between divalent cations and the highly acidic mucous glycoproteins may contribute substantially to the physical and rheologic properties of mucous secretions. For example, rat goblet cell intestinal mucus becomes less soluble in the presence of increasing concentrations of calcium (31). However, a similar effect in CF mucus secretions has not been described. Furthermore, calcium-mucin complex formation is an unlikely pathogenetic factor because calcium levels are not elevated in CF tracheobronchial secretions, cervical secretions, and meconium (29). The milky appearance of CF submaxillary saliva initially was considered to be evidence for calcium-glycoprotein interactions (32) but the insoluble material has been shown to be a complex of calcium and a low molecular weight phosphoglycoprotein, unrelated to the mucous glycoproteins (33).

Interactions of mucous glycoproteins with other cationic compounds in secretions such as the polyamines and their metabolites (29), lysozyme, lactoferrin, proline-rich polypeptides (34) and immunoglobulins (35) may well occur. However, the role which complexes of mucous glycoproteins and these substances play in determining the properties of mucus has not been fully explored. The possibility that these complexes contribute to abnormalities of mucus in CF deserves consideration.

Altered Physicochemical Properties of Mucous Glycoproteins: Mucous glycoproteins are responsible in large part for the viscoelastic properties of tracheobronchial secretions. Determinants of these properties include disulfide bonding (36) and a number of non-covalent interactions (37,38,39) between the large, elongated glycoprotein molecules, resulting in the formation of a gel matrix. Participation of other proteins in the formation of this matrix has been suggested (36).

Human respiratory tract mucous glycoproteins contain 75-80% carbohydrate, 20% or less peptide, and a variable amount of sulfate ester (40), at least some of which is present as D-galactose-6-SO_4 (41). Fucose, galactose, glucosamine, galactosamine, and N-acetylneuraminic acid are the only identifiable sugar residues and are arranged in a very heterogeneous group (41) of galactosaminyl-threonine (serine)-linked oligosaccharide chains. Nearly one-half of the peptide amino acids are threonine or serine residues, and at least 50% of these hydroxyamino acids are glycosylated (42). The molecular weight of reduced and carboxymethylated respiratory glycoprotein prepared from lung washings of an individual with cystic fibrosis is approximately 250,000 as determined by sedimentation equilibrium studies (42), but these glycoproteins behave as if they are much larger molecules on agarose gel filtration (40). In the non-reduced state they polymerize to form complexes with estimated molecular weights of 600,000 to 3,000,000 (35,43). It has been suggested that repulsive forces between sialic acid carboxyl and sulfate anionic sites on adjacent oligosaccharide chains are responsible for the extended conformation of these glycoproteins (44), but the contribution of anionic groups to the rheologic behavior of mucous glycoproteins currently is a subject of debate (45,46).

We have developed a system for purification of mucous glycoproteins in sputum or lung lavage fluids and their fractionation based on charge (40). Reduced and carboxymethylated glycoproteins containing all of the blood group substance, are recovered at the void volume of a BioGel A-5m column and separated into a series of components, ranging from sparsely acidic to highly acidic, on DEAE cellulose (Figure 2). These components have similar peptide and carbohydrate compositions. The major

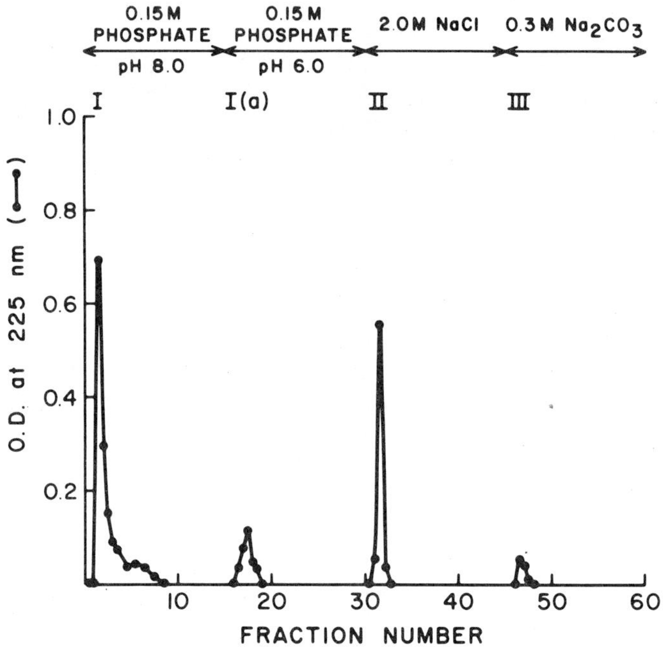

Archives of Biochemistry and Biophysics

Figure 2. Fractionation of reduced, carboxymethylated mucous glycoproteins from a tracheostomized subject with no evidence for lower respiratory tract disease on DEAE cellulose (12 × 0.75 cm column).

Glycoproteins were first purified by BioGel A-5m chromatography (40). Glycoproteins were applied to the column in 0.005M phosphate buffer (pH = 8.0) and eluted stepwise as indicated. Five mL fractions were collected. Less than 1% of material applied to the column was recovered in the run-through volume. Approximately 60% of material applied to the column was recovered in the four fractions. All four fractions contained blood group-active substance and were typical mucous glycoproteins with respect to sugar and amino acid composition (40).

sparsely acidic component (I) is a potent blood group antigen, has relatively weak influenza virus hemagglutination inhibition properties, and contains 1-2% sulfate. The major acidic component (II) is a less potent blood group antigen, a stronger inhibitor of virus hemagglutination, and is highly sulfated (6-7% by weight). Oligosaccharide chains of component II have a slightly longer mean chain length than those of component I. However, both components display a wide range of oligosaccharide chain lengths, from 1-20 sugar residues/chain. In general sulfated sugars are found in the long chains, and sialic acid resides in chains containing 4-7 sugar residues (47).

Cystic fibrosis tracheobronchial mucous glycoproteins differ from those of subjects with other hypersecretory states, e.g., chronic bronchitis, in that they contain predominently the highly acidic components as shown in Figure 3. Consequently, CF tracheobronchial secretions as a whole are more highly sulfated and exhibit stronger acidic properties. This observation has been made in our laboratory and by others (48). CF salivary glycoproteins are more potent inhibitors of influenza B virus hemagglutination than are salivary glycoproteins from control subjects (49), suggesting that they also have increased acidic properties. No unique mucous glycoprotein components have been found in CF secretions and none are missing. There is no evidence for alterations of the fucose and sialic acid contents of CF tracheobronchial mucous glycoproteins, as postulated a number of years ago for intestinal glycoproteins by Dische (50).

Explants of tracheal epithelium from subjects with CF also secrete large amounts of the highly sulfated glycoprotein components. This pattern reverts to a non-CF type with predominance of sparsely sulfated glycoprotein (component I) when secretion by explants is stimulated by adding methacholine to the culture medium during a 6 hour period (Table I). These results indicate that alterations in the distribution of mucous glycoprotein components from CF tracheobronchial epithelium cannot be attributed to excessive stimulation through cholinergic mechanisms.

The consequences of enhanced sulfation of CF respiratory tract mucous glycoproteins remain unclear . Evidence that sulfation of macromolecules decreases the water of hydration of these substances (51) suggests that increased sulfation and a relative paucity of water in CF respiratory tract secretions are related in some way.

A question of considerable importance is the relationship between extent of glycoprotein sulfation and the genetic defect in CF. It has been suggested, based on results of histochemical studies, that increased amounts of sulfated glycoprotein may reflect long standing irritation of the lung airways (52). Thus, the increased sulfation of CF tracheobronchial glycoproteins may be secondary to infectious bronchitis. On the other hand, two pieces of information suggest that increased secretion of highly

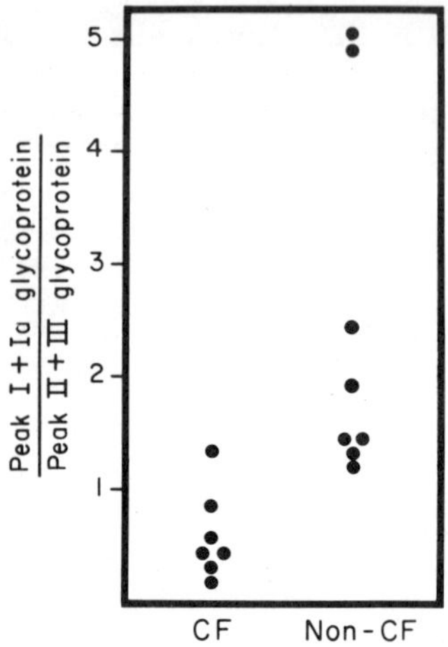

Archives of Biochemistry and Biophysics

Figure 3. Ratios of the amounts of tracheobronchial glycoprotein in sparsely acidic components (I, Ia) to amounts in highly acidic components (II, III) after fractionation on DEAE cellulose (see Figure 2). Secretions for these studies were obtained by bronchial lavage, performed by Robert E. Wood. Glycoproteins were quantitated by determining the total amount of fucose, galactose, galactosamine, glucosamine, and sialic acid in each fraction (40).

TABLE I

Distribution of Mucous Glycoprotein Components Secreted by Tracheal Explants of an O-secretor with CF, With and Without Methacholine Stimulation.

	No Methacholine			Methacholine[3]		
	I	II	III	I	II	III
% of Total Carbohydrate	34	49	16	57	32	11
Fucose[1]	3.3	4.0	--	1.2	0.9	--
Galactose[1]	6.1	6.3	--	2.7	3.1	--
Galactosamine[2]	1.0	1.0	--	1.0	1.0	--
Glucosamine[2]	2.6	5.1	--	2.3	3.2	--
N-acetylneuraminic acid[1]	0.8	0.3	--	0.3	0.3	--

1. Analyzed by gas liquid chromatography (40); residues/1 galactosamine.
2. Analyzed by amino acid analyzer (55); residues/1 galactosamine.
3. 50 ug/ml methacholine added to the culture medium for a 6 hour period.

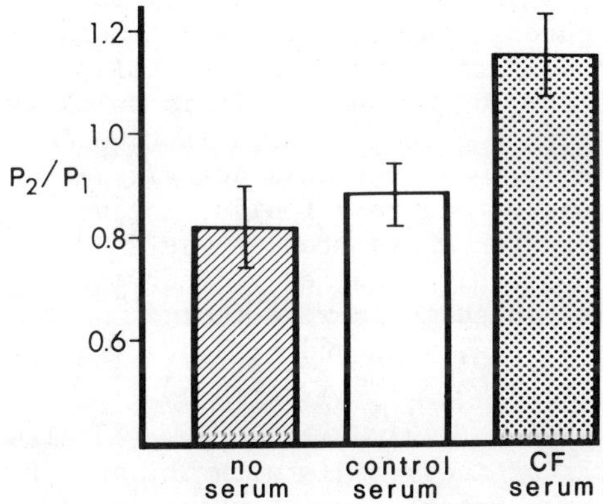

Figure 4. Ratio of $^{35}SO_4$ to ^3H in glycoproteins released from rabbit tracheal explants in the presence of no serum, 50% (v/v) normal serum, and 50% (v/v) CF serum. Data were generated and analyzed as outlined in legend for Figure 1. No serum vs. CF serum; $p < 0.02$. *Control serum vs. CF serum;* $p < 0.02$.

sulfated glycoproteins by CF respiratory epithelia may have a closer relationship to the gene defect. In double isotope studies, the ratio of $^{35}SO_4$ to ^{3}H-glucosamine incorporated into glycoproteins secreted by CF respiratory epithelial explants is significantly greater than the ratio for glycoproteins secreted by age-matched control explants (53). In addition, recent studies in our laboratory (Figure 4) suggest that brief exposure of rabbit tracheal explants to 50% CF serum effects the release of glycoproteins which are more highly sulfated than those released in the presence of 50% age and sex-matched control serum. CF tissues and/or body fluids may possess a property, independent of chronic infection, which modulates the process of glycoprotein sulfation, and which may play a role in the pathogenesis of the lung disease. It is unlikely that people with CF are missing a glycoprotein sulfatase activity, because we have been unable to detect this activity in both CF and non-CF tissue homogenates and body fluids, including purulent sputum. Substrate for these studies was $^{35}SO_4$-labeled respiratory mucous glycoprotein, labeled and harvested *in vitro*. Both intact and desialylated mucin were tested. Enzyme activity was assayed at several pHs, with and without divalent cations, and with and without Triton X-100 (54).

Summary: There are several, well documented alterations in the mucous secretions of individuals with CF, any of which alone or in combination, may interfere with mucous clearance from the lung or from passageways of other organs. It seems clear that additional studies to define basic aspects of normal mucous biochemistry, physiology and pharmacology must precede or parallel studies of CF mucous secretions if data pertinent to CF pathophysiology are to be generated. Particular needs include an understanding of the chemical determinants of physical and rheological properties of these secretions, and elucidation of mechanisms for control of sulfated glycoprotein biosynthesis.

Acknowledgements

This work was supported in part by Grants HL 14844 and HR 52957 from the United States Public Health Service, and a Grant from the Cystic Fibrosis Foundation of Cleveland.

Literature Cited

1. Wood, R.E., Boat, T.F., and Doershuk, C.F. Am. Rev. Resp. Dis. 113:833 (1976).
2. Davis, P., and di'Sant Agnese, P.A. New Eng. J. Med. 295:481 (1976).
3. Mayo, B.J., Klebe, R.J., Lankford, B.J., Morris, N.R., Barnett, D.R., and Bowman, B.H. Texas Reports on Biology and Medicine 34:107 (1976).
4. Conod, E.J., Conover, J., and Gaerlan, P. Pediat. Res. 11:45 (1977).
5. Mangos, J.A., and McSherry, N.R. Pediat. Res. 2:378 (1968).
6. Dearborn, D.G. In: Advances in Polyamine Research, Vol. II, Eds. D. Morris, R. Campbell, D. Daves, D. Bartos, and F. Bartos, Raven Press, New York (1978), pp. 273-279.
7. Bowman, B.H., Lankford, B.J., Fuller, G.M., Carson, S.D., Kurosky, A., and Barnett, D.R. Biochem. Biophys. Res. Commun. 64:1310 (1975).
8. Epstein, J., and Breslow, J.L. Proc. Natl. Acad. Sci. USA, 74:1676 (1977).
9. Rao, G.J.S., and Nadler, H.L. Pediat. Res. 9:739 (1975).
10. Shapira, E., Martin, C.L., and Nadler, H.L. J. Biol. Chem. 252:7923 (1977).
11. Davis, S.S. Bull. Physio-path. Resp. 9:47 (1973).
12. Potter, J.L., Spector, S., Matthews, L., and Lemm, J. Am. Rev. Resp. Dis. 99:909 (1969).
13. Khan, M.A., Wolf, D.P., and Litt, M. Biochim. Biophys. Acta 444:369 (1976).
14. Reid, L., and deHaller, R. Mod. Probl. Pediat. 10:195 (1967).
15. Zuelzer, W.W., and Newton, W.A. Pediatrics 4:53 (1949).
16. Reid, L. In: Ciba Foundation Study Group No. 32, Cystic Fibrosis, Eds. R. Porter and M. O'Connor, Little, Brown and Co., Boston (1968), pp.45-67.
17. Oppenheimer, E.H., and Esterly, N.R. Arch. Path. 96:149 (1973).
18. Bowman, B.H., Lockhart, L.H., et al. Science 164:325 (1969).
19. Bogart, B.I., Conod, E.J., and Conover, J.H. Pediat. Res. 11:131 (1977).
20. Czegledy-Nagy, E., and Sturgess, J.M. Lab. Invest. 35:58 (1976).
21. Boat, T.F., Kleinerman, J.I., Carlson, D.M., Maloney, W.H., and Matthews, L.W. Am. Rev. Resp. Dis. 110:428 (1974).
22. Baker, A.P. "GAP" Conference Reports, Cystic Fibrosis Foundation (1976).
23. Boat, T.F., Cheng, P.W., and Wood, R.E. Mod. Probl. Pediat. 19:141 (1977).
24. Baker, A.P., Hillegass, L.M., et al. Am. Rev. Resp. Dis. 115:811 (1977).

25. Baker, A.P., and Hillegass, L.M. Arch. Biochem. Biophys. 165:597 (1974).
26. Boat, T.F., and Cheng, P.W. In: Cystic Fibrosis, Projections into the Future, Eds., J.A. Mangos and R.C. Talamo, Stratton Intercontinental Medical Book Corp., New York, 1976, pp. 165-177.
27. Neutra, M., Grand, R.J., and Trier, J.S. Lab. Invest. 36:535 (1977).
28. Sturgess, J., and Reid, L. Clin. Sci. 43:533 (1972).
29. Dearborn, D.G. In: Cystic Fibrosis: Projections into the Future, Eds., J.A. Mangos and R.C. Talamo, Stratton Intercontinental Medical Book Co., New York, 1976, pp. 179-191.
30. Wood, R.E., Wanner, A., Hirsch, J., and Farrell, P.M. Am. Rev. Resp. Dis. 111:233 (1975).
31. Forstner, J.F., and Forstner, G.G. Pediat. Res. 10:609 (1976).
32. Gugler, E., Pallavicini, J.C., Swerdlow, A., and di Sant' Agnese, P.A. J. Pediat. 71:585 (1967).
33. Boat, T.F., Wiesman, U.N., and Pallavicini, J.C. Pediat. Res. 8:531 (1974).
34. Bailleul, V., Richet, C., Hayem, A., and Degand, P. Clin. Chim. Acta 74:115 (1977).
35. Degand, P., Roussel, P., Lamblin, G., Durand, G., and Havez, R. Bull. Physio-Path. Resp. 9:199 (1973).
36. Roberts, G.P. Arch. Biochim. Biophys. 173:528 (1976).
37. Shih, C.K., Litt, M., Khan, M.A., and Wolf, D.P. Am. Rev. Resp. Dis. 115:989 (1977).
38. Snary, D., Allen, A., and Pain, P.H. Biochem. Biophys. Res. Commun. 40:844 (1970).
39. Hill, H.D., Reynolds, J.A., and Hill, R.L. J. Biol. Chem. 252:3791 (1977).
40. Boat, T.F., Cheng, P.W., Iyer, R., Carlson, D.M., and Polony, I. Arch. Biochem. Biophys. 177:95 (1977).
41. Roussel, P., Lamblin, G., Degand, P., Walker-Nasir, E., and Jeanloz, R.W. J. Biol. Chem. 250:2114 (1975).
42. Cheng, P.W., and Boat, T.F. Fed. Proc. 35:1444 (1976).
43. Lopez-Vidriero, M.T., Das, I., and Reid, L.M. In: Respiratory Defense Mechanisms, Part I, Eds., J.D. Brain, D.F. Proctor, and L.M. Reid. Marcel Dekker, New York, (1977), pp. 289-356.
44. Gottschalk, A., and Thomas, M.A.W. Biochim. Biophys. Acta 46:91 (1961).
45. Meyer, F.A., King, M., and Gelman, R.A. Biochim. Biophys. Acta 392:223 (1975).
46. Deman, J., Mareel, M., and Brunyneel, E. Biochim. Biophys. Acta 297:486 (1973).
47. Cheng, P.W. Unpublished Observation.
48. Lamblin, G., Lhermitte, M., Lafitte, J.J., Filliat, M., Degand, P., and Roussel, P. Bull. Europ. Physiopath. Resp.

13:175 (1977).
49. Boat, T.F., Davis, J., Cheng, P.W., and Stern, R.C. Proceedings of the Fourth International Symposium on Glycoconjugates, In press.
50. Dische, Z., di Sant'Agnese, P.A., Pallavicini, C., and Youlos, J. Pediat. 24:74 (1959).
51. Ablett, S., Lillford, P.J., Baghdadi, S., and Derbyshire, W., In: Magnetic Resonance in Colloid and Interface Science, Eds., H. Resing, and C. Wade, A.C.S. Symposium Series 34 (1976), p. 344.
52. Jeffery, P.K., and Reid, L.M., In: Respiratory Defense Mechanisms, Part I, Eds., J.D. Brain, D.F. Proctor, and L.M. Reid, Marcel Dekker, New York, (1977), pp. 193-245.
53. Boat, T.F., and Cheng, P.W., In: Cystic Fibrosis: Projections into the Future, Eds., J.A. Mangos and R.C. Talamo, Stratton Intercontinental Medical Book Corporation, New York, (1976), pp. 165-177.
54. Jentoft, N., and Cheng, P.W. Unpublished observations.
55. Cheng, P.W., and Boat, T.F. Anal. Biochem. 85:276 (1978).
56. Boat, T.F., and Kleinerman, J.I. Chest 67:32 S (1975).

RECEIVED April 17, 1978.

6

Glycoprotein Storage Disease

J. NEVIN ISENBERG

Pediatric Gastroenterology Division, Department of Pediatrics—UCLA, Center for the Health Sciences, Los Angeles, CA 90024

I. Alpha-1-antitrypsin (αAT) is a serum glycoprotein that has attracted considerable attention from the medical community because of its association with two divergent disease processes - emphysema and cirrhosis. On routine serum protein electrophoresis, it is the predominant protein in the α_1 globulin area. Ironically (with regard to the current topic), the disease states are associated with a marked reduction in the protein appearing in this band. Thus, this routine clinical test can serve as a screening test for alpha-1-antitrypsin deficiency-type emphysema or cirrhosis. This subject has been reviewed in 1976 by Harvey L. Sharp, M.D., who first recognized the association between α1AT deficiency and childhood cirrhosis.[1] In the subsequent paragraphs, we will utilize the material presented there and update the current status with information published since then.

The protein has a molecular weight in the range of 50,000. It has approximately 12% carbohydrate which includes galactose, mannose, N-acetyl-glucosamine and sialic acid. The name, α1AT, comes from its electrophoretic position ($-\alpha_1-$) and from the fact that it is responsible for 90% of the serum trypsin inhibitory capacity. Since this potential may be less important physiologically than the inhibitory capacity for other proteases, it has also been referred to as α_1 protease inhibitor as the study of protease inhibitors (P_i) has increased in general. As it is an acute phase reactant, its serum levels and presumably synthesis increase during inflammation.

Serum contains about 2 mg of α1AT per ml which has the capacity to inhibit 1.1 mg of trypsin. Both quantity (immunodiffusion methods) and function (trypsin inhibitory capacity) are used to evaluate serum from patients with emphysema or cirrhosis. The complexities begin when the deficient serum is P_i typed. Using an acid starch gel electrophoresis, Fagerhol first appreciated the microheterogeneity of α1AT which results in three major and five minor protein bands as products of each

0-8412-0452-7/78/47-080-122$05.00/0

allele.[2] When this separation was followed by antigen-antibody crossed electrophoresis, the full range of 24 phenotypes for this genetic locus was appreciated. Since each allele is expressed by a gene product (with microheterogeneity) the inheritance pattern is regarded as autosomal codominant. P_i type "M" is the predominant phenotype and has moderate electrophoretic mobility. Phenotypes moving more rapidly are identified by letters which are before "M," while slower ones are designated by letters after "M." Reduced circulating α1AT has been associated with alleles Z (< 10%), P (12.5%), S (30%) and with P_i-- in which no gene product has been identified. The "M" allele is 50% (MM=100%). Disease states are associated, then, with diploid phenotypes giving low activity such as P_iZZ, P_iSZ, P_i--, etc.

Population genetic studies have shown the wide variation in allele distribution. The highest gene frequency for P_iZ is .026[3] in Swedes vs. .0116 in a study from St. Louis.[1] P_iS is the pathologic allele most frequent in Spain (.112 vs. .034 USA). Pathologic P_i types are very infrequent among non-whites. The P_i allele system seems closely linked to the G_m locus (IgG heavy chain marker) with placement on chromosome 2,8 or 12 likely.[4] In Sveger's study[3], 120 P_iZZ infants and 48 P_iSZ infants were detected from 200,000 live births, indicating that one infant in 1200 births had a potentially pathologic phenotype! The figure (proportionately lower) for the USA makes this inherited glycoprotein disorder among the most common genetic disorders. Prenatal diagnosis is not possible.

The Lung Disease

The postulated mechanism by which emphysema develops in the deficient phenotypes is thought to be the result of uninhibited proteolysis that destroys the fibrous network of the lungs. α1AT does inhibit the leukocyte proteases including elastase and collagenase and is small enough to leave the vasculature to diffuse to a site of inflammation to function as a protease inhibitor. The susceptibility of heterozygotes (e.g., P_iMZ) to environmental influences which would provoke lung damage has been disputed.[1] In a recent study heterozygous children compared with normals from the same families showed no differences except in forced flow pulmonary functions analyzed by matched pairs.[5] These abnormalities were of the type seen with α1AT deficiency emphysema and suggest a predisposition for the heterozygote. Only follow-up will clarify the significance of this finding.

The Liver Disease

When Sharp first associated childhood cirrhosis with α1AT deficiency (P_iZZ), the hepatocytes contained eosinophilic cytoplasmic granules that were shown to be PAS positive after diastase digestion of glycogen. These globules are most evident in periportal hepatocytes and become more prominent with age in those children affected. These granules are positive when sections are treated with fluorescein-labelled antibody to α1AT. These globules are found on close inspection in individuals with P_iZ phenotype whether or not cirrhosis is present. A single report has identified globules in the liver of a 79-year-old man with P_iMM phenotype that were PAS(+) and positive for α1AT by immunofluorescent and immunoperoxidase methodology.[6] The significance of α1AT globules in the absence of a pathologic phenotype is not clear at present.

This material appears in dilated portions of the endoplasmic reticulum (ER) and in no other organelle when specimens are examined with the electron microscope. It is usually amorphous with a fine granular or occasional fibrillar appearance. Both smooth and rough endoplasmic reticulum are involved but not the Golgi apparatus. Cells are not always uniformly involved and only 80% have any material.[7] These are concentrated in periportal areas. The globules vary in size from one to 40 microns in diameter, with 15-20 microns diameter particles predominating.

The liver injury begins at birth as evidenced by the frequency of cholestasis in newborns with P_iZZ or P_iSZ phenotypes.[1,3] However, some children escape this phase but demonstrate cirrhosis when liver biopsy is done as part of the clinical evaluation of hepatosplenomegaly. The cholestatic phase usually passes within six months. Although affected children then remain asymptomatic until cirrhosis becomes sufficiently advanced to result in portal hypertension with ascites and/or hematemesis, liver function tests show continued evidence for hepatocellular injury without cholestasis. Interestingly, however, hyperbilirubinemia may precede enzyme elevation in the newborn period!

While originally the liver disease presenting in early childhood was thought to be almost uniformly fatal in adolescence secondary to the consequences of advancing cirrhosis, others have had experiences to the contrary.[8] The Toronto group reported four of 18 patients with neonatal cholestasis who have no liver function or physical abnormality in mid-adolescence.[8] Early cirrhosis with or without neonatal cholestasis was a bad prognostic factor.[8] In Sveger's neonatal screening study, the P_iZZ infants were classified into three groups according to the severity of hepatic involvement.[3] Fourteen of 122 had prolonged obstructive jaundice, eight had a suspicion of liver disease by examination, and the remainder had no clinical evidence for liver disease. The P_iSZ children were all in the third category. Continued evaluation of the infants over their first six months of life showed that half of those in the third category with P_iZZ phenotypes had biochemical abnormalities indicating liver injury. This was not found in the SZ group.

Sass-Kortask could draw no prognostic significance from biopsies taken during the cholestatic phase. They saw equivalent degrees of injury in well survivors and early deaths.[8] French workers have formed a different conclusion on the basis of reviewing biopsies from fifteen P_iZZ children with neonatal cholestasis.[9] Those that showed improvement in liver injury had cholestasis and cellular injury, but only slight portal fibrosis. When fibrosis was extensive and bile ducts proliferated, early cirrhosis occurred. A third group, in which prolonged cholestasis was noted, had ductular hypoplasia. The follow-up period for most of the patients in the series is less than five years, which is insufficient to make a final judgement; but those in Groups II and III are progressing toward the complications of cirrhosis while those in Group I seem to be improving.[9] All had dPAS(+) globules whose size and number did not correlate with the histologic grouping.[9]

PAS(+) globules are also noted in emphysematous adults of P_iZZ phenotype.[1] Fibrosis, cirrhosis and hepatoma were noted. No α1AT globules are seen in the malignant cells.[10] Adults with phenotypes resulting in partial deficiencies had no greater incidence of liver disease than P_iMM controls.[11]

The Liver Storage Material

Clearly, the accumulation (storage?) of α1AT-like material in the liver seems well documented and is associated with cirrhosis. The mechanism by which the hepatic injury occurs is more difficult to envision than that proposed for the development of emphysema in deficient phenotypes. The liver disease is not merely a product of deficient serum α1AT since it seems to be most easily related to the P_iZ allele and not other deficient phenotypes. Further, the P_iZ protein retains full protease inhibitor capacity proportionate to the amount present.

An understanding of the mechanism for liver injury will very likely require a better understanding of how the α1AT-like material gets to the liver in the first place. It is synthesized by the liver and recently has been partially characterized as a product of human liver cell monolayer cultures. P_iM and P_iZ types were cultured. Only the P_iZ cells showed globules with α1AT immunofluorescent positivity. The P_iZ protein from the cells differed from that released to the culture medium and extracted from the supernatant.[12] This may become an important laboratory approach to understanding more about the pathogenesis of liver disease.

In his review, Sharp listed four hypotheses that might explain the accumulation of α1AT in the livers of P_iZZ individuals:

1) A point mutation leading to an amino acid substitution in the peptide which would interfere with subsequent transport from the liver; 2) A glycyl transferase deficiency resulting in an inadequately glycosylated protein that could not be transported; 3) P_iZZ is synthesized and secreted but more rapidly taken up by the liver; and 4) A pro- α1AT is the storage product because a cleaving enzyme is missing. It could be added that the amino acid substitution (1) might result in inadequate glycosylation (2) and secretory failure.

Many groups are working on isolation of α1AT from serum and liver. No one has found a markedly larger α1AT in the liver so that support for pro-α1AT is minimal.

Increased hepatic uptake is a known feature of asialoglycoproteins. Working with purified α1AT, Gan's group showed that removing sialic acid and enzymatically altering the underlying galactose residue, altered the circulation half-life but did not change binding to concanavilin A, trypsin, or chymotrypsin inhibitory activity or immunological reactivity.[13] The unaltered α1AT had a half-life of 18 hours. After neuraminidase treatment removing four of six sialic acid residues, the half-life was reduced to 30 minutes. With subsequent treatment of the galactose residues with galactose oxidase, the half-life was restored to 18 hours.[13]

Similarly, isolated M and Z α1AT were labelled with radioactive iodine and injected into P_iMM subjects. Simultaneous disappearance rate measurement showed the half-life of M equal to seven days and for Z five days. These differences are too small to explain the serum deficiency seen in P_iZZ.[14]

Many investigators have suggested that the α 1AT of P_iZZ is deficient in sialic acid, but the reported amount and composition of the carbohydrate on this variant protein shows many discrepancies from group to group. In 1975, the Swedish group reported isolation and characterization of the α1AT P_iZ from the inclusion bodies in a P_iZZ liver.[15] It contained no sialic acid, was virtually insoluble, failed to inhibit proteases but did show the same immunoreactivity as serum α1AT.[15] Its amino acid composition was similar to P_iM but the sugars were all reduced to the following percent of that found in P_iM: mannose - 50%, galactose - 25%, N-acetylglucosamine - 16%. These findings, in the investigator's view, suggested a potential substitution in the amino acid backbone leading to decreased solubility, ineffective transport through the ER and secondarily inadequate glycosylation.[15]

In the same period, Sharp joined forces with Glew and co-workers to pursue the latter's observation that sialyltransferase was deficient in the serum of patients with the P_iZZ phenotype. After more extensive studies, it appeared that the decreased serum* levels were in general related to advanced cirrhosis but not accompanied by decreased activity in liver homogenates. No kinetic differences in the enzyme were noted between the sialyltransferase from P_iMM and P_iZZ.[1]

In 1976, two groups reported an amino acid substitution in the protein backbone of the Z variant of α1AT.[16,17] Both groups made peptide maps from digests of the isolated α1AT made in two different ways and found differences in one or more peptides. Both agree that lysine is substituted for glutamic acid in the major variant peptide. The USA group believe that another glutamic acid in the M peptide is replaced by glutamine in the Z peptide. Similarly, an amino acid substitution was confirmed in a histidine containing peptide of the S-variant.[18] Valine was substituted again for glutamic acid.[18,19] The S and M variant had the same sialic acid and carbohydrate content.[19]

However, dissenting voices have been raised about the Z variant findings. Owen and Carrell could find no peptide from the Z variant in the expected part of the peptide map after tryptic digestion of both M and Z.[20] A sequential analysis of an M peptide not seen in the Z map enabled prediction where the variant peptide should be. Thus far, sequential analyses of numerous peptides from the digests has failed to show any amino acid differences.

Further information strengthening the argument for the potential accumulation of an asialo- α1AT comes from studies of Emerson which imply that the liver asialoglycoprotein binding protein (LABP) is present in the membranes of the Golgi apparatus and SER. Thus, P_iZ with exposed galactosyl residues secondary to incomplete sialylation would bind to the SER and be no further transported.[21] Comparing binding to LABP in vitro, native serum P_iZ binds more avidly than native M but much less well than completely desialylated M. This suggests that there is circulating P_iZ which has exposed galactosyl residues.

Circular dichroism has been applied to the analysis of α1AT from normal serum. Sixteen to 20% is α-helix, 30-50% in the β pleated sheet conformation and the remainder is aperiodic. This conformational profile remained stable over a pH range from 4.7 to 8.8, but was markedly changed by pH 2.5, guanidine hydrochloride and SDS. These findings implicate electrostatic interaction, hydrogen bonding and hydrophobic regions as vital to the three dimensional structure of α1AT.[22] Obviously, the application of this technique to the variant α1AT with single amino acid substitutions may yield information on conformational changes that could interfere with glycosylation. Its application to the problem of microheterogeneity will likewise be interesting.

Microheterogeneity

The factors responsible for the microheterogeneity in the products of a single gene are not resolved. In some glycoproteins the microheterogeneity has been attributed to a variable degree of sialylation. Using isoelectric focusing to analyze the DEAE cellulose - NaCl gradient column chromatographic separation of purified P_iM, five pooled fractions each showed up to three of the five bands present in the α1AT preparation before DEAE fractionation. The first pool eluted with a low NaCl concentration had the more cathodal isoproteins, while the last pool (hi concentration NaCl) had the more anodal ones. If this was due to sialylation differences, then Pool V should have been richer in sialic acid. However, all pools were homogenous by immunoelectrophoresis and had the same amino acid, hexose, hexosamine and sialic acid concentrations! When the pools were enzymatically desialylated the bands in each pool shifted cathodally but the number and relative positions of the isoproteins remained the same. This study casts doubt on uneven sialylation as an explanation for the microheterogeneity in α1AT.[23]

This ability to separate the various forms of the gene product will enable the ultimate identification of what is responsible for the microheterogeneity. Allen has used isoelectric focusing to study the microheterogeneity in P_iZZ children with cirrhosis and compared it to adults with emphysema and P_iZZ individuals with no clinically evident disease.[24] All bands inhibit trypsin and were antigenically α1AT by crossed immunoelectrophoresis. The cirrhotic children had higher levels of α1AT and increasing amounts with age of isoprotein in the Z2 (anodal) band region (isoprotein point = 4.534). Thus, it appears that there are differences even in the P_iZ product which might be used in conjunction with newborn screening to predict whether infants of P_iZZ phenotype will be at risk for liver or lung disease. Using this Z2 band difference as their marker, family studies showed that penetrance of the Z allele was of the order of 70%.[25] If these observations hold true, then it may be possible to better understand why some P_iZZ families have liver disease without lung involvement and why the converse exists. Up to this point, environmental factors (which would stimulate the synthesis of acute phase reactant) were being sought to help explain the onset of liver disease in the newborn. The majority of investigators had been unable to identify any infectious or toxic agent present in those children developing the cholestatic phase of the disease.[1] Furthermore, since the incidence of small for gestational age infants is increased in the P_iZZ phenotype, there is some evidence that the disease may begin in utero.[1,3]

Therapy

No therapy exists for this glycoprotein storage disease.

Factors that increase serum levels of α1AT in P_iMM do not alter the level of P_iZZ. Steroids have not been helpful. Several patients have received liver transplants with conversion of serum P_i phenotype from ZZ to MM. All died of transplantation complications. At autopsy, no PAS(+) globules were noted in the liver which speaks strongly against a peripheral desialylation mechanism for altering α1AT with resultant storage.

II. ASPARTYLGLUCOSAMINURIA

In the usual sense, storage disease implies accumulation of a macromolecule whose catabolism is interrupted by a deficiency in one or more of the hydrolytic enzymes involved in its breakdown. For glycoproteins, disturbances in the breakdown of the peptide backbone or carbohydrate side chains could result in such accumulation. However, deficiencies in proteases would affect all proteins and have far reaching, perhaps lethal, effects. In addition, glycoprotein accumulation might be the consequence of disturbances in the breakdown of the carbohydrate side chain. Yet, the carbohydrate side chains are shared by the glycolipids and abnormalities in their cleavage are discussed elsewhere in this symposium.

The bond unique to the glycoproteins is that between the first sugar moiety of the side chain and the amino acid residue in the peptide backbone. In man, the predominant bond in serum glycoproteins seems to be the N-glycosidic linkage between N-acetylglucosamine and asparagine. The lysosomal enzyme responsible for the hydrolysis of this bond is [2-acetamido-1-(β-L-aspartamido)-1,2-dideoxy-β-D-glucose aspartamido hydrolase (EC 3.5.1.26)] AADGase.

A deficiency in this enzyme was first demonstrated in two mentally retarded English siblings who had excreted large amounts of an unusual compound in the urine which was identified as 2-acetamido-1-(β-L-aspartamido)-1,2-dideoxy-β-D-glucose (AADG).[1] This was the initial recognition of the disorder, aspartylglucosaminuria (AGU), which has subsequently been diagnosed in Finland[2] and the United States.[3]

The Finnish workers have demonstrated the prevalence of this disorder in their country (1:26,000 live births) by screening residents of national institutions for the mentally retarded.[4] The coarse features, visceromegaly and vacuolated lymphocytes result in clinically confusing this disease with the mucopolysaccharidoses and mucolipidoses.[3] Mental retardation characterized by impaired speech and motor clumsiness predominates.[3,4] Connective tissue lesions including dyostosis multiplex, umbilical hernias and hypermobile joints are common.[3,4] Systolic heart murmurs are frequent. Our patient showed mitral insufficiency on heart catheterization suggesting that this may be another metabolic disorder directly influencing valvular function.[4] Despite a predisposition to recurrent infections, no deficits in leukocyte function have been demonstrated.[5]

Extensive light and electron microscopic studies have been carried out on various tissues from affected patients.[6,7,8] The major changes in all tissues are lysosomal. In addition to the readily discernible lymphocyte vacuoles on the peripheral smear, electron dense filamentous material is noted in the enlarged lysosomes of leukocytes.[8] Virtually every lysosome in the hepatic parenchymal cells is enlarged. Some PAS positive droplets are seen. Ultrastructurally, the lysosomes contain fine granular material, electron dense structures, and electron lucent lipid droplets.[6,8] Kupffer cell lysosomes are likewise involved.[6,8] Similar lysosomal changes have been noted in the apical portions of proximal tubule cells in the kidney[6], in the basal portion of jejunal enterocytes[8], and in fibroblasts from a skin biopsy.[8] The most striking changes have occurred in the neuronal cytoplasm from cerebral cortex, cerebellum and thalamic regions.[7] The cytoplasm is virtually filled with membrane bound vacuoles of either electron lucent granular material or electron dense membranous or granular material with occasional lipid droplets.[7] No single feature is pathognomonic for AGU although seldom is such a variety of lysosomal changes seen in other storage diseases.

The reduced activity of AADGase was first noted in a patient by assay of seminal fluid.[1] Markedly decreased activity was observed in brain, liver and spleen but not in kidney biopsy material.[9] Leukocyte homogenates are also deficient in AADGase.[8] Other lysosomal enzymes involved in the catabolism of the carbohydrate side chain have normal activity levels except for N-acetyl-β-glucosaminidase which is elevated.[8,9] Measurement of the enzyme in cultured fibroblasts permits detection of heterozygotes and enables prenatal diagnosis.[10] AADGase has been isolated from liver of normals and AGU enabling mixing experiments that rule out inhibitors as a cause for the reduced activity.[11,12] Only one isoenzyme has been noted.[11,12] The pH optimum (7.7) and Km (0.16 mM) have been determined on a 17,500 fold purification of AADGase that was inhibited by N-acetylcysteine but not other SH agents.[12]

Even from the first description of AGU, compounds other than AADG were appreciated in the urine.[1] Characterization of these plus the nature of the principal tissue storage compound has taken place. The only quantitatively important compound in tissue is AADG.[13] The higher molecular weight compounds containing asparagine but with additional carbohydrate residues have been termed glycoasparagines and fall into neutral and acidic groups when analyzed by high voltage electrophoresis at pH 2.0 and pH

5.3.[13] Five neutral and nine acidic glycoasparagines were separated and analyzed. Neutral glycoasparagines (N_1 - N_5) differed in the number of neutral sugar residues attached to AADG. The acidic glycoasparagines fell into three groups with one or two sialic acid residues; the sulphated compounds formed the third group. The neutral sugars identified were glucosamine > galactose >> mannose > glucose > fucose. Mannose was noted only in the neutral group, glucose only in acidic compounds and fucose in both.[14] Aula, working in conjunction with several Japanese biochemists, has reported the sequence of several glycoasparagines as shown in the table below[15,16] :

GLYCOASPARAGINE SEQUENCES IN THE URINE OF AGU

AADG

β-Gal-(1→4)-AADG

α-NANA-(2→3)-β-Gal-(1→4)-AADG

α-NANA-(2→4)-β-Gal-(1→4)-AADG

α-NANA-(2→?)-β-Gal-(1→?)-β-GlcNAc-(1→?)-β-Gal-(1→4)-AADG

α-Man-(1→6)-β-Man-(1→4)-β-GlcNAc-(1→4)-AADG

That these compounds are present in only small amounts in the tissues thus far studied makes the source of urine glycoasparagines open to speculation. They may have their origin in connective tissue. We have suggested that the AADG may serve as an acceptor compound for various transferases leading to synthesis of new glycoasparagines which are released from tissue sites after reaching some critical conformation.[8]

ACKNOWLEDGEMENT

To Harvey L. Sharp, M.D., for introducing me to both these diseases of glycoprotein metabolism.

LITERATURE CITED

I. Alpha-1-Antitrypsin

1. Sharp HL: Gastroenterology 70: 611 (1976).

2. Fagerhol MK: Scand J Clin Lab Invest 23: 97 (1969).

LITERATURE CITED (Cont.)

I. Alpha-1-Antitrypsin (Cont.)

3. Sveger T: N Engl J Med 294: 1316 (1976).

4. Third International Workshop - Human Gene Mapping: Cytogenetics and Cell Genetics 16: (1976).

5. Vance JC, Hall WJ, Schwartz RH, Hyde RW, Roghmann KJ, and Mudholkar GC: Pediatrics 60: 263 (1977).

6. Bradfield JWB and Blenkinsopp WK: J Clin Path 30: 263 (1977).

7. Yunis EJ, Agostini Jr RM, and Glew RH: Am J Pathol 82: 265 (1976).

8. Moro SP, Cutz E, Cox DW, Sass-Kortsak A: J Pediatr 88: 19 (1976).

9. Hadchouel M and Gautier M: J Pediatr 89: 211 (1976).

10. Blenkinsopp WK and Haffenden GP: J Clin Path 30: 132 (1977).

11. Fisher RL, Taylor L and Sherlock S: Gastro 71: 646 (1976).

12. Gautier M, Martin JP and Polim G: Biomedicine (Express) 27: 116 (1977).

13. Yu SD and Gan JC: Arch Biochem Biophys 179: 477 (1977).

14. Laurell CB, Nosslin B and Jeppsson JO: Clin Sci Molec Med 52: 457 (1977).

15. Jeppsson JO, Larsson C and Eriksson S: N Engl J Med 293: 576 (1975).

16. Jeppsson JO: FEBS Letters 65: 195 (1976).

17. Yoshida A, Lieberman J, Gaidulus L and Ewing C: PNAS 73: 1324 (1976).

18. Owens MC, Carrell RW and Brennan SO: BBA 453: 257 (1976).

19. Yoshida A, Ewing C, Wessels M, Lieberman J and Gaidulis L: Am J Hum Genet 29: 233 (1977).

20. Owens MC and Carrell RW: FEBS Letters 79: 245 (1977).

LITERATURE CITED (Cont.)

I. Alpha-1-Antitrypsin (Cont.)

21. Emerson WA: GAP Conference 1: 7 (1977).

22. Jirgensons B: BBA 493: 352 (1977).

23. Hercz A and Barton M: Can J Biochem 55: 661 (1977).

24. Allen RC and Hennigar GR: Am J Clin Path 67: 209 (1977).

25. Allen RC: GAP Conference 1: 16 (1977).

II. ASPARTYLGLUCOSAMINURIA

1. Pollitt RJ, Jenner FA and Merskey H: Lancet 2: 253 (1968).

2. Palo J and Mattsson K: J Ment Defic Res 14: 168 (1970).

3. Isenberg JN and Sharp HL: J Pediatr 86: 713 (1975).

4. Autio S: J Ment Defic Res (Monograph Series 1) 1972.

5. Fleisher TA, Isenberg JN and Sharp HL: J Pediatr 87: 833 (1975).

6. Arstila AU, Palo J, Haltia M, Riekkinen P and Autio S: Acta Neuropath (Berl) 20: 207 (1972).

7. Haltia M, Palo J and Autio S: Acta Neuropath (Berl) 31: 243 (1975).

8. Isenberg JN and Sharp HL: Human Pathol 7: 469 (1976).

9. Palo J, Riekkinen P, Arstila AU, Autio S and Kivimäki T: Acta Neuropath (Berl) 20: 217 (1972).

10. Aula P, Näntö V, Laipio ML and Autio S: Clin Gen 4: 297 (1973).

11. Savolainen H: Biochem J 153: 749 (1976).

12. Dugal B and Strømme J: Biochem J 165: 497 (1977).

13. Palo J, Pollitt RJ, Pretty KM and Savolainen H: Clin Chim Acta 47: 69 (1973).

14. Pollitt RJ and Pretty KM: Biochem J 141: 141 (1974).

LITERATURE CITED (Cont.)

II. ASPARTYLGLUCOSAMINURIA (Cont.)

15. Sugahara K, Funakoshi S, Funakoshi I, Aula P and Yamashina I: J Biochem 80: 195 (1976).

16. Akasaki M, Sugahara K, Funakoshi I, Aula P and Yamashina I: FEBS Lett 69: 191 (1976).

RECEIVED March 15, 1978.

Lysosomal Enzyme Deficiency Diseases—Glycoprotein Catabolism in Brain Tissue

ERIC G. BRUNNGRABER

Missouri Institute of Psychiatry and the Departments of Psychiatry and Biochemistry, University of Missouri—Columbia, St. Louis, MO 63139

The heteropolysaccharide chains of the glycoproteins in brain tissue provide the glycosidic linkages which serve as the substrates for a number of lysosomal glycosidases. The genetically-induced absence of a glycosidase, or a mutation which causes one or more of these glycosidases to become non-functional, produces a block in the catabolism of these protein-linked heteropolysaccharides. This pathological event, observed in the lysosomal enzyme deficiency diseases, causes an accumulation of undegraded glycopeptides or oligosaccharides, and an excessive urinary excretion of these substances.

About 65 percent of the glycoprotein-carbohydrate of brain tissue is released in the form of sialoglycopeptides which contain NeuNAc, Fuc, GlcNAc, Gal, and Man (1) after treatment of the glycoproteins with proteolytic enzymes, usually papain or pronase. Approximately 20-40 percent of these sialoglycopeptides contain, in addition, sulfate ester groups. All of these sialoglycopeptides appear to contain 3 Man residues per glycopeptide molecule (Table I). Two, three, and possibly four chains consisting of -GlcNAc-Gal-NeuNAc (Fuc) are attached to this trimannoside core. The sialoglycopeptides were partially resolved by column electrophoresis. Fraction I contains the sialoglycopeptide with the highest molecular size and negative charge. The molecular size

TABLE I
SIALOGLYCOPEPTIDES DERIVED FROM BRAIN GLYCOPROTEINS

Fraction:	I	II	III	IV	V	VI	VII
NeuNAc	4	4	3	3	2	1	1
Gal	4	4	4	3	3	2	2
Man	3	3	3	3	3	3	3
GlcNAc	6	6	6	6	5	4	4
Fuc	1	1	1	1	2	1-2	1-2

0-8412-0452-7/78/47-080-135$05.00/0

and charge decreased as the number and size of the -GlcNAc-Gal-NeuNAc (or Fuc) chains attached to the trimannoside core is reduced.

The composition of fractions VI and VII is of particular interest. These sialoglycopeptides account for about 15-20 percent of the glycopeptide-carbohydrate of whole rat brain. Two -GlcNAc-Gal-NeuNAc (or Fuc) chains are attached to the trimannoside core. The Man residues are sufficiently exposed so that these glycopeptides show a greater affinity to Con A-Sepharose columns (2) than the more acidic polysialoglycopeptides of fractions I to IV. These glycopeptides also have a greater ability to inhibit the precipitation of glycogen by Concanavalin A than the more negatively charged polysialoglycopeptides (3). It was also found (4) that these glycopeptides yield methylation products that resemble those obtained from a glycopeptide from transferrin:

(A)

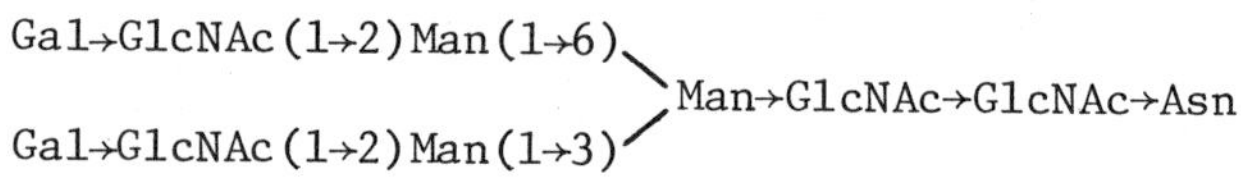

Sialoglycopeptides of fractions I to V, characterized by higher values for the ratios of NeuNAc/Man, Gal/Man, and GlcNAc/Man may possess structures which resemble those of the glycopeptides derived from fetuin (4), as suggested some years ago (5) and for which Krusius et al (4,6) have obtained some evidence.:

(B)

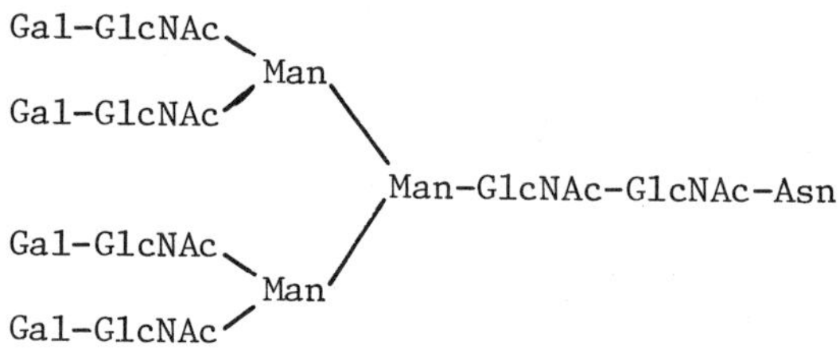

The brain sialoglycopeptides contained relatively high amounts of terminal nonsubstituted Gal and GlcNAc residues which suggested that the peripheral chains are often incomplete (6).

As we shall see, an accumulation of glycopeptides with structures corresponding to (A) have been shown to accumulate in fucosidosis, I-cell disease, and the GM1- and GM2-gangliosidoses. An accumulation of oligosaccharides or glycopeptides with 3 or 4 terminal -GlcNAc-Gal-NeuNAc (or fucose) branches has not yet been reported. Does this suggest the presence in tissues of a hitherto undescribed endo-β-glucosaminidase which splits the bond between the GlcNAc residue of the -GlcNAc-Gal-NeuNAc(or Fuc) branch chain linked to the trimannoside core? The existance of such an enzyme might explain the relatively large accumulation of oligosaccharides and glycopeptides with structure (A) in the liver, brain and urine in several of the lysosomal enzyme deficiency diseases, since this structure would be the accumulated end product of the catabolism of a wide variety of heteropolysaccharide chains.

Finne et al (7) demonstrated the presence of disialosyl (α-N-acetylneuraminyl-(2→8)N-acetylneuraminyl) groups in the sialoglycopeptides from brain. Thus, glycopeptides of fractions I and II, which contain 4 NeuNAc residues, may contain disialosyl groups, ensuring the availability of one or two of the 4 Gal residues as a terminal sugar or as a sugar that is penultimate to a terminal fucose or sulfate residue.

Approximately 20-25 percent of the glycoprotein-carbohydrate of rat brain consisted of mannoglycopeptides which contain 6 Man and 2 GlcNAc residues per glycopeptide molecule (8). Mannoglycopeptides, readily isolated since they bind strongly to Con A-Sepharose, contained terminal Man residues and would not be affected in the GM1-gangliosidosis and fucosidoses. They would make a major contribution to the accumulated oligosaccharides in mannosidosis.

When purified liver lysosomes were incubated in the presence of fetuin or orosomucoid (9), about one half of the peptide bonds were cleaved and most of the terminal NeuNAc residues were released. Release of Gal and GlcNAc was slower and did not exceed 30 percent of the theoretical yield. Release of Man was not detected, and branch points in which Man residues were involved appeared to resist attack.

Many studies have shown that purified β-galactosidase and β-N-acetylglucosaminidase can cleave terminally exposed carbohydrate groups of intact glycoproteins, and it appears likely that this process occurs in the lysosomes of the intact organism. There may be an exception, however. Highly purified α-fucosidase from rat liver catabolized the hydrolysis of 1-2, 1-3, and 1-4 fucosyl linkages in glycopeptides, but not in intact glycoproteins (10). Since fucose is in a terminally linked position, failure to cleave it will prevent the further degradation of any chains terminated by this sugar.

An endo-β-N-acetylglucosaminidase which hydrolyzes the di-N,N-acetylchitobiosyl bond in the linkage region in glycopeptides to produce oligosaccharides with GlcNAc in a reducing position and Asn-GlcNAc (Fig. 2) has been demonstrated in the hen oviduct (11) and methods for its assay have been described (12). Although most of the carbohydrate can be removed from native glycoproteins by the endo-glucosaminidase, denaturation of the glycoproteins was necessary to effect complete renewal (13). The activity of this enzyme accounts for the accumulation of oligosaccharides in some of the lysosomal enzyme deficiency diseases.

Oligosaccharides may also be produced by the action of 4-L-aspartylglycosylamine amidohydrolase, which cleaves the bond between GlcNAc and asparagine (14). The size and structure of the carbohydrate group did not appear to be an important factor in controlling activity, but the presence of amino acids in addition to the asparagine in the peptide portion of the glycopeptide molecule reduced the rate of this reaction. This enzyme is not active on intact glycoproteins (15).

Treatment of the delipidated brain tissue with proteolytic enzymes released all of the heteropolysaccharide chains of the glycoproteins, along with glycosaminoglycans and nucleic acids. The latter substances are removed by precipitation with cetylpyridinium chloride. Dialysis of the glycopeptide preparation separates the sialoglycopeptides, which are largely non-dialyzable, from the dialyzable, Con-A-binding, mannoglycopeptides. In the storage diseases, accumulated degradation products derived from the sialoglycopeptides will appear in the dialyzable glycopeptide fraction. Thus, an increase in the carbohydrate content of this fraction does not necessarily indicate an accumulation of the normal constituents of this fraction, but does indicate the appearance of abnormal degradation products (usually present only in traces) derived from the nondialyzable sialoglycopeptides.

Treatment of the delipidated tissue with proteolytic enzymes has the disadvantage of obscuring the evidence for the possible accumulation of glycopeptides or oligosaccharides in the tissue. These substances can be recovered quite simply by extracting the tissue with water or aqueous solvents. As we shall see, the latter procedure has the advantage of providing the accumulated oligosaccharides which have proved to be amenable for separation and structural analysis. A thorough study would allow for the isolation of these substances prior to subjecting the insoluble tissue residue to proteolysis.

GM2-gangliosidoses (deficiency of β-hexosaminidase).

Tingey (16) reported that the hexosamine level in the delipidated brain tissue from Tay-Sachs diseased patients was elevated. Hexosamine-containing substances could also be extracted from the residue with boiling water. While only 3-10 percent of the total brain hexosamine could be extracted from normal brain in this manner, the amount of the material extracted from Tay-Sachs brain was substantially elevated. An elevation of protein-bound hexosamine in Tay-Sachs disease was confirmed by others (17-21). Suzuki (20) found that the aqueous extract prepared from the brain of patients with Sandhoff's disease contained elevated levels of hexosamine while the increase in hexosamine concentration of defatted Tay-Sachs brain tissue and in aqueous extracts prepared from the residue was moderate.

Glycopeptides were recovered from the delipidated cerebral cortical tissue by treatment of the material with papain (22-24). The NeuNAc content of the dialyzable and non-dialyzable glycopeptides obtained from Tay-Sachs brains was normal. Thus, catabolism of these heteropolysaccharide chains differed from that of the gangliosides (Fig. 1): cleavage of the NeuNAc residues was not dependent on the prior removal of hexosamine. The concentration and carbohydrate composition of the non-dialyzable sialoglycopeptides did not differ from normal values. On the other hand, a four-fold elevation of hexosamine and mannose content in the dialyzable glycopeptides fraction was observed in both Tay-Sachs and Sandhoff's diseases.

The accumulated dialyzable glycopeptides did not bind strongly to Con A-Sepharose and presumably did not contain the terminal Man linkages which interact with the lectin. Thus, the accumulated material did not correspond to the mannoglycopeptides, also present in the dialyzable glycopeptide preparation. Probably breakdown products of the sialoglycopeptides, these accumulated glycopeptides presumably contained terminal GlcNAc residues.

Oligosaccharides which contained a trimannosyl core and GlcNAc at non-reducing termini were isolated from the liver of a case with Sandhoff's disease (25). Accumulation of this GlcNAc-containing oligosaccharide, or the excessive excretion in the urine, was not observed in Tay-Sachs disease (26). The oligosaccharides were isolated by homogenizing the liver in 10 volumes of chloroform-methanol-water (1:2:0.3,v/v/v). The material was then partitioned into an upper aqueous phase which was evaporated to dryness. Gangliosides were extracted from the dried residue by repeated extraction with dry chloroform-methanol. Glycopeptides were also isolated from delipidated tissue by means of proteolysis with papain.

Tsay and Dawson (27) minced brain tissue in water. The mixture was sonicated and centrifuged, and the water soluble material was fractionated by gel filtration. The structure of an isolated oligosaccharide was determined by treatment with the appropriate glycosidases, periodate oxidation, and GLC analysis. Its structure corresponded to that of the oligosaccharide previously isolated from the liver of Sandhoff patients (25). The amount isolated from brain was extremely small compared to the levels accumulated in the liver of these patients.

GlcNAc (β,1→2)Man(α,1→3)╲
                                                   Man(β,1→4)GlcNAc
GlcNAc (β,1→2)Man(α,1→6)╱

Whether the oligosaccharides in brain are a product of brain metabolism is not entirely clear. It is possible that incompletely degraded materials produced in the liver and organs can enter the circulation from which they may be taken up by brain cells. On the other hand, brain tissue does appear to contain heteropolysaccharide chains (Table I) with a carbohydrate composition which corresponds to that of the accumulated product.

The deficiency of hexosaminidase A in Tay-Sachs disease, and hexosaminidases A and B in Sandhoff's disease results in the accumulation of ganglioside GM2 (Fig. 1) and oligosaccharides (Fig. 2, reaction D) both of which contain terminally linked β-hexosamine residues. Both hexosaminidases A and B have been shown to act on terminal N-acetyl glucosamines of oligosaccharides derived from glycoproteins (28).

GM1-gangliosidosis, deficiency of β-galactosidase.

An oligosaccharide consisting of a degradation product derived from glycoproteins was isolated from the liver of patients with GM1-gangliosidosis (29):

$$\begin{array}{l}\text{Gal}(\beta,1\to4)\text{GlcNAc}(\beta,1\to2)\text{Man}(\alpha,1\to3)\searrow\\ \qquad\qquad\qquad\qquad\qquad\qquad\qquad\qquad\text{Man}(\beta,1\to4)\text{GlcNAc}\\ \text{Gal}(\beta,1\to4)\text{GlcNAc}(\beta,1\to2)\text{Man}(\alpha,1\to6)\nearrow\end{array}$$

The nondialyzable and dialyzable glycopeptides recovered after digestion of the delipidated residue from the cerebral gray matter showed an over two-fold elevation in Gal content (31,32). Man and GlcNAc were also elevated. These results suggested the accumulation of glycopeptides with exposed Gal residues. Degradation can proceed no further after the removal of the terminal NeuNAc and fucose residues. There was an especially marked accumulation of glycopeptides which were derived from the NeuNAc-rich, highly negatively charged, glycopeptides the sugar content of which was characterized by high values for the Gal/Man and GlcNAc/Man ratios (fractions I and II, Table I). In addition to these changes, electrophoretic analysis of the non-dialyzable glycopeptide preparation suggested that a large proportion of the sialoglycopeptides were deficient in -GlcNAc-Gal-NeuNAc branches which are attached to the trimannoside core. This change cannot be attributed to the absence of β-galactosidase, and may represent secondary changes due to the effects of the disease. In a second case, the concentration of dialyzable glycopeptides containing hexosamine, Gal and Man showed a 3-7 fold elevation which was dependent on the brain area studied. The accumulated glycopeptides did not bind to Concanavalin A-Sepharose (24). A six year-old boy was reported by Patel et al (33) to have a 2.5 fold elevation in glycoprotein-galactose in brain.

Tsay and Dawson (27) recovered an oligosaccharide from the brain of GM1-gangliosidosis patients the structure of which was identical to that recovered from liver and urine by Wolfe et al (29,30,31). Its structure resembled that of the oligosaccharides recovered from the brain of Sandhoff patients, except for the two terminally-linked Gal residues (Fig. 2). Only a relatively small amount of this material was recovered from brain; the accumulation of ganglioside GM1 and glycopeptides released by proteolysis was considerably greater.

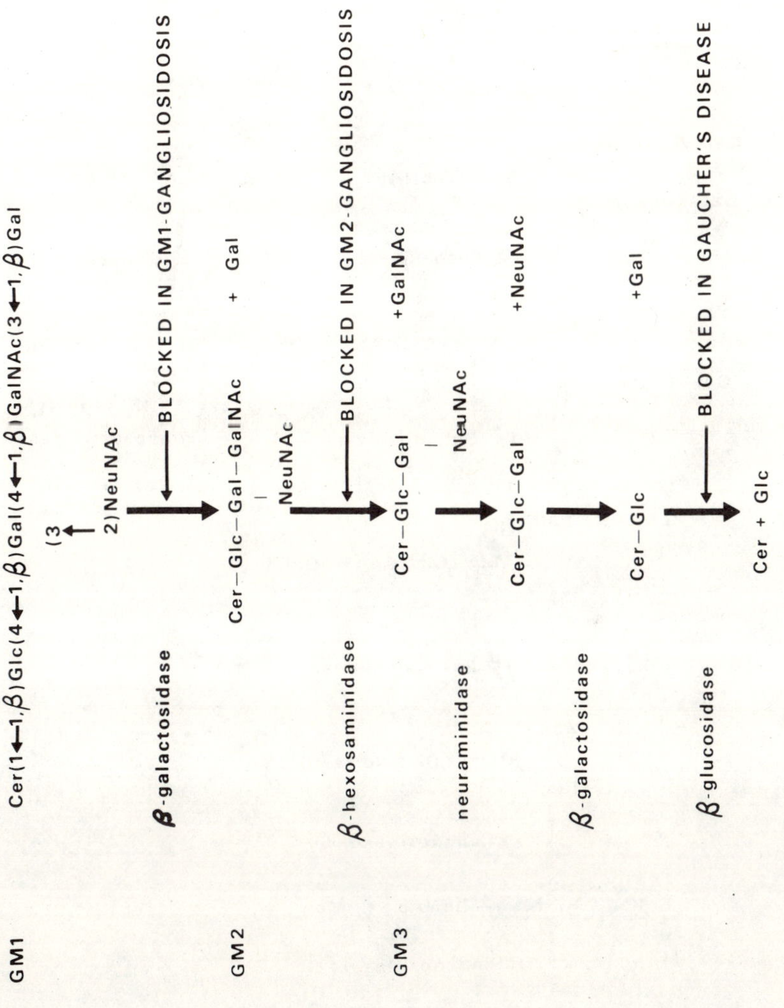

Figure 1. Block in ganglioside catabolism in GM-1-gangliosidosis, GM2-gangliosidosis, and Gaucher's Disease

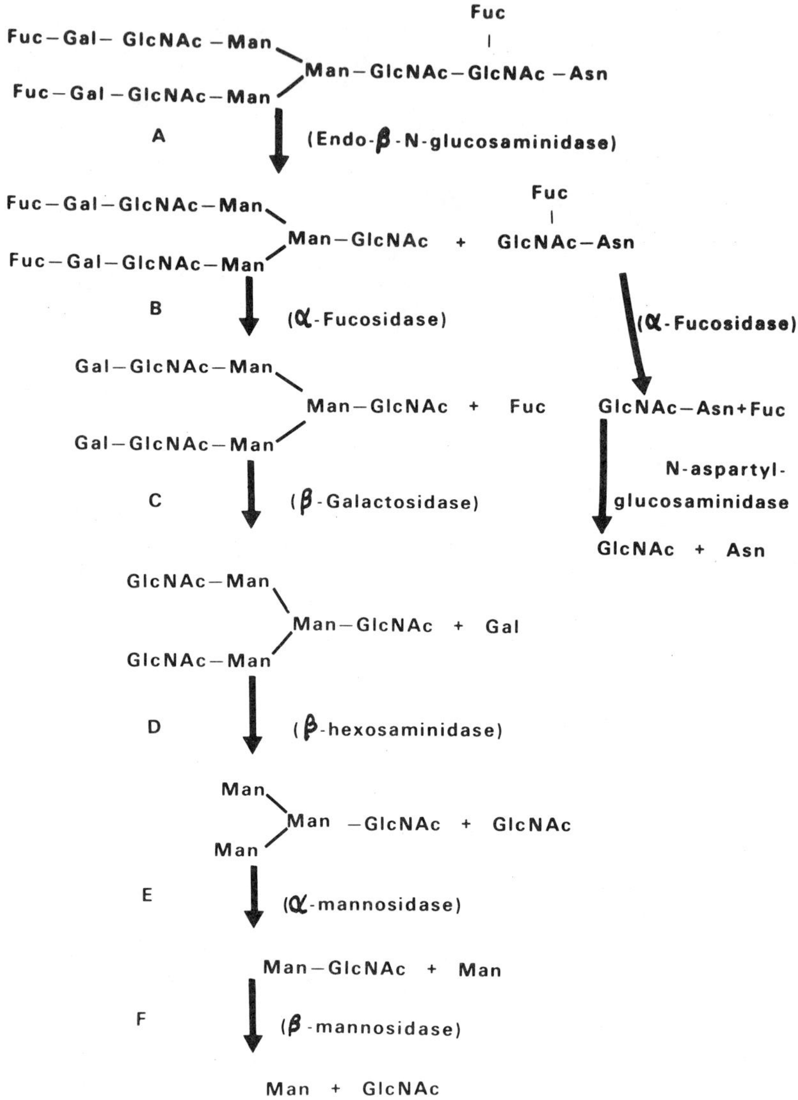

Figure 2. Probable catabolic pathway for the degradation of oligosaccharide chains derived from brain glycoproteins. Reactions B, C, D, and E are blocked in fucosidosis, GM1-gangliosidosis, GM2-gangliosidosis, and mannosidosis, respectively.

Gaucher's Disease (deficiency of glucocerebrosidase)

Whether partial degradation products derived from glycoproteins accumulate in the brain of the neuropathic forms of Gaucher's disease is not known. Kanfer et al (34) treated defatted liver tissue with proteolytic enzymes. Glycopeptides derived from glycoproteins were markedly elevated in the Gaucher liver; a 20 fold elevation in hexose, hexosamine, fucose and NeuNAc was recorded. The accumulated material contained Gal and GlcNAc as the major sugars; Glc and Man were present in traces. The disease is caused by deficit in glucocerebrosidase, a β-glucosidase. Although glucose-containing glycoproteins undoubtedly exist (1), there is no evidence for the accumulation of Glc-containing glycopeptides or oligosaccharides. Gaucher spleen appears to produce an excess of a glycoprotein activating factor, capable of activating the glucocerebrosidase from normal spleen (35-37). While the activator from Gaucher spleen was a glycoprotein, that from control spleens was not. The amino acid composition of the two activators also differed (38). The role of the glycoprotein activator remains obscure.

Metachromatic Leukodystrophy (MLD) and Multiple Sulfatase Deficiency (MSD).

The sialoglycopeptide preparation from brain contains molecules which carry sulfate-ester groups. Consequently, it appeared likely that partial degradation products derived from glycoproteins which bear sulfated oligosaccharide chains might accumulate in MLD, a disease caused by the absence of a functioning sulfatase A. However, sulfatides are the natural substrate for this enzyme. The hexosamine levels in the delipidated tissue residue from cases of MLD, in which it was established that sulfatase A is deficient, fell within the normal range (39,40).

The situation differs in MSD, a variant of MLD in which the tissues of the patients lack sulfatases A, B, and C. The nonlipid hexosamine in tissues from such patients show an elevation of protein-bound hexosamine (41), and it has been shown that this is due, at least in part, by the elevation of glycosaminoglycans. It is not known whether glycopeptides or oligosaccharides derived from glycoproteins accumulate in this disease.

It has been suggested (42) that O-sulfated glycosaminoglycans are the natural substrate for sulfatase B, since this enzyme is deficient in two diseases (MSD and Maroteaux-Lamy disease) in which O-sulfated glycosaminoglycans accumulate (mainly dermatan sulfate). However, co-cultivation of MSD fibroblasts with Sanfilippo A or Hunter cells did not correct for the degradation defect in glycosaminoglycan-SO_4-S^{35} metabolism. Sanfilippo and Hunter fibroblasts lack heparan sulfatase and dermatan sulfatase, but arylsulfatase B in these cells is normal or elevated (42). These findings led to the suggestion that sulfatase B may be involved in degrading sulfated glycopeptides (43).

Sulfatase B appears to correspond to N-acetylgalactosamine 4-sulfatase (44). The sulfatase responsible for the accumulation of protein-bound hexosamine still remains to be identified. Nor is it yet established whether sulfated glycopeptides derived from glycoproteins accumulate in MSD. The unknown sulfatase may be a component of sulfatase A or B, both of which exist in multiple forms, or a sulfatase which does not respond to the synthetic substrates commonly in use to detect sulfatase activity.

The mucopolysaccharidoses.

The defects in Hurler, Hunter, and Sanfilippo A and B syndromes, those forms of the mucopolysaccharidoses involving severe neurological deficits, are caused by nonfunctioning α-iduronidase, sulfoiduronate sulfatase, heparan-N-S-sulfatase, and α-N-acetylglucosaminidase, respectively. These linkages are, as far as is known, absent in glycoproteins. Consequently, no aberration in glycoprotein catabolism is expected, nor has such a defect been reported. Glycopeptides isolated from the brain of a Hunter patient showed no remarkable changes (45).

Diseases with a primary lesion in glycoprotein catabolism.

Mannose-rich oligosaccharides accumulate in the brain of mannosidosis patients (46-48):

Man-Man-GlcNAc

Man-Man-Man-GlcNAc

```
Man-Man
       \
        Man-GlcNAc
       /
    Man
```

The disease occurs also in Angus calves, and the defatted brain tissue residue of the afflicted animals showed a 4-6 fold elevation in Man content (49). Deficiency of α-mannosidase causes a block in the catabolism of the heteropolysaccharide chains: mannose-containing glycopeptides or oligosaccharides are derived from the sialoglycopeptide trimannoside core, as well as from the mannoglycopeptides.

In patients with a deficiency of α-fucosidase (fucosidosis), fucose-containing substances were shown to accumulate in the delipidated tissue, including that of the brain (50). Aqueous extracts of sonicated brain tissue yielded an oligosaccharide (27):

```
Fuc-Gal-GlcNAc-Man
                  \
                   Man-GlcNAc
                  /
Fuc-Gal-GlcNAc-Man
```

A disaccharide, fucosylated N-acetylglucosamine, was also detected in the brains of fucosidosis patients (Fig. 2).

Aspartylglucosaminuria is caused by a deficit in the enzyme 1-aspartamido-β-N-acetylglucosamine amidohydrolase. Aspartylglucosaminylamine and aspartyloligosaccharides accumulate in organs, including the brain (51,52).

The mucolipidoses are caused by a deficiency of α-neuraminidase (53) resulting in the excessive excretion of sialyloligosaccharides. Most of the oligosaccharides isolated had a core consisting of three mannose residues to which two GlcNAc-Gal-NeuNAc chains, or incomplete portions of these chains, were attached (54). Although not all of the linkages in the oligosaccharides of brain tissue have been definitively established, it appears likely that their structure will prove to be identical to that of one of the sialoglycopeptides isolated from urine (55):

```
NeuNAc(α,2→6)Gal(β,1→4)GlcNAc(β,1→2)Man(α,1→3)
                                               \
                                                Man(β,1→4)GlcNAc
                                               /
NeuNAc(α,2→6)Gal(β,1→4)GlcNAc(β,1→2)Man(α,1→6)
```

A sequence of the steps involved in the catabolism of the heteropolysaccharides of brain glycoproteins has been postulated (Fig. 2) based on the structures of the oligosaccharides which accumulate in the various lysosomal enzyme diseases discussed above.

Niemann-Pick's Disease

The primary lesion in the infantile, type A, form of Niemann Pick's disease is a deficiency of the enzyme, sphingomyelinase. While sphingomyelin is the primary storage product, many cases which have been examined have demonstrated an increase in gangliosides and cholesterol in brain tissue. Tingey (16) and Norman et al (56) have reported an elevation of protein-bound hexosamine as well. Brunngraber et al (57) noted a two fold elevation of glycoproteins-carbohydrate which was recovered as sialoglycopeptides upon proteolytic digestion of the delipidated tissue residue from gray matter. Mannoglycopeptide levels remained within the normal range. The composition of the sialoglycopeptides was altered, since the number of -GlcNAc-Gal-NeuNAc chains attached to the internal trimannoside core was reduced. It is possible, but not demonstrated, that the accumulated glycopeptides in Niemann-Pick disease may correspond to the sialylated forms of the oligosaccharides which accumulate in the GM1-gangliosidoses. It was suggested that sphingomyelin catabolism may be an early step in the degradation of plasma membranes, and a lesion in the catabolism of this phospholipid may have the secondary effect of impeding the degradation of other plasma membrane components.

Neuronal-Ceroid Lipofuscinoses (NCF).

A form of NCF may involve glycoprotein storage. Adelman et al (58) found a nearly two fold increase in the NeuNAc content of sialoglycopeptide preparations obtained from tissue sections of the cerebral cortex of a case with clinical features resembling those of Batten's disease.

Cell recognition of lysosomal enzymes.

Studies utilizing cultured fibroblasts derived from patients with lysosomal enzyme deficiency diseases have shown that these cells secrete lysosomal glycosidases into the medium, and are capable of re-assimilating these enzymes. Hickman and Neuman (59,60) have proposed that lysosomal enzymes are transported from their site of synthesis and glycosylation in the Golgi apparatus to the lysosome by this process of exocytosis and subsequent pinocytotic re-incorporation. Such a mechanism may be a part of a membrane recycling mechanism, since internalization of the enzyme is accomplished by an internalization of a patch of plasma membrane, forming pinocytotic vesicles destined to fuse with the lysosome (61).

Fibroblasts from patients afflicted with one of the lysosomal deficiency diseases are capable of assimilating the deficient enzyme which had been added to the culture medium, thereby correcting the metabolic defect. Uptake of lysosomal glycosidases is dependent on the recognition of the heteropolysaccharide chain of the glycoprotein enzyme by a receptor on the surface of the fibroblast. Thus, the terminal Gal residue appeared to be important for the recognition and uptake of α-N-acetylglucosaminidase (62) by fibroblasts of Sanfilippo B patients. Mannose residues appeared to be important for uptake of β-galactosidase by fibroblasts from patients with GM1-gangliosidosis (63).

Cultured fibroblasts appear to recognize phosphohexosyl groups in β-glucuronidase, β-hexosaminidase, and β-galactosidase (64,65) and α-L-iduronidase (66). The recognition of phosphohexosyl groups may be a general characteristic of pinocytosis of lyosomal glycosidases. These findings may be related to the earlier reports on the isolation of a phosphoglycoprotein from brain (67-69). Phosphorylated mannoglycopeptides were isolated from the phosphoglycoprotein, and the phosphate residue which is attached to one of the hydroxyl groups of mannose, was removed by the action of alkaline phosphatase (69). The relation of this phosphoglycoprotein to the lysosomal glycosidases remains to be established. If the capacity of the fibroblast to secrete and re-assimilate lysosomal enzymes is a specific example of a more general phenomena, phosphoglycoproteins and/or glycosidases may play an important role in the turnover of the plasma membrane.

Literature Cited

1. Brunngraber, E. G.; Hof, H.; Susz, J.; Brown, B. D.; Aro, A.; and Chang, I. Biochim Biophys Acta, 304: 781, 1973.
2. Krusius, T. FEBS Lett, 66: 86, 1976.
3. Hof, H. I.; Susz, J. P.; Javaid, J. I.; and Brunngraber, E. G. Neurobiology, 5: 347, 1975.
4. Krusius, T.; Finne, J.; and Rauvala, H. FEBS Lett, 71: 117, 1976.
5. Brunngraber, E. G. Adv Exptl Med and Biol, 25: 255, 1972.
6. Krusius, T.; and Finne, J. Eur J Biochem, 78: 369, 1977.
7. Finne, J.; Krusius, T.; and Rauvala, H. Biochem Biophys Res Communs, 74: 405, 1977.
8. Javaid, J. I.; Hof, H. I.; and Brunngraber, E. G. Biochim Biophys Acta, 404: 74, 1975.
9. Aronson, N. N., Jr.; and DeDuve, C. J Biol Chem, 243: 4564, 1968.
10. Opheim, D. J.; and Touster, O. J Biol Chem, 252: 739, 1977.
11. Tarentino, A. L.; and Maley, F. J Biol Chem, 251: 6537, 1976.
12. Tarentino, A. L.; and Maley, F. Anal Biochem, 77: 185, 1976.
13. Trimble, R. B.; and Maley, F. Biochem Biophys Res Communs, 78: 935, 1977.
14. Kohno, M.; and Yamashina, I. Biochim Bipphys Acta, 258: 600, 1972.
15. Fidler, I. J.; Montgomery, P. C.; and Cesarini, J. P. Cancer Res, 33: 1376, 1973.
16. Tingey, A. J Neurochem, 3: 230, 1959.
17. Stefanko, St.; Guminska, M.; and Pietrzykowa, B. Schweiz Arch Neurol Neurochir Psychiat, 90: 295, 1962.
18. Edgar, G. W. F. Proc of the Fifth Internatl Congr Neuropathology, Zurich, 1965. Amsterdam, Excerpta Med Found, 1966, p. 328.
19. Thieffry, S.; Bertrand, I.; Bargeton, E.; and Edgar, G.W.F. Rev Neurol, 102: 130, 1960.
20. Suzuki, Y.; Jacob, J. C.; Suzuki, K.; Kutty, K. N.; and Suzuki, K. Neurology, 21: 313, 1971.
21. Norman, R. M.; Tingey, A. H.; Newman, C. H. H.; and Ward, S. P. Arch Dis Child, 39: 634, 1964.
22. Brunngraber, E. G.; Haberland, C.; and Brown, B. D. Brain Res, 38: 151, 1972.
23. Brunngraber, E. G.; Brown, B. D.; and Aro, A. J Neurochem, 22: 125, 1974.
24. Brunngraber, E. G.; Davis, L. G.; Javaid, J. I.; and Berra, B. Adv Exptl Med and Biol, 68: 31, 1976.
25. Ng Ying Kin, N. M. K.; and Wolfe, L. S. Biochem Biophys Res Communs, 59: 837, 1974.
26. Wolfe, L. S.; and Ng Ying Kin, N. M. K. Adv Exptl Med and Biol, 68: 15, 1976.
27. Tsay, G. C.; and Dawson, G. J Neurochem, 27: 733, 1976.

28. Bearpark, T.; Bouquelet, S.; Fournet, B.; Montreuil, J; Spik, G.; Stirling, J.; Strecker, G. FEBS Lett, 84: 379, 1977.
29. Wolfe, L. S.; Senior, R. G.; and Ng Ying Kin, N. M. K. J Biol Chem, 249: 1828, 1974.
30. Ng Ying Kin, N. M. K.; and Wolfe, L. S. Biochem Biophys Res Communs, 66: 123, 1975.
31. Brunngraber, E. G.; Bérra, B.; and Zambotti, V. FEBS Lett, 34: 350, 1973.
32. Berra, B.; Di Palma, S.; and Brunngraber, E. G. Clin Chim Acta, 57: 301, 1974.
33. Patel, V.; Goebel, H. H.; Watanabe, I.; and Zeman, W. Acta Neuropathol, 30: 155, 1974.
34. Kanfer, J. N.; Stein, M.; and Spielvogel, C. Adv Exptl Med and Biol, 19: 225, 1972.
35. Ho, M. W.; and O'Brien, J. S. Proc Natl Acad Sci, US, 68: 2810, 1971.
36. Ho, M. W.; O'Brien, J. S.; Radin, N. S.; and Erickson, J. S. Biochem J, 131: 173, 1973.
37. Ho, M. W. Biochem J, 136: 721, 1973.
38. Peters, S. P.; Coyle, P.; Coffee, C. J.; Glew, R. H.; Kuhlenschmidt, M. S.; Rosenfeld, L.; and Lee, Y. C. J Biol Chem, 252: 563, 1977.
39. Haberland, C.; Brunngraber, E. G.; Witting, L. A.; and Daniels, A. Acta Neuropathol, 26: 93, 1973.
40. Federico, A.; Guazzi, G. C.; and Di Benedetta, C. Eur Neurol, 15: 163, 1977.
41. Eto, Y.; Meier, C.; and Herschkowitz, N. N. J Neurochem, 27; 1071, 1976.
42. Austin, J. Arch Neurol, 28: 258, 1973.
43. Eto, Y.; Wiesmann, U.; and Hierschkowitz, N. N. Arch Neurol, 30: 153, 1974.
44. Dorfman, A.; Arbogast, B.; and Matalon, R. Adv Exptl Med and Biol, 68: 261, 1976.
45. Roukema, P. A.; Overkerk, C. H. and Van Den Berg, G. Clin Chim Acta, 44: 277, 1973.
46. Norden, N. E.; Lundblad, A.; Svensson, S.; Ockerman, P. A.; and Autio, S. J Biol Chem, 248: 6210, 1973.
47. Norden, N. E.; Lundblad, A.; Svensson, S.; and Autio, S. Biochemistry, 13: 871, 1974.
48. Tsay, G. C.; Dawson, G.; and Matalon, R. J Pediat, 84: 865, 1974.
49. Norden, N. E.; Lundblad, A.; Ockerman, P. A. and Jolly, R. D. FEBS Lett, 35: 209, 1973.
50. Van Hoff, F.; and Hers, H. G. Eur J Biochem, 7: 34, 1968.
51. Palo, J.; and Savolainen, H. J Chromatog, 65: 447, 1972.
52. Palo, J. and Savolainen, H. Ann Clin Res, 5: 156, 1973.
53. Cantz, M.; Gehler, J.; and Spranger, J. W. Biochem Bipphys Res Communs, 74: 732, 1977.

54. Michalski, J. C.; Strecker, G.; and Fournet, B. FEBS Lett, 79: 101, 1977.
55. Strecker, G.; Peers, M. C.; Michalski, J. C.; Hondi-Assah, T.; Fournet, B.; Spik, G.; Montreuil, J.; Farriaux, J. P.; Maroteaux, P.; and Durand, P. Eur J Biochem, 75: 391, 1977.
56. Norman, R. M.; Tingey, A. H.; and Fowler, M. C. Proc of the Fifth Internatl Congr Neuropathol, Zurich, 1965, Amsterdam, Excerpta Med Found, 1966, p. 143.
57. Brunngraber, E. G.; Berra, B.; and Zambotti, V. Clin Chim Acta, 48: 173, 1973.
58. Adelman, L. S.; Young, E.; and Bass, N. H. Neurology, 24: 1045, 1974.
59. Hickman, S.; and Neufeld, E. F. Biochem Biophys Res Communs, 49: 992, 1972.
60. Neufeld, E. F.; Sando, G. N.; Sarvin, A. J.; and Rome, L. H. J Supramol Struct, 6: 95, 1977.
61. Brunngraber, E. G. Neurochemistry of the Aminosugars, Springfield, Ill. C. C. Thomas, Pub. 1978.
62. Von Figura, K.; and Kresse, H. Prot Biol Fluids, 22: 275, 1975.
63. Hieber, V.; Distler, J.; Myerowitz, R.; Schmickel, R. D.; and Jourdian, G. W. Biochem Biophys Res Communs, 73: 710, 1966.
64. Kaplan, A.; Fischer, D.; Achord, D.; and Sly, W. J Clin Invest, 60: 1088, 1977.
65. Kaplan, A.; Achord, D. T.; and Sly, W. S. Proc Natl Acad Sci US, 74: 2026, 1977.
66. Sando, G. N.; and Neufeld, E. F. Cell, 12: 619, 1977.
67. Davis, L. G.; Javaid, J. I.; and Brunngraber, E. G. FEBS Lett, 65: 30, 1976.
68. Davis, L. G.; Costello, A. J. R.; Javaid, J. I.; and Brunngraber, E. G. FEBS Lett, 65: 35, 1976.
69. Davis, L. G.; Brettschneider, I.; and Brunngraber, E. G. Fed Proc, 36: 750, 1977.

RECEIVED March 15, 1978.

8

Incorporation of Exogenous Enzymes into Lysosomes

A Theoretical and Practical Means for Correcting Lysosomal Blockage

P. G. PENTCHEV, J.A. BARRANGER, A. E. GAL, F. S. FURBISH, and R. O. BRADY

Developmental and Metabolic Neurology Branch, National Institute of Neurological and Communicative Disorders and Stroke, National Institutes of Health, Bethesda, MD 20014

The lysosomal storage disorders have served as useful prototypes for delineating the biochemical and cellular pathology of inherited metabolic disorders. Progress has led to an identification of most primary enzymatic lesions in these disorders and to a better understanding of the cellular manifestations resulting from these enzymopathies. Specific enzymatic tests are now available for the unequivocal diagnosis of most of these storage diseases. Reliable heterozygote detection and prenatal diagnosis has enabled genetic counseling to effectively reduce the incidence of many of these crippling diseases (1).

Investigators have also addressed considerable effort to the intriguing possibility of enzyme replacement as an effective therapeutic measure for the treatment of the lysosomal storage disorders. Although more than a decade of effort has been channelled in this direction, an unequivocal answer has yet to appear. An evaluation of the status of this research program can perhaps be best initiated by examining the cellular and biochemical bases for enzyme replacement. Remarkable insight into this phenomena was given in 1964, at a time when only one lysosomal storage disorder was recognized:

"Both in our pathogic speculations and in our therapeutic attempts it may be well to keep in mind that any substance which is taken up intracellularly by an endocytic process is likely to end up within lysosomes. This obviously opens up many possibilities for interaction including replacement therapy."(2).

The implication of the dynamic nature of lysosomal interactions is that one can, in theory, equate the ability of endogenous and exogenous enzymes to stimulate intralysosomal digestion (Figure I). It is now well documented that enzymes that are initially found in the extracellular environment can be incorporated into the lysosomal architecture of cells (Table I). It has

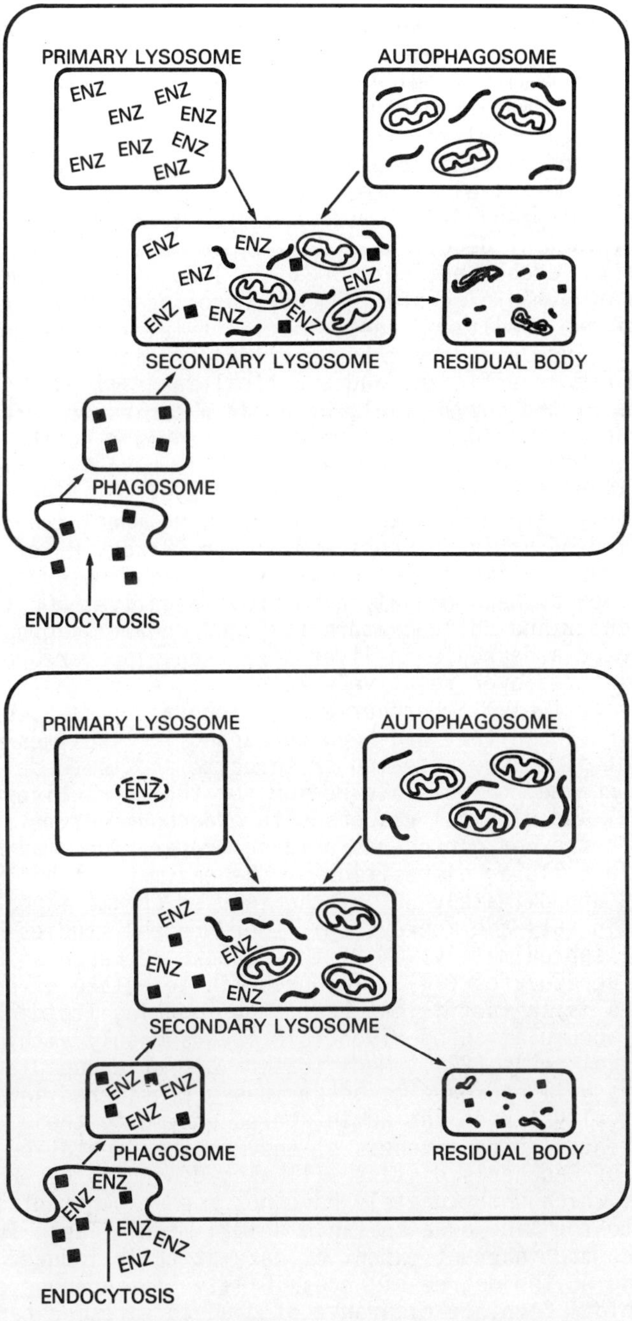

Figure 1. Representation of lysosomal interactions. (Top) *Normal lysosomal digestion.* (Bottom) *Lysosomal digestion stimulated by exogenous enzyme.*

also been shown that these "migrant" enzymes are functionally active in reversing previously blocked metabolic pathways and/or reversing abnormal morphology of lysosomes (Table II).

In 1966, it was predicted that Gaucher's disease might prove to be the ideal model for enzyme replacement therapy (15). The accumulation of glycolipid occurs predominantly within the cells of reticuloendethelial system. Consequently intravenously administered enzyme preparations were expected to have ready access to major pathological depots of stored lipid. Our enzyme infusion studies with Gaucher patients began in 1973 with the successful isolation of human placental glucocerebrosidase (16). A brief summary of the results of these initial trials is presented in Table III. Encouraged by these preliminary findings and having developed a more efficient and practical means of purifying large quantities of the enzyme (18), we began a series of infusions in patients using considerably more glucocerebrosidase than initially employed. The results of the latter trials have been disappointing. No significant changes in either the circulating or liver levels of glycolipid were observed following the intravenous administration of native glucocerebrosidase (Brady, R. *et al.*, unpublished observations). Concomittant clinical observations such as regression of hepatomegaly also prove negative. On the other hand, Belchetz and collaborators (19) and Beutler's group (20) have reported a decrease in liver size when glucocerebrosidase was administered over relatively long periods of time (13 and 3-1/2 months respectively). Furthermore, Belchetz, *et al.*, injected glucocerebrosidase that had been entrapped in liposomes while Beutler, *et al.*, used resealed erythrocyte ghosts to deliver the placental enzyme. A possible reason for the refractiveness of biochemical and clinical responses to exogenous enzyme in our trials with the new glucocerebrosidase preparations may be an unfavorable cellular disposition of the enzyme. It has been shown that approximately 30% of the infused enzyme is taken up by the liver in rats and monkeys (5). More recent studies have shown that approximately 90% of this uptake represents incorporation into hepatocytes (21). The predominate uptake of enzyme by hepatocytes is an inefficient route for inducing lipid clearance since the accumulation of glycolipid occurs mainly within reticuloendothetial cells (22). Modification and/or encapsulation of the enzymes will now have to be seriously considered in an attempt to specifically target the administered enzyme to these cells.

The eventual effectiveness of enzyme supplementation in the treatment of Gaucher's disease will also depend on several other parameters which unfortunately have not been experimentally evaluated due to the lack of a suitable animal model (Table IV). To begin with, the inherent extent of enzymatically induced clearance will depend on the degree of accessibility of exogenous enzymes to stored lipid. Complete clearance of glucocerebroside can be effected when specimens of Gaucher liver tissue are incubated with glucocerebrosidase *in vitro* (Figure II). However, the *in situ*

Table I. Demonstrated Lysosomal Uptake of Exogenously Administered Enzymes

Enzyme	Tissue	Reference
β-Glucuronidase	Cultured Fibroblasts	(3)
α-Glucosidase	Rat Liver	(4)
Glucocerebrosidase	Rhesus Monkey Liver	(5)
Hexosaminidase	Rat Liver	(6)

Table II. Experimental Models for Correction of Lysosomal Accumulation with Exogenous Enzymes

Enzyme Addition	Effect	Reference No.
A. Tissue Culture		
Invertase and β-glucosidase	Reversal of Sucrose and cellobiose accumulation in macrophases	7
Deficient lysosomal hydrolases:	Metabolic correction in patients' fibroblasts:	
Mucopolysaccharidases	Mucopolysaccharidoses	8
Arylsulfatase-A	Metachromatic leukodystrophy	9
Acid lipase	Wolman's Disease	10
α-Galactosidase	Fabry's Disease	11
β-N-Acetylhexosaminidase	Sandhoff Disease	12
B. Animals		
Invertase	Prevention of hepatic sucrose accumulation in rat	13
Dextranase	Reversal of hepatic dextran accumulation in rat	14

Table III. Initial Results of Enzyme Replacement Trials in Gaucher's Disease (*17*)

1. Intravenous infusions of highly purified human placental glucocerebrosidase were well tolerated by patients with no untoward immunological or pyrogenic response.

2. Exogenous enzyme was rapidly cleared from the circulation.

3. A clearance of the elevated glycolipid levels was observed in the blood stream.

4. Following enzyme infusion, lower levels of glycolipid were found in liver biopsy samples.

Table IV. Assessment of Lysosomal Responses in Evaluating the Effectiveness of Enzyme Replacement for the Treatment of Storage Disorders

1. The extent of enzymatically induced clearance of stored material.

2. The rate at which stored material accumulates.

3. The degree of lysosomal overloading in relationship to clinical and pathological manifestations.

4. The clinical consequence of reversing intralysosomal accumulation.

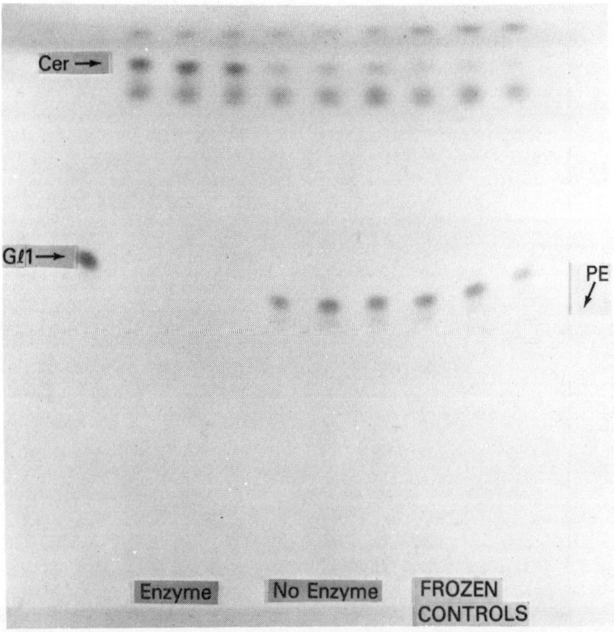

Figure 2. Clearance of glucocerebroside from Gaucher liver tissue by exogenous glucocerebrosidase in vitro. Fresh liver needle biopsy samples from a Gaucher patient were incubated in phosphate-buffered saline at pH = 6.0 for 12 hr at 37°C without and with 1000 units of highly purified glucocerebrosidase as indicated. Glucocerebroside in these preparations and in nonincubated samples was subsequently quantitated as previously described (23). Gl l = glucocerebroside; Cer-ceramide, PE = phosphatidyl ethanolamine.

Table V. Glycolipid Accumulation in Livers of Gaucher Patients

Patient	Sex	Type	Age	Residual Glucocerebrosidase Activity in Leukocytes	Liver Glucocerebroside
				(% of Control)	(mg/gram wet tissue)
1	F	Adult	44	16	0.25
2	M	Juvenile	15	21	0.70
3	F	Adult	54	30	1.63
4	F	Adult	15	29	4.30
5	M	Adult	30	- -	11.20
6	M	Adult	20	27	15.40
7	F	Adult	30	23	16.40
8	F	Adult	39	23	16.50
9	F	Adult	49	23	17.80
10	F	Adult	23	25	17.90
11	F	Adult	21	29	22.90
Control	F&M		20-35	100	0.03 - 0.07

accessability of lipid may be more restrictive and perhaps a limiting factor. Preliminary studies with mouse liver have shown that residual bodies have a lower tendency to fuse with lysosomes containing exogenous material (24). Secondly, the efficiency and course of enzyme replacement regimens will depend on the degree to which an induced clearance can be sustained. Obviously, a rapid rate of accumulation relative to clearance will limit the utility of this form of treatment. It may be useful to attempt to stabilize the exogenous enzyme in order to prolong its intra-lysosomal digestive effect. The half-life of human glucocerebrosidase in the liver of rats and Rhesus monkeys is approximately 8 hours (5). Generalized hypothermia (33°C) extends this half-life to over thirty hours (5). Lipids have also been shown to stabilize the enzyme *in vitro* (Furbish, F. and Brady, R., unpublished observations). Stabilization and protection of the enzyme may also be achieved by cross-linking or binding to matrices (25). Thirdly, the progression of primary lysosomal hypertrophies to clinical disease states remains obscure and speculative. The variability of biochemical and clinical manifestations in Gaucher's is illustrative of this gap in our knowledge. The range of elevated glucocerebroside in the liver of patients with Gaucher's disease may vary over 100-fold (Table V). Often little correlation was observed between the degree of glucocerebroside accumulation and the severity of clinical manifestations. Accordingly, a given level of clearance induced by enzyme supplementation may show considerable variation in terms of clinical benefits.

Finally, it has sometimes been tacitly assumed that an enzymatically induced reversal of lysosomal blockage will restore normality to the effected tissues. As already discussed, the reality of biochemically defined corrections has been well established with various models. However, descriptions of morphological correlates to biochemical correction are rather sparce in the literature. The classic experiments of Cohn *et al.* (7) reporting a visual observation of the reversal of lysosomal swelling in the sucrose-loaded macrophages upon the uptake by the cells of invertase remains to date the most descriptive accounting of this phenomena. The complexities of biochemical, morphological and physiological relationships in organisms are enormous. The present status of our knowledge is not sufficient to evaluate potentially corrective or simply preventive roles for enzyme replacement.

In conclusion, the realistic application of enzyme replacement as a therapeutic measure for the treatment of lysosomal storage disorders has yet to be established. Although the present state of the art is at a delicate embryonic stage of development, nature has extended a helpful hand in providing intracellular homing mechanisms. The continued creative and persistent efforts of investigators can likely complement this advantage towards a meaningful realization of this form of therapy.

REFERENCES

1. Brady, R., Metabolism 26: 329 (1977).

2. DeDuve, C., Fed. Proc. 23, 1045 (1964).

3. Lagunoff, D., Nicol, D., Prilzl, P., Lab. Invest. 29: 449 (1973).

4. Huijing, F., Waltuck, B., Whelan, W., in Birth Defects Original Article Series, Enzyme Therapy in Genetic Diseases, Vol. IX, No. 2, p. 191, (1973). Daniel Bergsma, Ed.

5. Pentchev, P., Kusiak, J., Barranger, J., Furbish, F., Rapoport,S., Massey, J., Brady, R., in Enzymes of Lipid Metabolism, Plenum Press, in Press.

6. Schlesinger, P., Rodman, J., Frey, M., Lang, S., Stahl, P., Arch. Biochem. and Biophys. 172: 606 (1976).

7. Cohn, Z., Ehrenreich, B., J. Exp. Med. 129: 201 (1969).

8. Neufeld, L., Shapiro, L., Ann. Rev. BIochem. 44: 357 (1975).

9. Porter, M., Fluharty, A., Kihara, H., Science 172: 1263 (1971).

10. Kyriakides, E., Paul, B., Balint, J., J. Lab. Clin. Med. 80: 810 (1972).

11. Dawson, G., Matalon, R., Li, Y., Pediat. Res. 7: 684 (1973).

12. Cantz, M., Kresse, H., J. Biochem. 47: 581 (1974).

13. Jacques, P., Bruns, G., Abs. 2nd Meet. Fed. Eur. Biochem. Soc. A36 (1978)

14. Colley, M., Ryman, B., Bioch. and Biophys. Acta. 451: 417 (1976).

15. Brady, R., N. Engl. J. Med. 275: 312 (1966).

16. Pentchev, P., Brady, R., Hibbert, S., Gal, A., Shapiro, D., J. Biol. Chem. 248: 5256 (1973).

17. Brady, R., Pentchev, P., Gal, A., Fed. Proc. 34: 1310 (1975).

18. Furbish, F., Blair, H., Shiloach, J., Pentchev, P., Brady, R. Proc. Natl. Acad. Sci. USA 74: 3560 (1977).

19. Belchetz, P., Crawley, J., Braidman, I., Gregoriadis, G. Lancet ii, 116 (1977).

20. Beutler, E., Dale, G., Guinto, E., Kuhl, W. Proc. Natl. Acad. Sci. USA, 74: 4620 (1977).

21. Furbish, F., Steer, C., Barranger, J., Jones, J., Brady, R. Biochem. Biophys. Res. manuscript submitted.

22. Brady, R., King, F. In Lysosomes and Storage Diseases (1973) H. G. Hers and F. Van Hoof, Editors, Academic Press, New York, p. 381.

23. Brady, R., Pentchev, P., Gal, A., Hibbert, S., Dekaban, A., N. Engl. J. Med. 291: 989 (1974).

24. Daems, W., in Enzyme Therapy in Lysosomal Storage Diseases. p. 3, (1974), J. M. Tager, G. S. Hooghwinkel and W. Daems, Editors, North-Holland Publishing Company.

25. Wold, F., in Birth Defects, Original Article Series, Vol. IX, No. 2, Enzyme Therapy in Genetic Diseases, p. 46. (1973). Daniel Bergama, Editor, Williams and Wilking Publishers, Baltimore, Md.

RECEIVED March 15, 1978.

9

Blood Coagulation—Initiation and Regulation by Limited Proteolysis

WALTER KISIEL

Department of Biochemistry, University of Washington, Seattle, WA 98195

I. Introduction

During the coagulation of blood, a number of plasma glycoprotein zymogens are converted to serine proteases in the presence of calcium ions and phospholipid. These activation reactions do not occur randomly but rather as a series of consecutive zymogen activation reactions where the product of one reaction is the catalyst for the next. The focal point of this cascade-like reaction is the formation of thrombin, the serine protease which converts fibrinogen to fibrin. *In vivo*, the formation of fibrin is thought to occur on the surface of platelets concurrent with their response to an injury in the vessel wall. Indeed, platelets constitute the primary defense in arresting the bleeding of ruptured blood vessels by their rapid adhesion to extravascular surfaces. The platelet adhesion reaction is followed by a morphological change in the platelet membrane which provides a catalytic phospholipid surface for optimal generation of thrombin. The ensuing fibrin serves to stabilize the hemostatic platelet thrombus by virtue of its mesh-like properties. Following tissue repair and vessel wall regeneration, the hemostatic plug is destabilized by the degradation of the fibrin network by plasmin arising from the fibrinolytic system.

The common names, as well as the Roman numeral designations, of the glycoproteins participating in coagulation are listed in Table I. Prothrombin, factor VII, factor IX, factor X, factor XI, factor XII, and prekallikrein exist in plasma as precursors to serine proteases. Also listed in Table I are the coagulation disorders associated with the deficiency of a particular coagulation factor. In the past decade, enormous progress has been made in the isolation and characterization of most of these proteins. The availability of homogeneous preparations of each zymogen has made is possible to establish unambiguously the mechanism whereby it is activated, and its proteolytic specificity following activation. The results of these studies have proven invaluable in our understanding of the molecular events in coagulation, and

0-8412-0452-7/78/47-080-160$05.00/0

TABLE I

The Proteins of Blood Coagulation

Common Name	Roman Numeral Designation	Associated Disorder
Fibrinogen	Factor I	Afibrinogenemia
Prothrombin	Factor II	Prothrombin deficiency
Tissue Factor	Factor III	--
Proaccelerin	Factor V	Parahemophilia
Proconvertin	Factor VII	ProSPCA deficiency
Antihemophilic Factor	Factor VIII	Hemophilia A
Christmas Factor	Factor IX	Christmas disease (hemophilia B)
Stuart Factor	Factor X	Stuart factor deficiency
Plasma Thromboplastin Antecedent (PTA)	Factor XI	PTA deficiency
Hageman Factor	Factor XII	Hageman trait
Fibrin-stabilizing Factor	Factor XIII	Factor XIII deficiency
Prekallikrein (Fletcher Factor)	--	Fletcher factor disease
HMW Kininogen (high molecular weight kininogen, Fitzgerald factor)	--	Fitzgerald trait

have greatly assisted the researcher in formulating a more intelligent approach to the understanding of the kinetics of the overall coagulation process *in vivo*. Furthermore, from such studies has emerged the concept that the principal mechanism for the initiation, propagation and regulation of the coagulation process involves the selective enzymatic cleavage of unique peptide bonds. In this article, the primary emphasis will be placed on more recent information regarding the mechanism of activation of each coagulation zymogen leading to fibrin formation. In addition, certain activated coagulation factors have been observed to carry out secondary proteolytic reactions in the test tube and these will be presented and discussed in relation to their potential regulatory relevance *in vivo*.

II. Molecular Events in Blood Coagulation

Two pathways apparently exist for the formation of factor Xa, the serine protease which converts prothrombin to thrombin. These pathways have been referred to as the intrinsic and extrinsic coagulation pathways. The intrinsic pathway includes those reactions that lead to factor Xa formation utilizing factors present only in plasma. These factors include factor XII, factor XI, factor IX and factor VIII. Presently, it is not clear whether high molecular weight kininogen and prekallikrein are also required for this pathway inasmuch as plasma lacking these proteins clots poorly upon experimentally initiating the intrinsic system. The extrinsic pathway of factor Xa generation involves the participation of plasma factor VII and tissue factor, a specific lipoprotein released from tissue surrounding the damaged blood vessel wall. These pathways clearly can be differentiated in the test tube depending on the initiating stimulus provided. Factor Xa formed by either the intrinsic or extrinsic pathway appears biochemically indistinguishable. Following its formation, factor Xa then triggers a sequence of reactions which results in the formation of insoluble fibrin. It is perhaps noteworthy to mention that each of these pathways is thought to be triggered by their initial interaction with elements of the vascular subendothelium. The rate and intensity of factor Xa formation *in vivo* conceivably is dictated by the pathway followed which in turn is related to the severity of the vascular trauma. Thus, vascular trauma which involves a break in the continuity of the vessel with concomitant release of tissue factor necessitates immediate formation of factor Xa which presumably occurs via extrinsic clotting. On the other hand, relatively mild vascular trauma in which the vessel does not lose its continuity, may call for a relatively slow, more controlled formation of factor Xa. Under these circumstances, factor Xa could be formed via the intrinsic pathway initiated by the interaction of factor XII with a subendothelial collagen surface.

A. The Activation of Factor X by the Intrinsic Pathway

The sequence of events which lead to the formation of factor Xa by the intrinsic pathway of blood coagulation is schematically represented in Figure 1. The first step in these cascading reactions is the activation of factor XII. Factor XII has been isolated from human and bovine plasma and characterized in regard to its size, structure and composition (1,2). Factor XII isolated from either source appears to be a single-chain glycoprotein with a molecular weight of about 75,000. Factor XIIa has also been isolated from bovine plasma by Fujikawa et al. (3). Bovine factor XIIa is composed of a heavy and a light chain held together by disulfide bonds. The light chain has a molecular weight of 28,000 and contains an amino-terminal sequence of Val-Val-Gly-Gly (3). The amino-terminal sequence of the heavy chain (mol wt 46,000) is Thr-Pro-Pro-Trp-Lys- and is identical to that observed for the single-chain, precursor factor XII. The The carboxyl-terminal residue in the heavy chain is arginine. Thus, the conversion of bovine factor XII to factor XIIa is due to the cleavage of a single internal arginyl-valine peptide bond. Factor XIIa exhibited amidase activity toward arginine-containing substrates and is readily inhibited by the serine protease inhibitor, diisopropyl phosphofluoridate (DFP). This inhibitor is bound exclusively to the light chain of factor XIIa indicating that the active site serine residue is located in this chain. Furthermore, the active site region in the light chain of factor XIIa contains an Asp-Ala-Cys-Gln-Gly-Asp-Ser-Gly-Gly- sequence which is typical of the other plasma serine proteases (2).

The mechanism whereby factor XII is activated physiologically is presently not clear. Moreover, the protease(s) which carries out this reaction has not been identified. The activation of factor XII has been thought to occur *in vivo* following its interaction with a charged surface which presumably is subendothelial collagen. The activation of factor XII has been investigated in the test tube utilizing glass or kaolin as the negatively charged surface. Incubation of single-chain factor XII with kaolin produces a [factor XII-kaolin] complex which does not have enzymatic activity towards small substrates. In addition, this complex is not inhibited by DFP suggesting that factor XII is not converted to an active enzyme simply by its interaction with the kaolin surface. However, the [factor XII-kaolin] complex will react with factor XI and prekallikrein converting these zymogens to active enzymes (3). Whether factor XII was converted to factor XIIa during this incubation is not known.

Griffin and Cochrane (4) and Chan et al. (5) have reported that human factor XII is activated in the presence of kallikrein, HMW kininogen and kaolin. These studies suggest that factor XII interacts stoichiometrically with HMW kininogen on the kaolin surface and is subsequently activated by kallikrein. From this information, it is conceivable that upon interaction with a charged surface in the presence of HMW kininogen, factor XII undergoes a conformational change which permits its active site to react slowly and preferentially with prekallikrein. The relatively small amount of kallikrein formed can reciprocally activate substantial amounts of factor XII. Once formed in sufficient levels, factor XIIa can then catalyze the activation of the next zymogen in the cascade, namely factor XI.

Factor XI is converted to factor XIa by factor XIIa, a reaction apparently augmented by HMW kininogen and kaolin (6). Factor XIIa is presently the only coagulation protease known which can activate factor XI in the test tube. The apparent obligatory role of factor XIIa in the activation of factor XI is interesting in view of the fact that patients with factor XII deficiency have no demonstrative bleeding tendencies, yet patients with factor XI deficiency often bleed considerably following injury or minor surgery. One possible explanation for this phenomenon is that factor XII-deficient plasma contains sufficient levels of factor XII to initiate coagulation. Assuming the coagulation process is propagated by an amplification mechanism, the smallest level of enzyme required would be in the initial reactions. An alternative explanation is that other heretofore unrecognized factors may play a major role in the physiological activation of factor XI. In this case, proteolytic cleavage of factor XI may be a secondary reaction catalyzed by factor XIIa and that other proteins such as prekallikrein and plasminogen proactivator constitute more specific primary substrates for factor XIIa.

Factor XI has been isolated from human and bovine plasma. Human factor XI is a glycoprotein with a molecular weight of about 124,000. It is composed of two identical polypeptide chains (mol wt about 60,000) held together by disulfide bonds. During the activation of factor XI an internal peptide bond in each of the two chains is cleaved by factor XIIa (7). This gives rise to two heavy chains (mol wt 35,000) and two light chains (mol wt 25,000) in factor XIa. The amino-terminal amino acid sequence of the light chains in factor XIa is Ile-Val-Gly-Gly. This sequence shows homology with other plasma serine proteases involved in coagulation. The light chains also contain the active site serine which reacts quantitatively with DFP. Thus, factor XIa is a serine protease containing two active sites per molecule of enzyme, and represents the first case of a serine protease with two catalytic sites (7).

The next zymogen to be activated in the intrinsic pathway is factor IX. Human and bovine factor IX have been isolated and well characterized in the past few years (8,9). Bovine factor IX is a single-chain glycoprotein with a molecular weight of 55,000. Human factor IX is also a glycoprotein with a molecular weight of 57,000 and possesses an amino-terminal amino acid sequence nearly identical to that of bovine factor IX (9).

Factor IX is one of the four coagulation factors which require vitamin K for its biosynthesis. The other factors include prothrombin, factor X and factor VII. Vitamin K participates in the biosynthesis of these proteins in the post-ribosomal carboxylation of specific glutamic acid residues present in the amino-terminal portion of these proteins (10). The new amino acid, γ-carboxyglutamic acid, is essential for calcium ion binding by these proteins and their subsequent interaction with acidic phospholipid membranes. The γ-carboxyglutamic acid residues in these proteins tend to occur as adjacent residues which facilitate the coordination of one calcium ion per pair of γ-carboxyglutamic acid residues (11).

Factor IX is activated by factor XIa in the presence of calcium ions in a two-step process. In the first step, single-chain factor IX is converted to a two-chain molecule by the cleavage of an internal Arg-Ala peptide bond. The product of this reaction has no enzymatic activity and consists of a light chain (mol wt 16,000) with an amino-terminal sequence of Tyr-Asn-Ser-Gly, and a heavy chain (mol wt 38,000) with an amino-terminal sequence of Ala-Gly-Thr-Ile. This intermediate is then converted to factor IXa in a rate-limiting step by limited proteolysis. In this reaction, a specific Arg-Val bond in the amino-terminal region of the heavy chain is cleaved by factor XIa with the concomitant release of an activation peptide. Thus, the factor IXa formed in this reaction is composed of two chains held together by disulfide bonds. The heavy chain (mol wt 27,300) contains the active site serine residue and contains an amino-terminal sequence of Val-Val-Gly-Gly. The light chain (mol wt 16,600), which originates from the amino-terminus of the zymogen, contains the γ-carboxyglutamic acid residues and the amino-terminal sequence of Tyr-Asn-Ser-Gly (12).

Factor IXa is the serine protease which converts factor X to factor Xa by limited proteolysis in the intrinsic pathway. In this reaction, factor IXa forms a macromolecular complex with an additional protein, factor VIII, in the presence of phospholipid and calcium ions. Factor VIII is the plasma protein that is absent in individuals with classic hemophilia, the most prevalent coagulation disorder known. Factor VIII has been isolated from human and bovine plasma and exhibits a molecular weight of about one million (13,14). These preparations possess potent coagulant activity as well as von Willebrand platelet aggregating activity. Treatment of the bovine factor VIII preparation with catalytic amounts of bovine thrombin results in a 25-fold increase in factor VIII procoagulant activity in the presence of $CaCl_2$. Gel filtration of this thrombin-treated preparation of factor VIII results in the separation of factor VIII coagulant activity and the platelet aggregating activity (15). However, no detectable protein was associated with the factor VIII coagulant activity. This has led to speculation that the factor VIII represents only a small fraction of the total protein in the factor VIII preparations described (15).

The role of factor VIII in the activation of factor X appears to be one of a regulatory nature. Conceivably, factor VIII could augment this reaction by promoting the formation of a more reactive conformation in either factor IXa or its substrate, factor X.

The final step in this pathway is the activation of factor X. Factor X is a glycoprotein with a molecular weight of 55,000 and one of the most well characterized proteins in blood coagulation. Indeed, the total amino acid sequence of bovine factor X has been completed (16). It consists of a heavy chain (mol wt 39,300) and a light chain (mol wt 16,500) held together by disulfide bonds. The heavy chain contains 307 amino acid residues with two glycosyl moieties linked to Asn-35 and Thr-300 (16). The heavy chain also contains the active site serine located in position 233. The light chain contains the vitamin K-dependent γ-carboxyglutamic acid residues.

The conversion of bovine factor X to factor Xa by factor IXa occurs as a result of the cleavage of a single peptide bond between Arg-51 and Ile-52. This results in the release of an activation peptide (mol wt 10,500) from the amino-terminal region of the heavy chain analogous to that observed in the activation of factor IX. Factor Xa is a serine protease which, in the presence of factor V, phospholipid and calcium ions, catalyzes the conversion of prothrombin to thrombin.

B. The Activation of Factor X by the Extrinsic Pathway

The proposed series of reactions which result in the activation of factor X by the extrinsic pathway are depicted in Figure 2. In these reactions, plasma factor VII, in the presence of calcium ions, forms a complex with a lipoprotein released from damaged tissue. This complex subsequently converts factor X to factor Xa which appears indistinguishable from that observed following intrinsic activation of factor X (17). Present evidence suggests that factor VII is converted to factor VIIa in order to catalyze this reaction. The lipoprotein serves as a high molecular weight cofactor which is absolutely essential for the expression of factor VIIa proteolytic activity.

Factor VII is a trace plasma glycoprotein. Its low concentration has virtually precluded the isolation of this protein from human plasma. Factor VII has been isolated from bovine plasma as a single-chain protein with a molecular weight of 45,500 (18,19). In the test tube, single-chain factor VII can be converted to a two-chain protein which exhibits 30-40 times the coagulant activity of the single-chain form. The activation of factor VII, that is, the formation of two-chain factor VII, has been observed following the incubation of factor VII with factor XIIa (20), factor XIa (21), kallikrein (21), factor Xa (22), and thrombin (22). Factor VIIa is composed of a heavy chain (mol wt 34,000) and a light chain (mol wt 23,000) held together by disulfide bonds. The active site serine is located in the heavy chain which contains the amino-terminal sequence of Ile-Val-Gly-Gly (20). The light chain contains the γ-carboxyglutamic acid residues and arises from the amino-terminal portion of the single-chain zymogen (20). The carboxyl-terminal residue of the light chain is arginine. Thus, factor VIIa is a serine protease arising from the proteolysis of a single internal Arg-Ile peptide bond. Factor VIIa has an absolute requirement for tissue factor in order to cleave the Arg-51-Ile-52 peptide bond in the heavy chain of factor X.

The physiological mechanism for the activation of factor VII is presently not known. While several serine proteases that participate in the intrinsic pathway will convert factor VII to factor VIIa *in vitro*, it is not known to what extent they participate in the activation of factor VII *in vivo*. In addition, factor X is the only protein substrate known for factor VIIa and the proteolytic range of factor VIIa has not been investigated.

While considerable information has accumulated on the molecular properties of factor VII, comparatively little is known regarding its interaction with purified tissue factor and the interaction of factor X with this [factor VII-tissue factor] complex. The tissue factor apoprotein, the protein component of the tissue factor lipoprotein, has been isolated from several tissues. The most highly characterized apoprotein preparation appears to be obtained from human brain microsomes (23). Tissue factor apoprotein prepared in this study is a single-chain glycoprotein containing about 6% carbohydrate and exhibited a molecular weight of 52,000 by SDS gel electrophoresis. The tissue factor apoprotein exhibits no coagulant activity in the absence of phospholipid. Addition of mixed phospholipids to the apoprotein, however, partially restores the tissue factor coagulant activity.

The role of tissue factor appears to be similar to that of factor VIII in that it serves to enhance the limited proteolysis of factor X by factor VIIa several orders of magnitude. No enzymatic activity has been observed associated with homogeneous preparations of tissue factor. Moreover, incubation of pure bovine factor VII with crude, highly active preparations of tissue factor does not result in the formation of two-chain, activated factor VII (21).

C. Prothrombin Activation by Activated Factor X

Prothrombin is converted to thrombin by factor Xa in the presence of calcium ions, phospholipid and factor Va (Figure 3). Prothrombin and its mechanism of activation has been intensively investigated over the last two decades, and the details of this complex reaction are reasonably clear (24).

Prothrombin is a glycoprotein with a molecular weight of

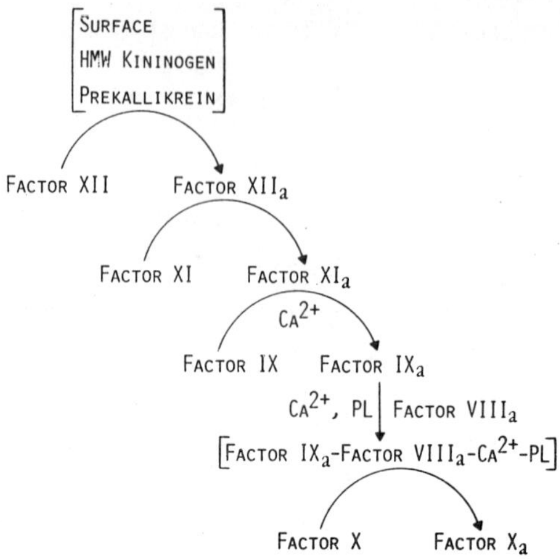

Figure 1. Abbreviated mechanisms for the activation of factor X by the intrinsic pathway

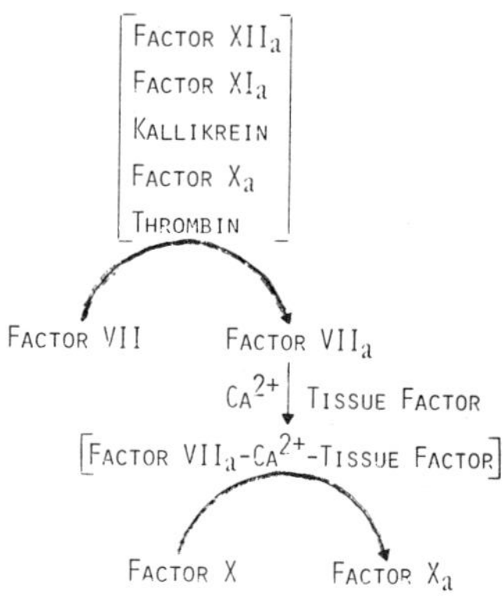

Figure 2. Abbreviated mechanisms for the activation of factor X by the extrinsic pathway

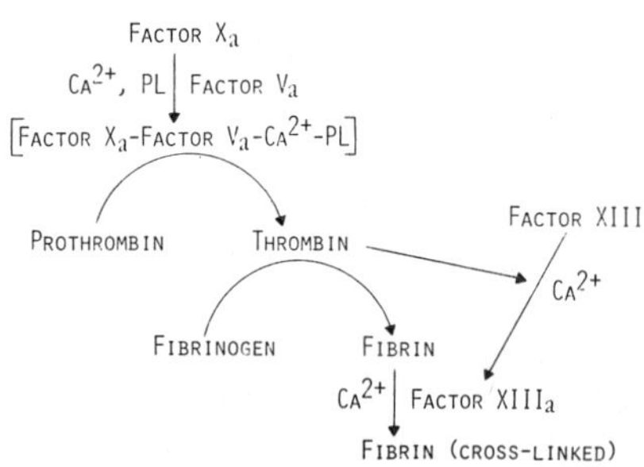

Figure 3. Mechanisms for the formation of cross-linked fibrin following the activation of factor X

70,000. It is one of the four vitamin K-dependent clotting factors and contains 10 γ-carboxyglutamic acid residues. Its concentration in plasma far exceeds that of any other coagulation zymogen. The amino acid sequence of bovine prothrombin has been determined (25) and the sequence of human prothrombin is nearing completion. Bovine prothrombin contains 582 amino acid residues with 12 disulfide bridges. The three carbohydrate chains of bovine prothrombin are N-glycosidically linked to asparagine residues at positions 77, 101 and 376 (25).

The conversion of prothrombin to thrombin occurs as a two-step process on the surface of a macromolecular aggregate commonly referred to as the "prothrombinase complex." This complex consists of factor Xa, factor Va, phospholipid and calcium ions. Pre-treatment of factor Xa with DFP obviates the conversion of prothrombin to thrombin indicating that factor Xa is the active principal in this complex. Factor V, like factor VIII and tissue factor, appears to function as a high molecular weight regulatory cofactor in this reaction. Its ability to potentiate this reaction is increased several fold by pre-treatment with factor Xa or thrombin (26).

The first step in the activation of prothrombin is the cleavage of a peptide bond between Arg-274 and Thr-275 by factor Xa. This reaction releases a large inactive glycopeptide fragment (mol wt 33,500) from the amino-terminal region of prothrombin and yields a single-chain thrombogenic protein called intermediate II. This protein has no coagulant activity and has a molecular weight of 39,000. Intermediate II is subsequently cleaved by factor Xa at a peptide bond between Arg-323 and Ile-324 to yield two-chain thrombin. Thrombin is composed of a heavy chain (mol wt 32,000) and a light chain (mol wt 5,700) held together by a disulfide bond. The heavy chain of thrombin also contains the active site serine residue.

D. Formation of Cross-linked Fibrin

The conversion of fibrinogen to fibrin monomer by thrombin is the penultimate step in the blood coagulation scheme. Fibrinogen is a large glycoprotein (mol wt 340,000) composed of three pairs of non-identical polypeptide chains. These chains, designated α, β and γ, are held together by several disulfide bonds. The molecular weights for the α, β and γ chains of human fibrinogen are 63,500, 56,000 and 47,000, respectively (27). Thus, the subunit structure for fibrinogen may be expressed as $\alpha_2\beta_2\gamma_2$.

During the conversion of fibrinogen to fibrin, four specific arginyl-glycine peptide bonds in the α and β chains are cleaved by thrombin with the concomitant release of four fibrinopeptides. Two fibrinopeptides A are released from the amino-terminus of the α chains and two fibrinopeptides B are released from the amino-terminal end of the β chains. Subsequent to the release of these fibrinopeptides, the fibrin monomers rapidly aggregate in an orderly process to form the "soluble" fibrin polymer. This fibrin polymer is soluble in denaturants such as urea or guanidine-hydrochloride, and is readily attacked by the fibrinolytic protease, plasmin.

Conversion of the "soluble" fibrin polymer to an "insoluble" fibrin clot is the terminal reaction in blood coagulation. This reaction is catalyzed by activated factor XIII in the presence of calcium ions. In this reaction, intermolecular peptide crosslinkages are formed by factor XIIIa between an amino group from a specific lysine residue on one fibrin monomer and a glutaminyl residue of an adjacent fibrin monomer (28). Two ε(γ-glutamyl) lysine peptide bonds are rapidly formed between intermolecular γ chains. Upon longer incubation, crosslinkages are also observed between intermolecular α chains. The total number of crosslinkages appears to be about six per mole of fibrin, two of which involve crosslinkages with the γ chains and four with the α chains. The presence of these crosslinkages results in a fibrin polymer which is insoluble in denaturing solvents and is quite resistant to plasmin digestion.

Factor XIIIa, the transglutaminase which catalyzes the crosslinking reactions in fibrin, circulates in plasma in a precursor form. In the presence of calcium ions, factor XIII is converted by thrombin to factor XIIIa by limited proteolysis. Factor XIII is a glycoprotein composed of two A chains (mol wt 75,000) and two B chains (mol wt 88,000). The native protein is a tetramer with a molecular weight of 320,000 (29). Factor XIII is activated by thrombin in a two-step process. In the first step, thrombin cleaves an arginyl-glycine bond in the amino-terminal region of the A chains, giving rise to an inactive tetrameric intermediate and two peptides (mol wt 4,000). In the next step, which requires calcium ions, the tetrameric intermediate dissociates into a catalytic dimer (A'_2) and an inactive dimer of B chains (30). Although the A' chains are presumably identical, the active cysteine residue is found in only one of the catalytic subunits. Thus, factor XIIIa appears to be an example of the "half-of-the-site reactivity" enzymes (31).

E. Mechanism Involving Proteolysis in the Termination of Coagulation

Under normal physiological conditions, the activation of a zymogen through proteolysis is essentially an irreversible process. Accordingly, several control mechanisms are available to regulate and terminate coagulation from spreading throughout the circulation. Perhaps the principal mechanism for the regulation of the activated coagulation factors is their neutralization by serine protease inhibitors found in the blood. Indeed, about 10% of the protein in plasma is serine protease inhibitors (32).

Another important mechanism for the control of coagulation involves the inactivation of specific clotting factors by limited proteolysis. Although the details of these reactions are just beginning to emerge, it appears that thrombin is central to this process. Excess thrombin formation is carefully controlled by several negative feedback reactions either initiated or directly catalyzed by thrombin. As pointed out earlier, proteolytic modification of factor V and factor VIII by thrombin enchances the coagulant activity of these proteins several fold. Continued incubation of these proteins with thrombin, however, results in the loss of their coagulant activity (15,33). Thrombin also catalyzes the rapid cleavage of a peptide bond between Arg-156 and Ser-157 in prothrombin giving rise to a non-thrombogenic fragment (mol wt 23,000) and a thrombogenic intermediate I (mol wt 55,000). Intermediate I is loosely associated with the "prothrombinase complex" and consequently cleaved by factor Xa at a much slower rate than prothrombin. Hence, the overall effect is to reduce thrombin formation as a result of these reactions.

Recently, a new vitamin K-dependent protein has been discovered and tentatively identified as Protein C (34). Protein C is a glycoprotein composed of a heavy chain (mol wt 40,000) and a light chain (mol wt 22,000) held together by disulfide bonds. Incubation of Protein C with thrombin results in the cleavage of a single peptide bond between Arg-14 and Ile-15 in the heavy chain with the concomitant formation of activated Protein C (35). Activated Protein C is a serine protease and the active site serine residue is located in the heavy chain. The amino acid sequence surrounding the active site serine residue in activated Protein C is homologous to the active site regions of the serine proteases involved in blood coagulation. Activated Protein C markedly prolongs the clotting time of plasma, and this anticoagulant effect was totally obviated by prior incubation of the enzyme with DFP. Recent evidence indicates that activated Protein C, in the presence of calcium ions and phospholipid, is a potent inactivator of factor V and factor VIII (35,36), and that these inactivation reactions were absolutely dependent on the enzymatic activity of activated Protein C. Thus, it appears that the level of thrombin activity is carefully regulated by negative feedback reactions which involve the limited proteolysis of prothrombin, factor V and factor VIII either directly catalyzed by thrombin or indirectly through the formation of activated Protein C.

Literature Cited

1. Griffin, J. H., and Cochrane, C. G. (1976) Methods Enzymol. 45, 56-65.
2. Fujikawa, K., Walsh, K. A., and Davie, E. W. (1977) Biochemistry 16, 2270-2278.
3. Fujikawa, K., Kurachi, K., and Davie, E. W. (1977) Biochemistry 16, 4182-4188.

4. Griffin, J. H., and Cochrane, C. G. (1976) Proc. Natl. Acad. Sci. USA 73, 2554-2558.
5. Chan, J. Y. C., Habal, K. M., Burrowes, C. E., and Movat, H. Z. (1976) Thromb. Res. 9, 423-430.
6. Wiggins, R. C., Bouma, B. N., Cochrane, C. G., and Griffin, J. H. (1977) Proc. Natl. Acad. Sci. USA 74, 4636-4640.
7. Kurachi, K., and Davie, E. W. (1977) Biochemistry 16, 5831-5839.
8. DiScipio, R. G., Hermodson, M. A., Yates, S. G., and Davie, E. W. (1977) Biochemistry 16, 698-706.
9. Fujikawa, K., Thompson, A. R., Legaz, M. E., Meyer, R. G., and Davie, E. W. (1973) Biochemistry 12, 4938-4945.
10. Stenflo, J., Fernlund, P., Egan, W., and Roepstorff, P. (1974) Proc. Natl. Acad. Sci. USA 71, 2730-2733.
11. Furie, B. C., Mann, K. G., and Furie, B. (1976) J. Biol. Chem. 251, 3235-3241.
12. Fujikawa, K., Legaz, M. E., Kato, H., and Davie, E. W. (1974) Biochemistry 13, 4508-4516.
13. McKee, P. A., Anderson, J. C., and Switzer, M. E. (1975) Ann. N.Y. Acad. Sci. 240, 8-33.
14. Legaz, M. E., Weinstein, M. J., Heldebrant, C. M., and Davie, E. W. (1975) Ann. N.Y. Acad. Sci. 240, 43-61.
15. Vehar, G. A., and Davie, E. W. (1977) Science 197, 374-376.
16. Titani, K., Fujikawa, K., Enfield, D. L., Ericsson, L. H., Walsh, K. A., and Neurath, H. (1975) Proc. Natl. Acad. Sci. USA 72, 3082-3086.
17. Fujikawa, K., Coan, M. H., Legaz, M. E., and Davie, E. W. (1974) Biochemistry 13, 5290-5299.
18. Kisiel, W., and Davie, E. W. (1975) Biochemistry 14, 4928-4934.
19. Radcliffe, R., and Nemerson, Y. (1975) J. Biol. Chem. 250, 388-395.
20. Kisiel, W., Fujikawa, K., and Davie, E. W. (1977) Biochemistry 16, 4189-4194.
21. Kisiel, W., unpublished data.
22. Radcliffe, R., and Nemerson, Y. (1976) J. Biol. Chem. 251 4797-4802.
23. Bjorklid, E., and Storm, E. (1977) Biochem. J. 165, 89-96.
24. Suttie, J. W., and Jackson, C. M. (1977) Physiol. Reviews 57, 1-70.
25. Magnusson, S., Peterson, T. E., Sottrup-Jensen, L., and Claeys, H. (1975) in Proteases and Biological Control, eds. Reich, E., Rifkin, D. B., and Shaw, E. (Cold Spring Harbor Laboratory, New York), Vol. 2, pp. 123-149.
26. Smith, C. M., and Hanahan, D. J. (1976) Biochemistry 15, 1830-1838.
27. McKee, P. A., Rogers, L. A., Marler, E., and Hill, R. L. (1966) Arch. Biochem. Biophys. 116, 271-279.
28. Lorand, L., Downey, J., Gotoh, T., Jacobsen, A., and Tokura, S. (1968) Biochem. Biophys. Res. Commun. 31, 222-230.

29. Schwartz, M. L., Pizzo, S. V., Hill, R. C., and McKee, P. A. (1973) J. Biol. Chem. 248, 1395-1407.
30. Chung, S. I., Lewis, M. C., and Folk, J. E. (1974) J. Biol. Chem. 249, 940-950.
31. Levitzki, A., Stallcup, W. B., and Koshland, D. E. (1971) Biochemistry 10, 3371-3378.
32. Heimburger, N., Haupt, H., and Schwick, H. (1971) in Proc. Int. Res. Conf. Proteinase Inhibitors, eds. Fritz, H., and Tschesche, H. (W. de Gruyter, Berlin), pp. 1-21.
33. Colman, R. W. (1969) Biochemistry 8, 1438-1445.
34. Stenflo, J. (1976) J. Biol. Chem. 251, 355-363.
35. Kisiel, W., Canfield, W. M., Ericsson, L. H., and Davie, E. W. (1977) Biochemistry 16, 5824-5831.
36. Vehar, G., unpublished results.

RECEIVED March 15, 1978.

Infection and Degenerative Diseases

Introduction

DON M. CARLSON

Department of Biochemistry, Purdue University, West Lafayette, IN 47907

Introductory Remarks

I wish to express my appreciation to Dr. Walborg for the invitation to chair this session of the Symposium and to acknowledge the efforts of the late Dr. Ward Pigman, my dear friend and colleague, who was instrumental in the original planning of the Symposium. There are five very timely and interesting presentations scheduled for this session and my introductory remarks will be brief.

Until recently there was limited interest in the biochemistry of complex carbohydrates. However, the excitement created by the probability that cell-surface glycoproteins, glycolipids and glycosyltransferases are involved in such biological phenomena as cell adhesion, cancer, diabetes, hemostasis, and cholera has attracted workers from all biological and medical disciplines. The number of metabolic diseases and dysfunctions related to glycoprotein and glycolipid metabolism is constantly growing. Certainly a better understanding of the chemistry and the mechanisms of biosynthesis of these complex molecules is needed to provide information leading to the control of these diseases.

In a Symposium dedicated to the memory of Dr. Ward Pigman we should reflect upon major past accomplishments in the field of complex carbohydrates. Two major discoveries opened up new horizons for biological research. First, the discovery of sugar nucleotides and their central role in glycoside biosynthesis. Second, the discovery of sialic acid. Others have presented elaborate biosynthetic schemes and structural aspects of complex carbohydrates, allowing me to skip details here and continue with the important aspects of this Symposium, the research presentations.

Summing Up

Before beginning the arduous task of summing up the presentations of this Symposium, I want to mention an organization which originated under the guidance and initiative of Dr. Ward Pigman. The Society for Complex Carbohydrates evolved from the 4-G's (Glycoprotein-Glycosaminoglycan-Glycolipid-Group), which was originated by Ward as a mechanism to periodically assemble researchers in these areas (relatively few at that time) to discuss current research interests. Dr. Pigman was the first President of the Society for Complex Carbohydrates, now an international organization. The Society meets annually, usually in September, for its national meeting. The Fourth International Symposium on Glycoconjugates, held in Woods Hole last fall, was sponsored by the Society. This ability and enthusiasm of Dr. Ward Pigman in organizing conferences, symposia and other meetings is exemplified by the excellent presentations of this Symposium.

When Dr. Walborg asked me if I would "sum up" this Symposium, I hesitated for two reasons. First, I was not well-informed in each of the diverse subjects to be presented, and second, I would be forced to remain inside and miss the lovely weather of Southern California. However, I could not decline an offer to sum up a Symposium dedicated to the memory of Dr. Ward Pigman As the hour is late, my remarks will be brief and I apologize for not mentioning all presentations, but the 10 minutes scheduled for this summation would allow 24 seconds per presentation, assuming equal time for each.

Since this Symposium was scheduled and sponsored partly by the Division of Carbohydrate Chemistry, I will dwell briefly on the chemistry, structures and biosynthesis of the oligosaccharide moieties. In the past few years we have progressed from using relatively insensitive, non-specific methodologies for determining sugar compositions and structures to the sophisticated procedures of today. Many of these were outlined by Dr. Sweeley, including GC/MS, HPLC, ^{1}H-NMR, and ^{13}C-NMR. For studies on the peptide portion of glycoproteins, it is now fashionable to isolate mRNA, prepare cDNA by reverse transcriptase and sequence the DNA. Several "signal peptides" have been detected in this manner.

Many glycosyltransferases are now available in relatively pure form and a better understanding of the regulatory mechanisms of glycoprotein and glycolipid biosynthesis is evolving. Studies on α_1-antitrypsin are especially fascinating. The abnormal glycoprotein apparently is stored in the RER as a result of improper "processing" of the polymannose chains (sometimes referred to as the simple-type chain). This processing may be necessary to the clearing or transporting of the glycoprotein to the golgi membranes, where completion of the complex-type chain occurs. The inability to "process" this glycoprotein may result from a single amino acid change in the peptide chain.

The binding and transport of toxic substances such as ricin, abrin, cholera toxin and so forth appears related to cell surface carbohydrates as receptors. Comparison of normal and transformed cells with regards to cell surface glycoproteins and glycolipids continues in many laboratories. There is little doubt that these efforts will be rewarded by contributing to the ultimate description of cancer on a molecular basis.

Interaction of complex carbohydrates and glycosyltransferases may possibly explain the phenomenon of cellular adhesion. Dr. Roseman proposed this hypothesis, and has emphasized that it remains a hypothesis. For many, the cell surface is the "workshop". We are continually probing its surface with unsophisticated approaches and in insulting ways but gradually are making progress at understanding this complex membrane that shelters our life's processes.

When confronted by Dr. Walborg regarding the summation, I asked him what he expected of me. He said, "After listening to the current status of studies on complex carbohydrates, just give some thoughts on where we are going." I can only say that with the intense interest and rapid progress in this field as indicated by presentations in this Symposium, my thoughts are that we should be back to the laboratory. Thank you.

RECEIVED May 18, 1978.

10

Biosynthesis of Murine Leukemia Virus Membrane Proteins and their Assembly into Virus Particles

R. B. ARLINGHAUS, E. C. MURPHY, JR., R. B. NASO, L. J. ARCEMENT, and W. L. KARSHIN

Biology Department, University of Texas System Cancer Center, M. D. Anderson Hospital and Tumor Institute, Houston, TX 77030

The cellular plasma membrane forms the surface layer of C-type RNA tumor virus particles. The cell's plasma membrane is also involved in one of the important phases of the life cycle of RNA tumor viruses, that is the budding of the complete virus particles from the infected cell. C-type RNA tumor viruses contain a core structure surrounded by a lipid envelope through which spikes protrude. These spikes are believed to be composed, in mouse systems, of the viral glycoprotein gp69/71 (1). Two non-glycosylated proteins termed p12(E) and p15(E) are also associated with the virion envelope. Their roles are not definitely known but recent studies suggest that gp69/71 and p15(E) exist as a complex and this complex forms the basic unit of the protruding spike (1, 2, 3). The core of these viruses is composed of four internal proteins termed p30, p15, pp12 (phosphoprotein) and p10 (4). The latter two have been shown to be associated with the viral genomic RNA (5, 6). Viral p30 is thought to be the capsid protein of the core (4) whereas viral protein p15 (a hydrophobic protein) could be involved in the association of the core with the lipid envelope.

The virus also contains a DNA polymerase in its core. This polymerase can readily copy RNA into DNA with an appropriate primer. It is present in virions in amounts estimated to be 50-100 molecules per virus particle (7) whereas the internal structural proteins are thought to be present in approximately 5,000 copies per virion (8). The envelope proteins [gp69/71, p15(E)] are also initially present in relatively large amounts but gp69/71 appears to be lost from the virion probably because of proteolytic degradation that releases the mature glycoprotein into the surrounding medium (9).

0-8412-0452-7/78/47-080-180$05.25/0

These viral proteins are encoded by three different genes within the 35S (≃3 x 10^6 daltons or 10,000 bases) viral genome. Each viral particle is thought to contain two identical copies of the RNA genome. The gene coding for the four internal core proteins has been termed 'gag' or group antigens; that coding for the viral RNA-directed DNA polymerase is termed 'pol' or reverse transcriptase (RT); the envelope proteins are coded within the envelope gene or 'env' (10).

The purpose of this report is to summarize the events leading to the formation of the mature viral envelope proteins and their assembly into virions. We also will propose a model for virion assembly and budding of the virus particle from the cell.

The Virion Proteins of Rauscher Murine Leukemia Virus

Rauscher murine leukemia virus (R-MuLV) coded proteins can be resolved by electrophoresis in gels of polyacrylamide under denaturing conditions (Fig. 1). The major mature proteins that can be seen are p30, p15, p12(E) and p10; minor amounts of gp69/71 and p15(E) are also present in the virion. A substantial amount of an uncleaved precursor to the core proteins, termed $Pr65^{gag}$, is also present in virus. The viral phosphoprotein pp12 is not solubilized by this procedure but it is readily seen in virus denatured with 6 M quanidine-HCl and fractionated by molecular seive chromatography in 6 M guanidine.

The radioactive virus, whose proteins are shown in Fig. 1, was prepared by incubating infected cells with ^{14}C-arginine and lysine for a 48 hr interval. Under these conditions much of the viral gp69/71 is lost from the virus particle (9). This loss of viral gp69/71 may correspond to the low amount of p15(E) that is present in mature virus particles (see below).

Murine C-type viruses appear to contain a complex made up of gp69/71 and p15(E) (1, 2, 3). The gp69/71-p15(E) complex can be dissociated with sulfhydril reagents such as mercaptoethanol (1, 2, 3). This complex may be the basic unit of the surface structure of the virus. Viral gp69/71 has been shown to be on the surface of the virus by several methods. One involves lactoperoxidase-catalyzed iodination of the virion surface components (11, 12). Another involves sodium borohydride reduction following galactose oxidase treatment (13). A third procedure concerns the use of immunoelectromicroscopy using monospecific antisera raised against purified gp70 (14), and most important, anti-gp70 can neutralize infectivity (15, 16). The surface location of p15(E) is indicated by the finding that antisera to p15(E) can precipitate intact virus (17, 18, 19) unlike antisera prepared against viral internal components. Thus, it is quite evident that both p15(E) and gp69/71 are present in the same place in the virion which is consistent with their being complexed together.

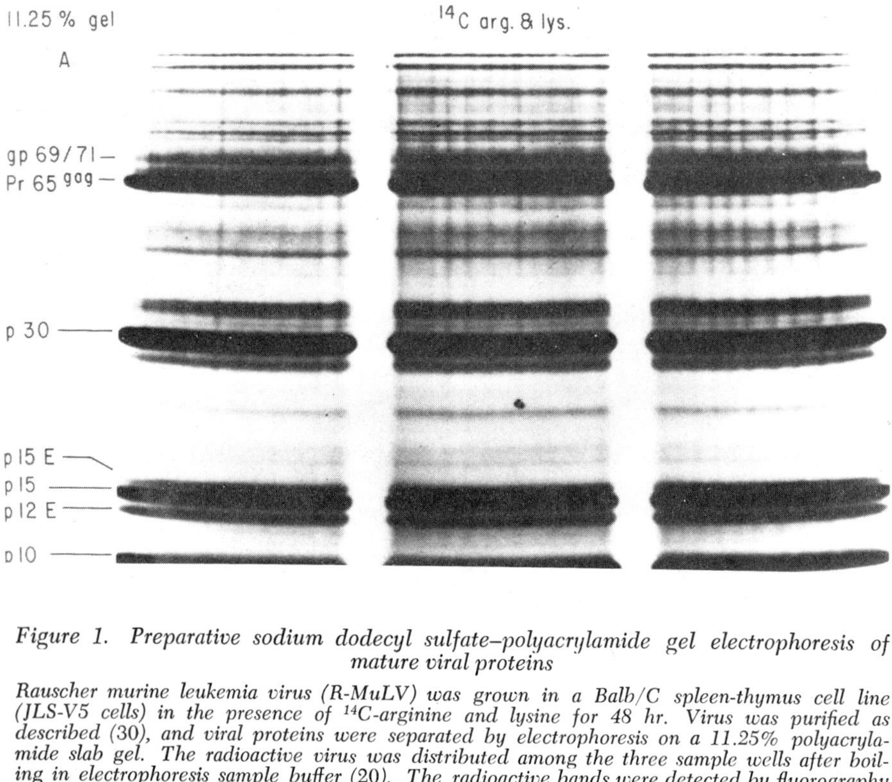

Figure 1. Preparative sodium dodecyl sulfate–polyacrylamide gel electrophoresis of mature viral proteins

Rauscher murine leukemia virus (R-MuLV) was grown in a Balb/C spleen-thymus cell line (JLS-V5 cells) in the presence of ^{14}C-arginine and lysine for 48 hr. Virus was purified as described (30), and viral proteins were separated by electrophoresis on a 11.25% polyacrylamide slab gel. The radioactive virus was distributed among the three sample wells after boiling in electrophoresis sample buffer (20). The radioactive bands were detected by fluorography using pre-flashed x-ray film so as to obtain a linear response to the amount of radiactivity (26).

Intracellular Viral Specific Glycoproteins

Virus infected and uninfected NIH Swiss mouse embryo fibroblasts (JLS-V16 cells) were incubated for 5-1/2 hr with ^{14}C-glucosamine to detect viral glycoproteins. The cytoplasmic extract, treated with NP-40 and sodium deoxycholate detergents (20), was mixed with an optimal concentration of antiserum raised in rabbits against purified R-MuLV that had been disrupted by the above detergents (21). Figure 2 shows the results of fractionation of the immunoprecipitates on a 10% sodium dodecyl sulfate (SDS) polyacrylamide gel. Figure 2 lane A shows the radioactive glycoproteins were not observed in uninfected cells in the area where two broad bands termed gPr90^{env} and gp69/71 were seen in infected cells (Fig. 2, lane B). These results indicate that two major glycoproteins are present in viral infected cells but not in uninfected cells. One of the glycoproteins corresponds in molecular weight to the mature viral glycoprotein gp69/71. The other we have shown to be a precursor polyprotein containing gp69/71 and the non-glycosylated protein p15(E) (see below and references 22, 23).

Since both gp69/71 and gPr90^{env} readily incorporate radioactive glucosamine, it was of interest to test radioactive labeling with other sugars. Of particular interest is fucose since it is often a terminal sugar in glycoproteins (24). Fucosyl tranferases are also present in plasma membranes of the cell (25) thereby restricting fucosylation to the latter stages of glycoprotein processing. With these points in mind, we then incubated infected cells with ^{3}H-labeled fucose and precipitated the viral glycoproteins either with anti-RLV serum (Fig. 2, lane C) or anti-gp69/71 (Fig. 2, lane D). The results showed that gPr90^{env} did not incorporate fucose whereas the mature viral glycoprotein (gp69/71) did. Lanes G and H of Fig. 2 also show that virion gp69/71 also incorporates radioactive fucose as well as glucosamine. Thus, gPr90^{env} appears to contain glucosamine but lacks fucose and distinguishes it clearly from the mature gp69/71.

In a parallel experiment virus-infected cells were incubated with radioactive glucosamine for 5-1/2 hr and the cytoplasmic extract was divided into two parts. One portion was immunoprecipitated with anti-gp69/71 (Fig. 2, lane E) and the other portion was isolated by anti-R-MuLV immunoprecipitation (Fig. 2, lane F). Again the major viral glucosamine-containing glycoproteins recognized by either sera were gp69/71 and gPr90^{env}. We note that anti-R-MuLV sera also precipitated a minor glucosamine-containing glycoprotein that migrates just slower than gPr90^{env} (Fig. 2, lanes B and F). This glycoprotein may be related to 'gag' gene products because it is not recognized by anti-gp69/71 but it is seen with anti-RLV. Anti-RLV serum contains antibodies to 'gag' and 'env' determinants but not to 'pol' determinants (26). Infrequently another protein was also detected in the anti-R-MuLV precipitate from glucosamine labeled infected cells. It comigrates with $Pr200^{gag-pol}$ which is the precursor to the virion reverse transcriptase (26, 27, 28). Kabat and co-workers have reported that Friend leukemia virus infected cells contain 'gag' gene product-related glycoproteins of 95,000 and 200,000 mol. wt. (29). The latter were found in addition to the viral 'env' gene-related glycoproteins. The role of the 'gag' gene-related glycoproteins in the virus life cycle is unknown.

To show the specificity of immunoprecipitation, R-MuLV infected cells were pulse-labeled with ^{14}C-amino acids for 10 min and parallel cultures were pulse-labeled for 10 min and then incubated in complete growth medium for 60 min (chase). Cytoplasmic extracts were precipitated with monospecific antiserum directed against R-MuLV gp69/71 and p30 (22). Anti-gp69/71 precipitated $gPr90^{env}$ almost exclusively in pulse-labeled cells (Fig. 3, lane A) and both $gPr90^{env}$ and gp69/71 in chase incubated cells (Fig. 3, lane B). Anti-p30 serum, however, did not recognize $gPr90^{env}$ in pulse-labeled cells (Fig. 3, lane C) but precipitated mainly precursors $Pr80^{gag}$ and $Pr65^{gag}$ as well as $Pr200^{gag-pol}$. After chase incubation (Fig. 3, lane D), analysis of the anti-p30 precipitate showed that $Pr80^{gag}$ disappeared and $Pr65^{gag}$ was reduced in its intensity while p30 appeared as a major band. No gp69/71 or $gPr90^{env}$ were precipitated by antisera directed against p30.

The kinetics of viral glycoprotein synthesis were studied by pulse-labeling cells with ^{14}C-amino acids and chasing replicate cultures for 2, 4 and 6 hr periods. The results clearly showed that gp69/71 was not labeled in a 10 min pulse-labeling (Fig. 4, lane B) but became radioactive between 60 and 120 min of the chase incubation (Fig. 4, lane E) while at the same time there was a slight decrease in the intensity of $gPr90^{env}$. Chase incubations of up to 6 hr still showed a significant amount of $gPr90^{env}$ but the ratio of gp69/71 to $gPr90^{env}$ reached a value greater than one whereas after 2 hr of chase it was less than one.

As previously reported, precipitation of cellular extracts with antiserum to disrupted virus revealed other viral proteins (e.g. p30 and p15) that appeared upon chase at the expense of $Pr80^{gag}$ and $Pr65^{gag}$ (20, 21, 22, 30) (Fig. 4, lanes A, D, G, and J). Viral p12(E) also appeared during long chase periods (2 to 6 hr) whereas only p15(E) was seen in short chase periods (1 hr).

Presence of gp69/71 and p15(E) Peptide Sequences in $gPr90^{env}$

We have previously shown that $gPr90^{env}$ and gp69/71 contain many tryptic peptides that comigrate on ion exchange columns (22). However, the patterns obtained were quite complex, to simplify the complexity of the tryptic digest profiles and the subsequent comparisons of the peptide sequences, we elected to tag the viral proteins with radioactive tyrosine (23, 30). Figure 5 shows ion-exchange profiles in which ^{14}C-tyrosine-labeled $gPr90^{env}$ tryptic peptides were mixed with tryptic peptides of ^{3}H-tyrosine labeled

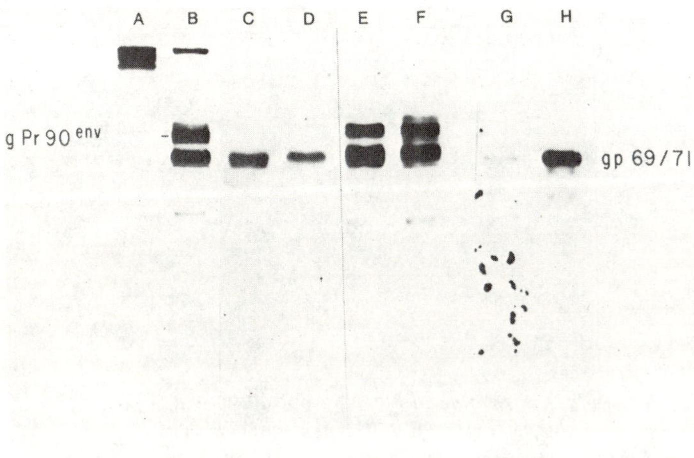

Figure 2. Polyacrylamide gel electrophoresis of R-MuLV specific glycoproteins from infected cells and virus.

Cytoplasmic extracts from uninfected (lane A) and R-MuLV-infected NIH Swiss mouse embryo fibroblasts (JLS-V16 cells) (lane B) were labeled for 5½ hr with ^{14}C-glucosamine and the anti-R-MuLV immunoprecipitates were analyzed as in Figure 1. Cytoplasmic extracts from R-MuLV-infected JLS-V15 cells were labeled with ^{3}H-fucose were immunoprecipitated with anti-gp69/71 serum (lane C) or anti-R-MuLV serum (lane D). R-MuLV-infected cells were labeled for 5½ hr with ^{14}C-glucosamine and the anti-gp69/71 immunoprecipitate (lane E) or the anti-R-MuLV immunoprecipitate (lane F) also were analyzed. Columns G and H are ^{3}H-fucose and ^{14}C-glucosamine-labeled virus, respectively, obtained form growth medium from R-MuLV-infected JLS-V16 cells following a 5½ hr labeling period.

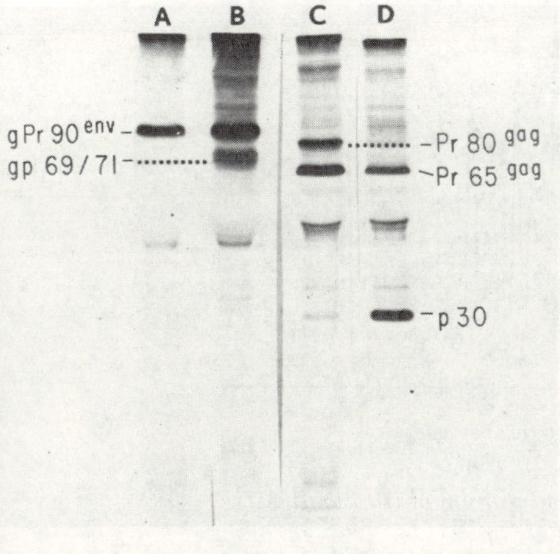

Figure 3. Analysis of intracellular R-MuLV specific proteins precipitated by anti-gp69/71 and anti-p30 sera.

Cytoplasmic extracts from R-MuLV-infected JLS-V16 cells labeled with ^{14}C-labeled amino acids were prepared and immunoprecipitation was performed with either anti-gp69/71 or anti-p30 serum. Electrophoresis on 10% gels and fluorography were as in Figure 1. Column A, anti-gp69/71 immune precipitate from a 10-min pulse labeling; column B, anti-gp69/71 immune precipitate from a 10-min pulse labeling with a 60 min chase; column C, anti-p30 immune precipitate from a 10-min pulse labeling; column D, anti-p30 immune precipitate of a 10-min pulse labeling with a 60-min chase incubation.

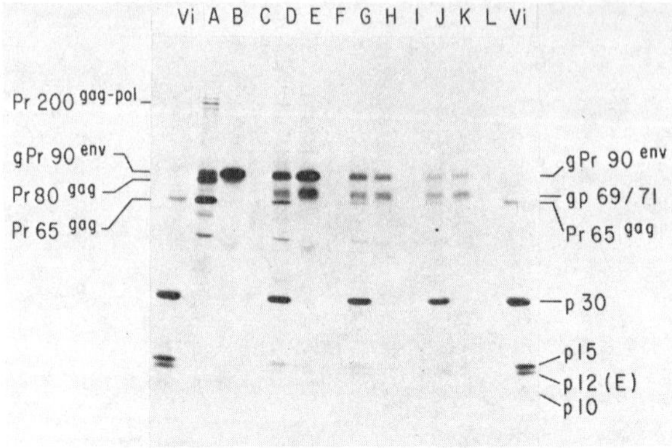

Figure 4. Formation of gp69/71 and gPr90^{env} in a pulse–chase experiment in R-MuLV-infected cells.

Cytoplasmic extracts from infected JLS-V16 cells were prepared, and immunoprecipitation was performed as described. Electrophoresis on 6–12% gradient gels and fluorography were carried out as in Figure 1. Columns Vi are ^{14}C-labeled amino acid marker virus; columns A, B, and C, 10 min pulse with ^{14}C-labeled amino acid protein hydrolysate; columns D, E, and F, 2-hr chase incubation of the 10-min pulse labeling; columns G, H, and I, 4-hr chase; columns J, K, and L, 6-hr chase; columns A, D, G, and J, anti-R-MuLV precipitated; columns B, E, H, and K, anti-gp69/71 precipitated; and columns C, F, I, and L, precipitated with anti-MuLV absorbed with excess R-MuLV proteins.

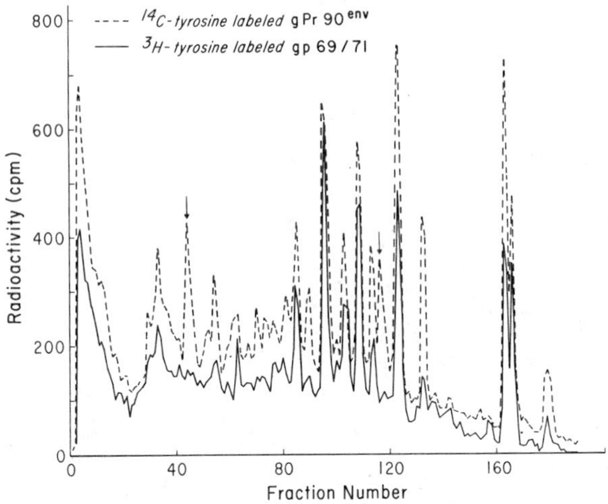

Figure 5. Ion exchange chromatography of tryptic digests of tyrosine-labeled gPr90^{env} and gp69/71. The arrows show the two additional tyrosine-containing tryptic peptides in gPr90^{env} that are not found in gp69/71.

gp69/71. The results clearly show that $gPr90^{env}$ shares many tyrosine-containing tryptic peptides with gp69/71. Figure 5 also shows that there are at least two additional tyrosine-containing tryptic peptides in $gPr90^{env}$ that are not found in gp69/71 (see arrows in Fig. 5). An analysis of a ^{14}C-tyrosine-labeled p15(E) tryptic digest mixed with a ^{3}H-tyrosine-labeled $gPr90^{env}$ digest (Fig. 6) clearly showed that these two additional tryptic peptide fractions are derived from p15(E). We conclude from these results that p15(E) and gp69/71 are cleavage products of $gPr90^{env}$.

The Gene Order of the Proteins Within $gPr90^{env}$

The order of gp69/71 and p15(E) within $gPr90^{env}$ was examined by incubating infected cells with radioactive amino acids in the presence and absence of pactamycin, a drug that selectively inhibits the initiation of translation at 5×10^{-7} M, but not elongation (31, 32). In the presence of pactamycin, one can preferentially label the C-terminal ends of nascent polypeptide chains. In the case of $gPr90^{env}$, we could determine which protein is at the C-terminus. Results of such studies (23) showed that pactamycin treatment for 30 to 60 seconds before and during pulse-labeling reduced the amount of radioactive gp69/71 formation in a chase incubation, whereas radioactive p15(E) was increased relative to that found in the absence of the drug. The results were quantitated yielding a ratio of p15(E) to gp69/71 of 0.39 in the control and 0.76 in the presence of pactamycin. These results suggest that p15(E) is C-terminal in $gPr90^{env}$.

Viral Protein p12(E)

We have identified a new viral protein, termed p12(E), in virus particles. It can be distinguished from phosphoprotein p12 (pp12) on several grounds. First p12(E) is not phosphorylated (23), it migrates in the void volume fraction of a 6 M guanidine-HCl column whereas pp12 elutes in the 12,000 mol. wt. range (23), and thirdly it has a peptide map much different from pp12 (23); and lastly, p12(E) and pp12 can be resolved by 2-dimensional gels consisting of isoelectric focusing in the first dimension and sodium dodecyl sulfate (SDS) polyacrylamide gel electrophoresis (PAGE) in the second (23). We emphasize the point that isolation of p12(E) from virus by SDS-PAGE often yields p12(E) contaminated with fragments of p15 and p30. We have found it necessary to use a two-step procedure involving guanidine-HCl agarose column chromatography followed by SDS-PAGE in order to obtain a relatively pure preparation of p12(E).

The role of p12(E) in virus structure is not known. It is tempting to propose that it is a degradation product of p15(E) and is the result of a proteolytic cleavage of the gp69/71-p15(E) complex resulting in the loss of gp69/71 from the virion (4, 9) and the generation of an uncomplexed p12(E) remaining in the

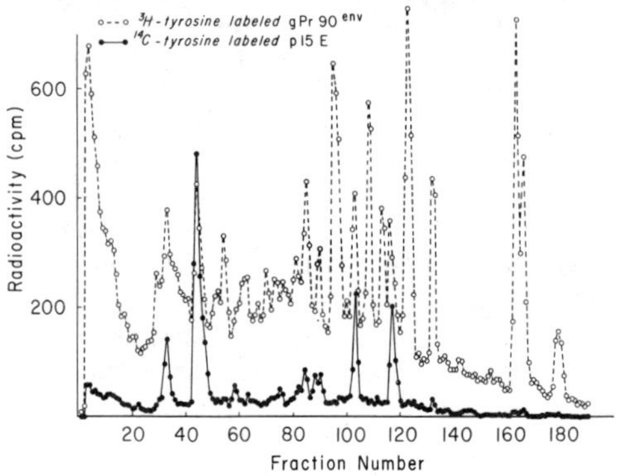

Figure 6. Ion exchange chromatography of tryptic digests of tyrosine-labeled $gPr90^{env}$ and p15(E). The p15(E) used in this experiment was purified from virus by sodium dodecyl sulfate-polyacrylamide gel electrophoresis as described in the legend to Figure 1.

virus envelope. We propose that this proteolytic cleavage occurs after the virus is released from the cell into the culture medium. We have shown that as R-MuLV particles age in the culture medium in which they were released, such particles show a steady loss of gp69/71 (Arlinghaus, unpublished results).

Cell-Free Synthesis of an Unglycosylated Precursor to gp69/71 and p15(E)

Cell-free translation of genomic RNAs from murine and avian RNA tumor viruses has shown that the full-length 35S RNA can code for the core protein precursor $Pr65^{gag}$ (28, 33, 34, 35. 36, 37) as well as for a larger polyprotein containing both 'gag' and 'pol' gene products (28, 34, 38, 39). The 'env' gene product has never been observed to be coded for by full length virion genomic RNA although the information for this gene product is definitely encoded in the viral genome.

In order to determine whether the viral envelope proteins could be synthesized from intracellular RNA from viral infected cells, we isolated 18-25S RNA and 25-35S RNA from a total RNA extract by sucrose density gradient centrifugation. These RNA classes were again fractionated on separate sucrose gradients and individual fractions were analyzed in a mRNA-dependent cell-free amino acid incorporation system (28, 34). The results of this experiment, displayed in Fig. 7 (lanes E and F) showed that only RNA of about 20-22S in size, can be translated into a polypeptide precipitable by antiserum to gp69/71. This polypeptide had an apparent molecular weight of 68,000 to 70,000. In some experiments this polypeptide migrated as two bands (a doublet). Polypeptides both larger and smaller than the 68,000 to 70,000 were also recognized by anti-gp69/71 serum. However, the smaller polypeptides were found in all RNA fractions and the larger ones were found in considerably smaller quantities than the 68,000 to 70,000 mol. wt. polypeptide. The identity of these smaller and larger polypeptides is unknown. Immunoprecipitation of the cell-free translation product of the 25S to 35S RNA class with anti-gp69/71 indicated that the 68,000 to 70,000 mol. wt. polypeptide was absent (results not shown). Hence, a 20-22S RNA seems to be the sole mRNA for an envelope-related protein in R-MuLV infected cells.

The translation product of each of the RNA fractions in the 18-25S and 25-30S size classes was also tested by immunoprecipitation with anti-p30 serum. The results showed that the 'gag' or core protein precursor $Pr65^{gag}$ and the RT precursor $Pr200^{gag\text{-}pol}$ are synthesized exclusively from RNA which is greater than 28S in size (results not shown). Tryptic peptide analysis of this 65,000 mol. wt. product showed that it had methionine-containing tryptic peptides chracteristic of authentic $Pr65^{gag}$. We have previously established that $Pr65^{gag}$ is the major product and $Pr200^{gag\text{-}pol}$ is a minor product in cell-free protein synthesizing systems of 35S R-MuLV genomic RNA translated (34).

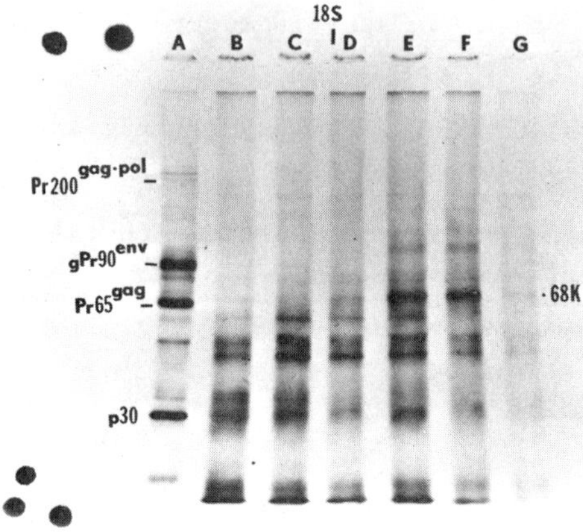

Figure 7. Sodium dodecyl sulfate slab gel electrophoresis of the cell-free product produced from 20-22S mRNA from R-MuLV infected JLS-V5 cells.

Total cellular nucleic acid was extracted from R-MuLV-infected JLS-V5 cells as described (28), and RNA was selectively obtained by washing the nucleic acid alcohol precipitate with 3M LiCl–5mM EDTA=(pH 6) which removes DNA. The RNA was fractionated by velocity sedimentation on a sucrose gradient, and the 18S-to-25S RNA fraction was pooled and again separated on a sucrose gradient. Aliquots of the fractions (1 μg of RNA) were analyzed in a mRNA-dependent protein synthesizing system (28, 34). The protein products were immunoprecipitated with anti-gp69/71 serum and analyzed on a 6–12% gradient slab-gel of polyacrylamide, as in Figure 1. Lane A is a pulse–chase experiment from R-MuLV-infected JLS-V16 cells showing authentic viral specific protein precursors obtained with anti-R-MuLV serum (20). Lanes B–G represent aliquots across a gradient in which 18S ribosomal RNA sediments to the middle of the centrifuge tube (lane D). Lane B represents the top two fractions, lane G represents the bottom two fractions, and fractions C, D, E, and F represent the in-between fractions.

Peptide Maps of the Envelope-related Protein Synthesized From Intracellular 20-22S mRNA

If the 68,000 to 70,000 mol. wt. polypeptide coded for by 20-22S mRNA is related to R-MuLV envelope precursor gPr90^{env}, it should possess tryptic peptides characteristic of gp69/71 and p15(E). Therefore, ^{35}S-methionine-labeled 68,000 to 70,000 mol. wt. polypeptide product of 20-22S mRNA was digested with trypsin and mixed with authentic gPr90^{env} tryptic peptides labeled with ^{3}H-methionine. The mixture was fractionated by ion exchange chromatography (Fig. 8). A nearly perfect co-elution of tryptic methionine-containing peptide was observed. The one exception being a minor peak (Fraction 45) in the *in vitro* synthesized product that was not observed in the authentic gPr90^{env}. The peptide peak eluting at Fraction 76 is characteristic of p15(E) (20, 22) as is the peptide peak at Fraction 83 if the tryptic peptides are not oxidized with performic acid. The peak at Fraction 30 is characteristic of gp69/71 (Karshin and Arlinghaus, unpublished results). We conclude from these experiments that the 68,000 to 70,000 mol. wt. polypeptide termed Pr68^{env}, is identical to gPr90^{env} in overall peptide content and Pr68^{env} probably represents the unglycosylated form of that molecule.

Translation of Fractionated R-MuLV Genomic RNA

Since a 20-22S intracellular mRNA species codes for Pr68^{env}, it was of interest to determine whether RNA from virus particles had a similar coding capacity. Viral RNA preparations obtained from virus, that has remained 24 hr in the culture fluid, have in addition to 35S RNA a range of fragments of the full length genome (35S) whose sedimentation rate varies down to 4S. Translation of various size classes of virion RNA fractionated on a sucrose gradient indicated that a significant amount of a polypeptide of the size of Pr68^{env} and precipitable by anti-gp69/71 is coded for by viral RNA of less than full length size. It is necessary to obtain the tryptic peptide map of this polypeptide to determine that it is a 'env' gene product and not a 'gag' gene related protein that is artifactually seen in the anti-gp69/71 precipitate.

Model for the Synthesis and Processing of R-MuLV Proteins

Our results and those of others (40, 41, 42, 43) indicate that there are at least two size classes of mRNAs that code for R-MuLV proteins. The full-length genome size 35S R-MuLV contains genes termed 'gag' or core proteins, 'pol' or the DNA polymerase and 'env' or the envelope proteins. However, only the 'gag' and 'pol' genes appear to be translated from 35S genomic RNA (28, 34, 39). Our findings indicate that a smaller size class of mRNA (≃22S) codes for the envelope proteins. Studies in the avian system indicate the gene order within the genome of non-defective avian sarcoma viruses (another C-type RNA tumor virus) is: 5'-'gag-pol-env-src'-poly(A)-3' (44). Given the same gene order in the R-MuLV genomic RNA (e.g. 'gag-pol-env'), our results indicate that only the first two genes are translated from virion RNA. Furthermore, our results suggest that only one ribosome of 15 to 25, that has completed the translation of the 'gag' gene, proceeds into the 'pol' gene (26, 28, 34) (Fig. 9). This latter process we have termed read-through (28, 34) and the product of this read-through event is a 200,000 mol. wt. polyprotein termed $Pr200^{gag-pol}$. It contains peptide sequences and antigenic determinants of the 'gag' protein and the viral DNA polymerase (26, 27, 28, 34). We have shown it to be the precursor of the virion reverse transcriptase (26, 27).

The detection of both RT-specific and core protein-specific antigenic determinants in a single precursor polyprotein ($Pr200^{gag-pol}$) raised the following issue concerning the translation of the viral genome. If the synthesis of the 'gag' and 'pol' gene products proceeds by synthesis and cleavage of this common precursor only, then this would entail the synthesis of equimolar amounts of 'gag' and 'pol' gene products. Virus particles contain much less polymerase than structural proteins (7, 8). Moreover, measurement of the steady state amount of RT-specific sequences in virus-infected cells has indicated that these sequences are present in much smaller amounts than the viral structural proteins (7, 45). This indicates that if RT polypeptide sequences are indeed made in equimolar amounts compared to the core proteins, they must be quickly degraded. We have been able to shed more light on this point by comparing the amount of RT precursors to the amount of core protein precursors after pulse-labeling of infected cells (26, 27). These conditions should more clearly approximate the rate of synthesis of these polypeptides. Our measurement revealed that 1/25 to 1/10 as much RT-specific precursors are present in the cell, after a pulse-labeling, as core protein precursors (26). Moreover, intermediate RT precursors appear to be formed by cleavage of preformed $Pr200^{gag-pol}$, while formation of core protein precursors is only compatible with this mechanism if very fast or nascent chain cleavage of $Pr200^{gag-pol}$ takes place, which is not supported by our studies with amino acid analogs or protease inhibitors (26). This makes it unlikely that the pathways of synthesis of 'gag' and 'pol' gene products only involve the cleavage of a common precursor.

Three possiblities are theoretically compatible with our findings. First, the mRNA (35S in size) coding for $Pr200^{gag-pol}$ is a minor RNA species compared to a smaller size mRNA involved in the synthesis of 'gag' gene products. Second, there is only one species of mRNA coding for both 'gag' and 'pol', but a translational control mechanism allows the synthesis on the mRNA of more 'gag' than 'pol' gene products. A third possibility is that two classes of similar size mRNA (e.g. 35S) exist in the infected cell. One codes for 'gag' and the other codes for 'gag-pol'. A

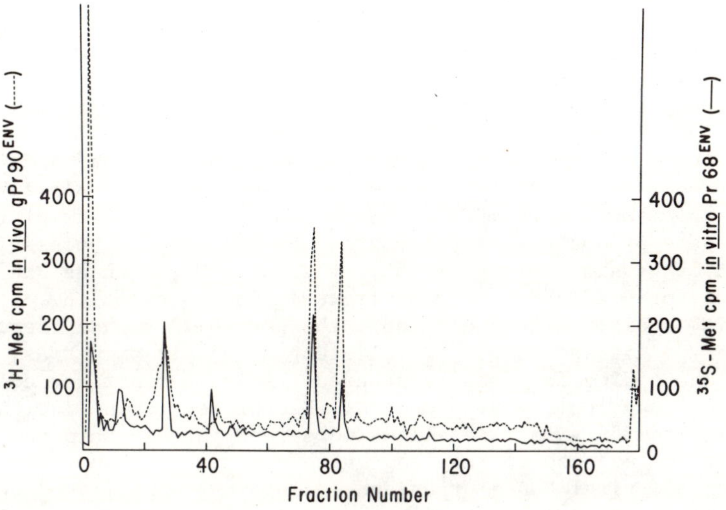

Figure 8. Ion exchange chromatography of tryptic digests of authentic gPr90^{env} and the 68,000 mol wt polypeptide (pr68^{env}) made from ≃ 22S mRNA from R-MuLV-infected JLS-V5 cells.

Pr68^{env} was labeled with ^{35}S-methionine in the cell-free mRNA-dependent protein synthesizing system (as in Figure 7). gPr90^{env} was isolated from R-MuLV-infected JLS-V16 cells labeled with ^{3}H-methionine. The proteins were purified by immunoprecipitation with anti-gp69/71 and sodium dodecyl sulfate gel electrophoresis. The bands were digested with TPCK-trypsin and fractionated on a cation exchange column (30).

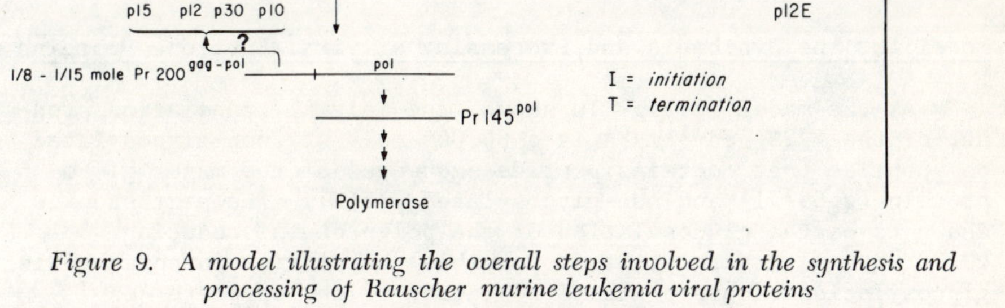

Figure 9. A model illustrating the overall steps involved in the synthesis and processing of Rauscher murine leukemia viral proteins

consideration of other available information favors the second or third possibility. Thus, the major species of viral mRNA in murine and avian RNA tumor virus-infected cells has a size of 35S, but species of 21S and 16S have also been detected (40, 41, 42, 43). Moreover, further studies indicate that the 35S RNA is the main species present in polyribosomes precipitable with anti-p30 mono-specific serum (46). Studies in avian systems also suggest that the full-length genomic RNA is the mRNA for 'gag' and 'pol' gene products (38, 47). These observations make it unlikely that there is a small size mRNA (≃16S) on which the majority of the 'gag' gene products are synthesized and disfavor the possibility that a large size species (≃35S) of viral mRNA, such as one able to code for $Pr200^{gag-pol}$, constitute a minor population.

It is difficult to eliminate the possibility that there are indeed two classes of 35S or full-length genomic size RNAs, a major one coding for 'gag' gene products and a minor one coding for 'gag' and 'pol' gene products. Recent results do indicate that stored 35S viral RNA allows more read-through into the 'pol' gene at the expense of the 'gag' gene precursors than fresh batches 35S viral RNA (Murphy and Arlinghaus, unpublished results). This result is consistent with the idea that some structural alteration(s) of 35S RNA genomic is responsible for $Pr200^{gag-pol}$ synthesis. This alteration could be brought into play by interaction of the 35S viral RNA with a second molecule. A candidate for this second molecule would be a RNA since pronease-treated viral 35S RNA does not alter the frequency of $Pr200^{gag-pol}$ synthesis over non-treated RNA (Murphy and Arlinghaus, unpublished results).

Our results indicate that the envelope gene is probably never translated from full length genomic size RNA but that a ≃22S intracellular mRNA serves as a mRNA for the envelope protein. The polarity of this mRNA is unknown but a similar viral envelope mRNA from avian sarcoma virus infected cells has the same polarity as viral genomic RNA (47, 48, 49). Therefore, we would assume that ≃22S mRNA composed of sequences from the 3' one-third of R-MuLV genomic RNA is the mRNA for the envelope proteins. We would also assume that this RNA is capped and contains some untranslated sequences found at the 5' end of the 35S genomic RNA (48).

Model for the Synthesis and Processing of Viral Envelope Proteins

As the model in Fig. 10 shows, the initial translation product of the ≃22S 'env' mRNA is a 68,000 mol. wt. non-glycosylated polypeptide that contains peptide sequences of the major glycoprotein (gp69/71) and non-glycosylated p15(E). The next step is shown to be the glycosylation of the polyprotein producing $gPr90^{env}$. Alternately, it is possible that inside infected cells glycosylation occurs prior to formation of the complete 'env' gene product which would result in the immediate formation of gPr90^{env}. If this were the case, Pr68^{env} would only be seen under conditions of cell-free translation and not in whole cells. However, it has not been established whether or not some Pr68^{env} is formed in infected cells. Cleavage and further glycosylation (fucose residues are added) of gPr90^{env} yields gp69/71 and p15(E) which probably exists as a complex held together by disulfide bonds and possibly additional non-covalent bonds. The basic unit of the protruding spike-glycoprotein knob is believed to be this gp69/71-p15(E) complex. The association of such basic units would then result in formation of the spike-knob structure that traverses and projects through the lipid envelope of the virus (Fig. 11).

The role of viral protein p12(E) in the structure of the virion is unknown. Since, it does not appear to form a complex with gp69/71 (50), we assume that it may be an artifact of aged virus resulting from the proteolysis of the spike-knob structure which in turn leads to the loss of the glycoprotein gp69/71 plus a small portion of the N-terminal end p15(E) from the virion.

Virus Assembly

Cleavage of viral polyproteins constitutes a predominant feature in the assembly of many animal and bacterial viruses (for reviews see 51, 52, 53, 54). RNA tumor viruses are no exception. We have shown that actinomycin D reduced the rate of cleavage of Pr65^{gag} (55) and have suggested that the rate of cleavage of Pr65^{gag} to the mature viral-proteins is catalyzed by the viral genomic RNA. Recent experiments have also shown that the cleavage rate of Pr200$^{gag-pol}$ is also affected by inhibiting viral RNA synthesis (Kopchick and Arlinghaus, unpublished results). We note that the cleavage of Pr80^{gag} to Pr65^{gag} is not affected under these conditions nor is the cleavage of gPr90^{env}. However, cleavages of Pr65^{gag} and Pr200$^{gag-pol}$ do occur (but at a slower rate) in the absence of viral RNA and, interestingly, the reverse transcriptase can be incorporated into virions that are deficient of viral RNA (56). Based on a number of studies from my own laboratory (55), we have proposed a model for the assembly of R-MuLV particles (Fig. 11). In this model, Pr65^{gag} rather than the mature viral proteins constitute the initial assembly unit. Units of Pr65^{gag} assemble, presumably at the cell membrane, into a structure that remains open on the inside, due to the electrostatic repulsion of the basic portion (p10) of Pr65^{gag}, which is then cleaved to form the viral proteins including p10. Viral p10 in turn is found in the virion in association with the viral RNA (5, 57). This model also proposes that Pr200$^{gag-pol}$ is incorporated into the virion by means of its similarity in structure to Pr65^{gag}. In fact, Pr200$^{gag-pol}$ contains all the antigenic determinants and peptide sequences that are found in Pr65^{gag} plus those of the reverse transcriptase. Thus, Pr200$^{gag-pol}$ would substitute for small number (30-50) of the Pr65^{gag} molecules resulting

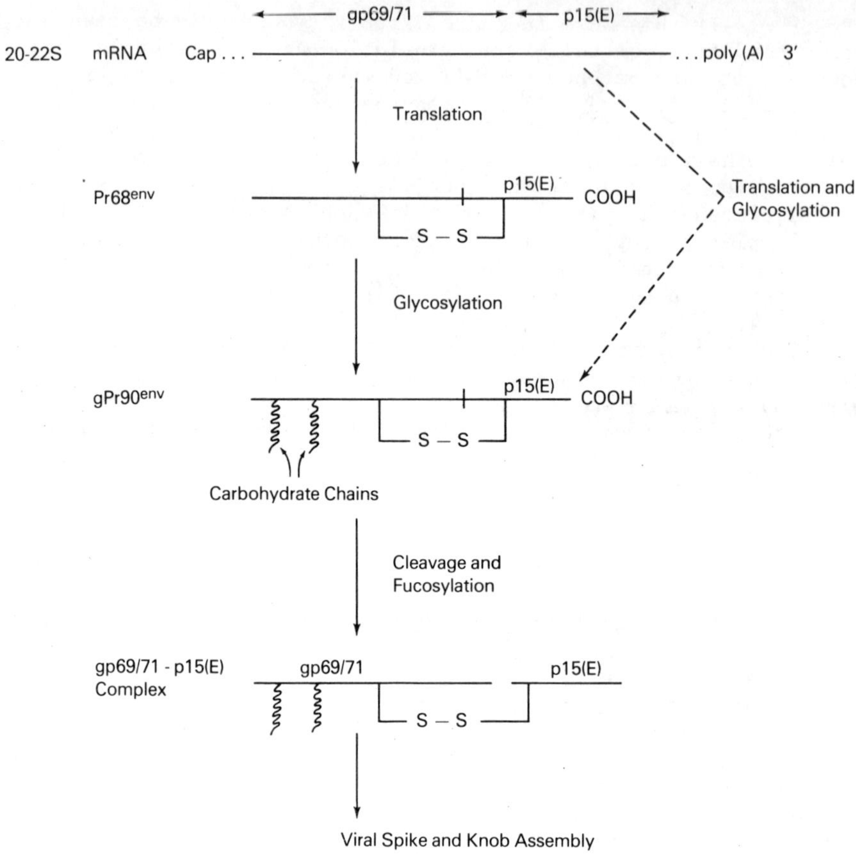

Figure 10. Synthesis of Rauscher murine leukemia virus envelope proteins

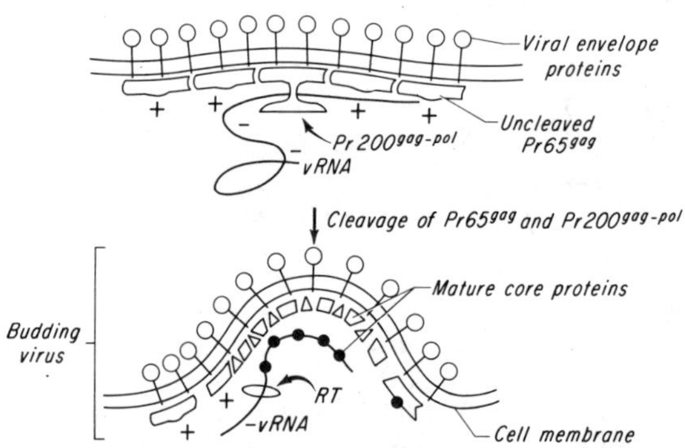

Figure 11. A model showing the assembly and budding of Rauscher murine leukemia virus particles

in the inclusion of RT in the virion.

A substantial amount of cleavage of $Pr65^{gag}$ (and $Pr200^{gag-pol}$) is proposed in the model in order to allow virus release. Viral RNA, due to charge neutralization, catalyzes the cleavage of $Pr65^{gag}$ and $Pr200^{gag-pol}$. This results in a more rapid rate of assembly of RNA-containing viral cores, and, if precursor cleavage is viewed to play a part in the budding process, in a faster rate of release of RNA-containing virus particles. We further propose that host cell RNA can substitute to a certain degree for viral RNA in catalyzing the cleavages of $Pr65^{gag}$ and $Pr200^{gag-pol}$, but host cellular RNA is packaged to a much lesser extent than viral RNA since virus isolated after actinomycin D treatment exhibited a lower ratio of RNA to protein than control preparations of virus (55). This suggests the existance of specific recognition by the assembled viral structures of viral genomic RNA. Whether the basis for this recognition lies in the specific interaction between some of the viral core proteins, such as pp12, and the viral RNA (6) remains to be determined.

In the above mentioned model of assembly, we have postulated that cleavages of $Pr65^{gag}$ and $Pr200^{gag-pol}$ is probably required for virus budding and release. There is a possibility, however, that $Pr65^{gag}$ and $Pr200^{gag-pol}$ may be cleaved not during budding but after release of the virus from the cell. For many years, C-type particles have been observed to undergo a post-release maturation process in which the core center of the virion collapses into a condensed structure that is clearly different from the organized concentric coils observed in the core of budding particles (58, 59, 60). The significance of this maturation process remains uncertain. The recent results of Yoshinaka and Lufteg (61) suggest that there is a positive correlation between the amount of $Pr65^{gag}$ in viral cores and the amount of immature cores. However, it remains to be shown whether or not the cleavage of $Pr65^{gag}$ can occur in released virions and if so that the changes in released virions (immature to mature) are related to precursor cleavage. A number of lines evidence suggest that cleavage of $Pr65^{gag}$ and $Pr200^{gag-pol}$ occur mainly prior to viral release (62) and prior to the immature to mature virus transition that takes place in the culture fluid.

Literature Cited

1. Pinter, A. and Fleissner, E. Virology (1977) 83, 417.
2. Leamnson, R. N., Shander, M. H. M. and Halpern, M. S. Virology (1977) 76, 437.
3. Witte, O. N., Tsukamoto-Adey, A. and Weissman, I. L. Virology (1977) 76, 539.
4. Bolognesi, D. P., Montelaro, R. C., Frank, H. and Schafer, W. Science (1978) 199, 183.
5. Davis, J., Scherrer, M., Tsai, W. P., and Long, C. J. Virol. (1976) 18, 709.

6. Sen, A., Sherr, C. J., and Todaro, G. J. Cell (1976) 7, 21.
7. Panet, A., Baltimore, D., and Hanafusa, T. J. Virol. (1975) 16, 146.
8. Stromberg, K., Hurley, N. E., Davis, N. L., Rueckert, R. R., and Fleissner J. Virol. (1974) 13, 513.
9. Bolognesis, D. P., Langlois, A. J., and Schaffer, W. Virology (1975) 58, 550.
10. Baltimore, D. Cold Spring Harbor Symp. Quant. Biol. (1974) 39, 1187.
11. Kennel, S. H., Del Villano, B. C., Levy, R. L., and Lerner, R. A. Virology (1973) 55, 464.
12. Witte, O. N., Weissman, I. L., Kaplan, H. S. Proc. Natl. Acad. Sci. (1973) 70, 36.
13. McLellan, W. and August, J. T. J. Virol. (1976) 20, 627.
14. Leamnson, R. N. and Halpern, M. S. J. Virol. (1976) 18, 956.
15. Hunsmann, G., Moennig, V., Pister, L., Seifert, E., and Schaefer, W. Virology (1974) 62, 307.
16. Ikeda, H., Pincus, T., Yoshiki, T., Strand, M., August, J. T., Boyse, E., and Mellars, R. C. J. Virol. (1974) 14, 1274.
17. Famulare, N. G., Buchhagen, D. L., Klenk, H. D., and Fleissner, E. J. Virol. (1976) 20, 501.
18. Ihle, J. N., Domotor, Jr., J. J., Bengali, K. M. J. Virol. (1976) 18, 124.
19. Ihle, J. N., Hanna, Jr., M. G., Schafer, W., Hunsman, G., Bolognesi, D. P., Huper, G. Virology (1975) 63, 60.
20. Arcement, L. J., Karshin, W. L., Naso, R. B., Jamjoom, G. A. and Arlinghaus, R. B. Virology (1976) 69, 763.
21. Naso, R. B., Arcement, L. J., and Arlinghaus, R. B. Cell (1975) 4, 31.
22. Naso, R. B., Arcement, L. J., Karshin, W. L., Jamjoom, G. A., and Arlinghaus, R. B. Proc. Natl. Acad. Sci. USA (1976) 73, 2325.
23. Karshin, W. L., Arcement, L. J., Naso, R. B., and Arlinghaus, R. B. J. Virol. (1977) 23, 787.
24. Kraemer, P. "Biomembranes" Vol. 1, p. 67, Ed. Mauson, L. A. (1971).
25. Bosmann, H. B., Hagopien, A., and Eylar, E. H. Arch. Biochem. Biophys. (1968) 128, 470.
26. Jamjoom, G. A., Naso, R. B., and Arlinghaus, R. B. Virology (1977) 73, 11.
27. Kopchick, J. J., Jamjoom, G. A., Watson, K. F., and Arlinghaus, R. B. Proc. Natl. Acad. Sci. USA (1978) In press.
28. Murphy, Jr., E. C., Kopchick, J. J., Watson, K. F., and Arlinghaus, R. B. Cell (1978) 13, 369.
29. Evans, L. H., Dreslar, S., and Kabat, D. J. Virol. (1977) 24, 865.
30. Arcement, L. J., Karshin, W. L., Naso, R. B., and Arlinghaus, R. B. Virology (1977) 81, 284.
31. Taber, R., Rekosh, D., and Baltimore, D. J. Virol. (1971) 8, 395.

32. Butterworth, B. E., and Rueckert, R. R. J. Virol. (1972) 9, 823.
33. Naso, R. B., Arcement, L. J., Wood, T. G., Saunders, T. E., and Arlinghaus, R. B. Biochim. Biophys. Acta (1975) 383, 195.
34. Murphy, Jr., E. C., and Arlinghaus, R. B. Virology (1978) In press.
35. Kerr, I. M., Olshevsky, U., Lodish, H., and Baltimore, D. J. Virol. (1975) 18, 627.
36. Pawson, T., Martin, G. S., and Smith, A. E. J. Virol. (1976) 19, 950.
37. Von Den Helm, K., and Duesberg, P. H. Proc. Natl. Acad. Sci. (1975) 72, 614.
38. Paterson, B. M., Marciani, D. J., and Papas, T. S. Proc. Natl. Acad. Sci. USA (1977) 74, 4951.
39. Philipson, L., Andersson, P., Olshevsky, U., Weinberg, R., and Baltimore, D. Cell (1978) 13, 189.
40. Fan, H. and Baltimore, D. J. Mol. Biol. (1973) 80, 93.
41. Gielkens, A. L. J., Van Zaane, D., Bloemers, H. P. J., and Bloemendal, H. Proc. Natl. Acad. Sci. USA (1976) 73, 356.
42. Shanmugam, G., Bhadure, S. and Green, M. Biochim. Biophys. Res. Commun. (1974) 56, 697.
43. Shincarial, A. and Joklik, W. Virology (1973) 56, 532.
44. Duesberg, P. H., Wang, L. H., Mellan, P., Mason, W. S., and Vogt, P. K. "ICN-UCLA Symposia on Molecular and Cellular Biology" Vol. IV, pp 107. Edited by D. Baltimore, A. S. Huang, and C. F. Fox. Academic Press, Inc., New York, New York. 1976.
45. Chen, J. H. and Hanafusa, H. J. Virol. (1974) 13, 340.
46. Mueller-Lantsch, N. and Fan, H. Cell (1976) 9, 579.
47. Haywood, W. S. J. Virol. (1977) 24, 47.
48. Mellon, P. and Duesberg, P. H. Nature (1977) 270, 63.
49. Stacey, D. W., Allfrey, V. G., and Hanafusa, H. Proc. Natl. Acad. Sci. USA (1977) 270, 631.
50. Montelaro, R. C., Sullivan, S. J., and Bolognesi, D. P. Virology (1978) In press.
51. Casjens, S. and King, J. Annu. Rev. Biochem. (1975) 44, 555.
52. Hershko, A. and Fry, M. Annu. Rev. Biochem. (1975) 44, 775.
53. Korant, B. D. "Cold Spring Harbor Symposium on Control of Cell Proliferation" Vol. 21, p. 621. Cold Spring Harbor Laboratory, Cold Spring Harbor, New York, 1975.
54. Showe, M. H. and Kellenberg, E. "Control Processes in Virus Multiplication" p. 407. Cambridge University Press, Cambridge, England. 1975.
55. Jamjoom, G. A., Naso, R. B., and Arlinghaus, R. B. J. Virol. (1976) 19, 1054.
56. Levin, J. G., Grimley, P. M., Ramseur, J. M., and Berezesky, I. K. J. Virol. (1974) 14, 152.
57. Fleissner, E. and Tress, E. J. Virol. (1973) 12, 1612.
58. Fleissner, E., Ikeda, H., Tung, J. S., Vitetta, E., Tress, E., Hard, Jr., W., Stockert, E., Boyse, E. A., Pincus, T., and

O'Donnell, P. Cold Spring Harbor Symp. Quant. Biol. (1975) 39, 1057.
59. Dalton, A. J. and Haguenau, F. "Ultrastructure of Animal Viruses and Bacteriophages" pp. 255. Eds. A. J. Dalton and F. Haguenau. Academic Press, New York, New York. 1973.
60. deHarven, E. Adv. Virus Res. (1974) 19, 221.
61. Yoshinaka, Y. and Luftig, R. B. Proc. Natl. Acad. Sci. USA (1977) 74, 3445.
62. Jamjoom, G. A., Ng, V. L., and Arlinghaus, R. B. J. Virol. (1978) 25, 408.

RECEIVED March 15, 1978.

11

Glycoprotein Inhibitor of Lipoprotein Lipase from Aortic Intima—Its Possible Role in Atherosclerosis

PREMANAND V. WAGH

Veterans Administration Hospital, 300 East Roosevelt Road, Little Rock, AR 72206

Early observations on the presence of glycoproteins in the mammalian arterial wall have been reported (Buddecke, 1960; Muller-Spreer et al, 1960; Berenson and Fishkin, 1962). Although considerable research interest is centered upon the relationship of glycoproteins in the aortic wall to atherosclerosis and ageing, little is known in regard to the structure and function of these macromolecules. Several soluble glycoproteins have been purified from bovine aorta after extraction of the tissue with neutral buffers (Radhakrishnamurthy et al, 1964, Maier and Buddecke, 1971).

The concentration of arterial glycoproteins in atherosclerosis and in ageing is a subject of question and is being actively investigated. The wet weight of the aorta has been found to be increased in atherosclerosis and among the aged (Manley and Mullinger, 1967). Our data (Wagh et al, 1973), shown in Table I, provide evidence that the glycoprotein concentration increased significantly in atherosclerotic tissue as expressed by an increase in total neutral sugars, hexosamine and sialic acid. These results suggest that the increased glycoprotein concentration may have a functional role in the formation of atherosclerotic plaque.

Several functions have been ascribed to arterial glycoproteins such as transplantation rejection and maintenance of structural integrity (Anderson, 1976). Furthermore, Ishii (1971) demonstrated that preparations of crude glycoprotein obtained from various organs including the aorta of dog inhibited dextran sulfate released lipoprotein lipase (LPL) activity of human plasma. LPL catalyzes the hydrolysis of the triglyceride moiety of chylomicrons and very low density lipoproteins (Eisenberg and Levi, 1975), and is normally located on the surface of endothelial cells where it is physiologically active in the normal clearance of lipoprotein-bound triglycerides (Robinson, 1970).

Since the surface of the intimal layer of aortic wall is susceptive to continuous hemodynamic insult and thus may be involved in the genesis of atherosclerosis and other physiological events, we focussed our attention here for the purpose of isolation of glycoprotein. One of the shortcomings in the isolation of purified glycoprotein from the intimal region has been to separate it from the rest of the aortic wall. A new procedure (dermatome procedure) was developed in our laboratory (Roberts et al, 1974) that rapidly and reliably separates the intima in sufficient quantities from both porcine and human aorta.

Studies on Porcine Intimal Glycoprotein (Lipolipin). We reported on the purification and characterization of a novel glycoprotein from the intimal region of porcine aorta (Wagh and Roberts 1972). Swine aorta was chosen because in this species, the aorta is relatively free from atherosclerotic lesions and the material is readily available. The purification procedure involved extraction of intimal tissue with neutral buffer, $(NH_4)_2SO_4$ precipitation between 40-90% saturation and two DEAE-cellulose chromatographic steps. The purified glycoprotein was homogeneous upon analytical ultracentrifugation (4.86 S) and by polyacrylamide disc-gel electrophoresis. Generally, 40 mg of purified glycoprotein was obtained from 2 Kg of wet weight of thoracic aorta. Due to the presence of equimolar glucose and galactose and the absence of hydroxylysine in the molecule, it was suggested that this glycoprotein was unique in its characteristics and therefore of a new type. The isoelectric point of the purified glycoprotein was 4.3 as observed by analytical isoelectric focusing (Baig and Ayoub, 1976).

Subsequent studies (Roberts and Wagh, 1976) revealed that sodium dodecyl sulfate - polyacrylamide gel electrophoresis of the native and its S-carboxyamidomethyl derivative at different polyacrylamide concentrations did not affect the molecular weight (72,000 daltons) indicating the absence of subunits. The carboxy-terminal amino acid was found to be serine. Attempts to determine the identity of the amino acid indicated that the amino group was not free. The glycoprotein did not contain an alkali-labile (O-glycosidic) carbohydrate-peptide linkage as tested by β-elimination reaction. The release of monosaccharides from the intact glycoprotein as a function of time was studied employing mild acid hydrolysis (0.5 M HCl, 80°C) and also by the use of neuraminidase, α-D- and β-D-glucosidases and β-D-N-acetylglucosaminidase. From the observations on the release of monosaccharides and analogy with standard features determined by other investigators on soluble aortic glycoproteins (Radhakrishnamurthy et al, 1964; Radhakushnamurthy and Berenson, 1966; Klemer and Nager, 1967; Maier and Buddecke, 1971), a prediction was made as to the general features of the carbohydrate moiety of the glycoprotein (Fig. 1). This postulated structure must still withstand results of future investigations including methylation studies, oxidation with periodate and the use of other specific glycosidases.

The purified porcine intimal glycoprotein was tested for its LPL inhibitory activity by employing post-heparin dog plasma as the source of enzyme and Ediol (stabilized coconut oil emulsion) as the triglyceride substrate. Because it inhibited post-heparin

TABLE I
Concentration of Component Sugars of Normal and Atherosclerotic Aortae

Component	Normal	Sclerotic
Wet weight (mg/cm^2)	195.7	250.4 (a)
DDAF (mg/cm^2)	39.3	36.9 (b)
Total carbohydrate (mg/cm^2)	2.10	2.30 (a)
Hexose (μg/cm^2)	870	978 (a)
Hexosamine (μg/cm^2)	651	698 (c)
Uronic acid (μg/cm^2)	297	308 (NS)
Sialic acid (μg/cm^2)	285	317 (b)
Total carbohydrate (μg/mg DDAF)	55.4	63.7 (a)
Hexose (μg/mg DDAF)	23.1	27.1 (a)
Hexosamine (μg/mg DDAF)	17.0	19.4 (a)
Uronic acid (μg/mg DDAF)	7.9	8.5 (NS)
Sialic acid (μg/mg DDAF)	7.5	8.8 (a)

The concentration of components are expressed as weight per unit area and weight per unit dry, defatted, ash-free residue (DDAF). All values represent averages from 15 samples of aortae obtained from individuals of 70.3 ± 2.8 (mean ± S.E.M.) years of age. Paired t test: (a) = P<0.01; (b) = P<0.05; (c) = P<0.02; (NS) = not significant. From Wagh et al (1973), with permission from Elsevier/North-Holland Biomedical Press.

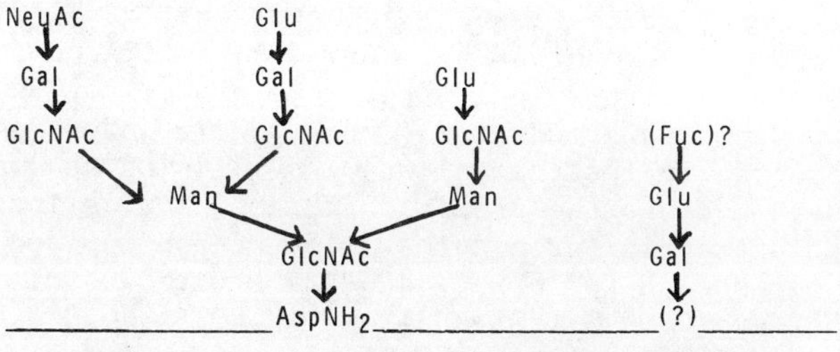

Biochimica et Biophysica Acta

Figure 1. A predicted structure for the carbohydrate moiety of porcine intimal glycoprotein. NeuAc, N-*acetyl-neuraminic acid; GlcNAc,* N-*acetyl-*D-*glucosamine. (From Roberts and Wagh [1976])*

plasma LPL, the molecule was named "lipolipin" (from lipoprotein lipase inhibitor) (Wagh, 1975).

The inhibition of LPL activity at various lipolipin concentrations and a triglyceride concentration of 2.5 mM is shown in Fig. 2. Inhibition of LPL increased exponentially with lipolipin concentration. At the triglyceride concentration used, the reaction velocity was reduced 50% when the concentration of lipolipin was 1.5×10^{-5} M. To determine the nature of inhibition, we measured the velocity of lipolysis at various concentrations of triglyceride and the inhibitor. The inhibition was observed to be non-competitive (Fig. 3). Further studies (unpublished data) in our laboratory indicated that porcine lipolipin inhibits purified bovine milk LPL (Egelrud and Olivecrona, 1972) when radioactive triglyceride emulsion stabilized by gum arabic (Hernell et al, 1975) was used as the substrate. However, the mechanism by which lipolipin inhibits LPL activity is not understood due to complex nature of the enzyme reaction.

The stability of lipolipin under various conditions was examined by electrophoresis on polyacrylamide gel. Lipolipin is unstable after approximately 2 weeks when stored in neutral Tris·HCl buffer or as a freeze-dried powder. The instability is indicated by additional bands in the gel and the loss of its inhibitory property (Wagh, 1975).

Studies on Human Intimal Glycoprotein. We extended our investigation of LPL inhibitor to human atherosclerotic intima (Wagh and Olivecrona, 1978). Atherosclerotic human aortae (Grade III-IV) were obtained at necropsy from male veterans 60-80 years of age. Intimal layer was separated by the "dermatome procedure" and intima powder was prepared as described previously (Roberts et al, 1974). The yield of intima powder was 1.5 - 2.0 g per 100 g of wet weight of aortae.

Purification of human LPL inhibitor demanded that the original procedure for porcine tissue (Wagh and Roberts, 1972) be modified (Wagh and Olivecrona, 1978). Briefly, intima powder (5 g) was extracted twice with 20 vol of acetone at 4°C for 30 min. The acetone powder was dried at 25°C for 30 min under vacuum. This step resulted in the removal of 0.75 g lipids and was found to be necessary because the presence of these lipids interfered in subsequent extraction procedures. The acetone powder was extracted two times each for a period of 24 h with 20 vol of extraction buffer (0.05 M Tris·HHCl - 1 mM EDTA - 0.3 M NaCl, pH 7.4). The ammonium sulfate precipitate of the extract between 40-80% saturation was collected, dissolved in 5 mM Tris·HCl containing 1 mM EDTA and 50 mM NaCl, pH 7.4 buffer and dialyzed exhaustively against the same buffer. The dialyzed solution was designated as the ammonium sulfate fraction. This fraction upon chromatography on a DEAE-cellulose column resulted in three peaks which contained material that inhibited LPL activity (Fig. 4).

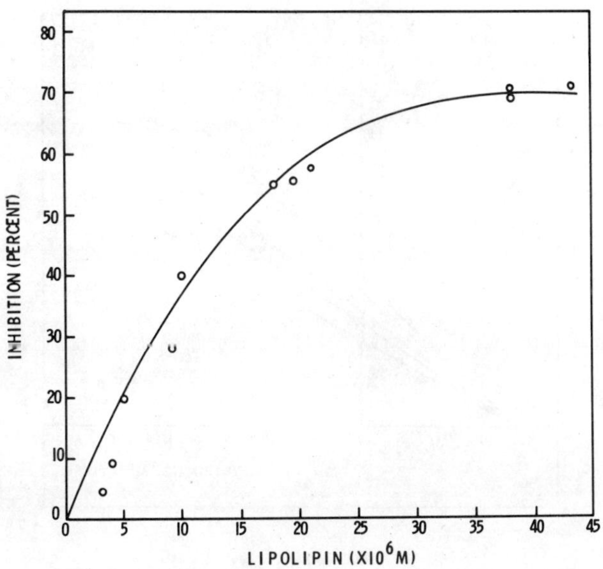

Advances in Experimental
Medicine and Biology

Figure 2. Inhibition of LPL as a function of lipolipin concentration. Incubation mixture contained 0.2 mL of post-heparin plasma, 0.5 mL of buffer (0.2M NH_4OH–NH_4Cl at a pH of 8.6) containing various amounts of lipolipin and 0.1 mL of Ediol (20 micromoles triglyceride per milliliter). Final concentration of triglyceride was 2.5mM. Percent inhibition was calculated as the decrease in LPL activity as related to assay mixtures without added lipolipin. (From Wagh [1975])

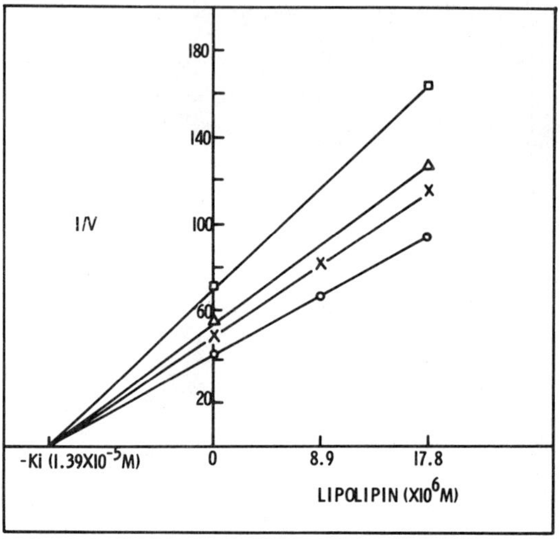

Advances in Experimental
Medicine and Biology

Figure 3. Dixon plot of LPL activity. Incubation mixture contained 0.2 mL of post-heparin plasma, 0.5 mL of buffer (0.2M NH_4OH–NH_4Cl at a pH of 8.6 containing various amounts of triglycerides). Velocity is expressed as micromoles of free fatty acids liberated per milliliter of post-heparin plasma per minute. Final concentrations of triglyceride were: (□—□) 0.7mM; (△—△) 1.6mM; (×—×) 2.5mM; and (○—○) 4.3mM. (From Wagh [1975])

Peak 1 contained large amounts of hemoglobin which seriously interferred with the LPL assay. The protein material in peak 2 eluted at the same ionic strength of the gradient as that for porcine lipolipin. Its homogeneity was assessed by polyacrylamide gel electrophoresis, analytical isoelectric focusing and sedimentation velocity measurement. Electrophoretic analysis as observed by scanning the gel stained with Coomassie Brilliant Blue dye at 520 nm revealed one major and three minor components. The major component constituted approximately 80% of the total protein. The electrophoretic mobility of the major component after staining for both the protein and carbohydrate (periodic acid-Schiff stain) was similar to that of highly purified porcine lipolipin. Analytical isoelectric focusing in a pH 4-6 ampholyte system resulted in a pattern of one major and two minor protein bands as visualized by Coomassie Blue staining. The major component electrofocused at pH 4.3. Upon ultracentrifugation, this material resolved in a major species which sedimented at 4.75 S. The total carbohydrate content of the peak 2 material was 2.55%. The carbohydrate contained hexosamine, glucose, galactose, mannose, fucose and sialic acid in a molar ratio of 3:1:2:2:1:1, respectively with sialic acid taken as unity. This material eluted from Bio-Gel P-150 in the same position as described for porcine lipolipin (Roberts and Wagh, 1976).

A partially purified preparation was employed for the kinetic studies. Instead of chromatography on DEAE-cellulose column, a batch-wise DEAE-cellulose fractionation method was used (Wagh and Olivecrona, 1978). The preparation contained all the protein in peak 2 with some overlapping proteins contributed from peak 3 (approximately 20%) but was free from proteins in peak 1 (Fig. 4). This material inhibited the activity of milk LPL against long-chain triglycerides under all conditions that we have studied (Wagh and Olivecrona, 1978). The release of free fatty acids was linear with time both with and without inhibitor, indicating that the inhibitor did not cause a progressive inactivation of the enzyme with time. Inhibition of both the basal activity and the serum-stimulated activity was the same for upto a 7-fold increase of enzyme concentration. It was observed that the inhibition was non-competitive with respect to serum (Fig. 5). When the amount of triglyceride substrate was increased at a constant level of serum, activity in the absence of inhibitor first increased but then decreased at high concentrations of triglyceride; however, the inhibition was relieved at high levels of triglyceride substrate (Fig. 6). Therefore, the inhibition may be competitive with respect to triglyceride substrate.

Concluding Remarks: I should like to end by posing four questions to which we presently do not have answers or our knowledge is at best incomplete. First, are porcine and human intimal lipolipins similar? There is some evidence that both of these molecules share certain characteristics in common, e.g., molecular

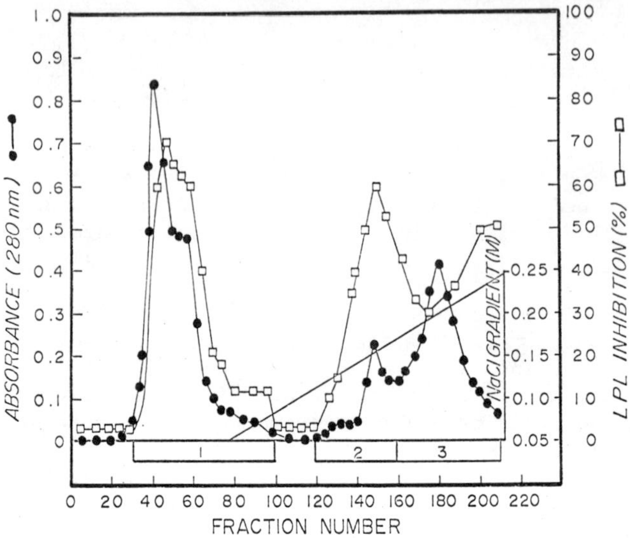

Atherosclerosis

Figure 4. Chromatography of ammonium sulfate fraction on DEAE-cellulose. The ammonium sulfate fraction was chromatographed on a DEAE-cellulose column. The elution was initiated with 5mM Tris · HCl — 1mM EDTA–0.05M NaCl, pH = 7.4 (standard buffer). The elution was initiated with 450 mL of standard buffer and 0.25M NaCl in the same buffer. Fractions of 6 mL were collected. The flow rate was 53 mL/hr. (From Wagh and Olivecrona [1978])

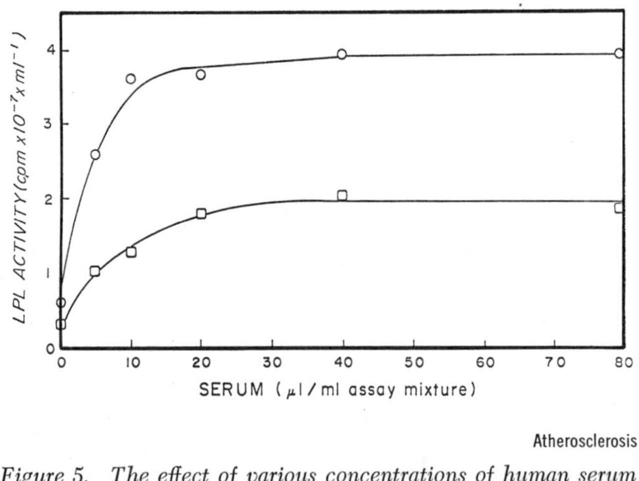

Atherosclerosis

Figure 5. The effect of various concentrations of human serum on LPL inhibition. (○—○) control; (□—□) 356 µg inhibitor per mL assay. For details see Ref. Wagh and Olivecrona (1978).

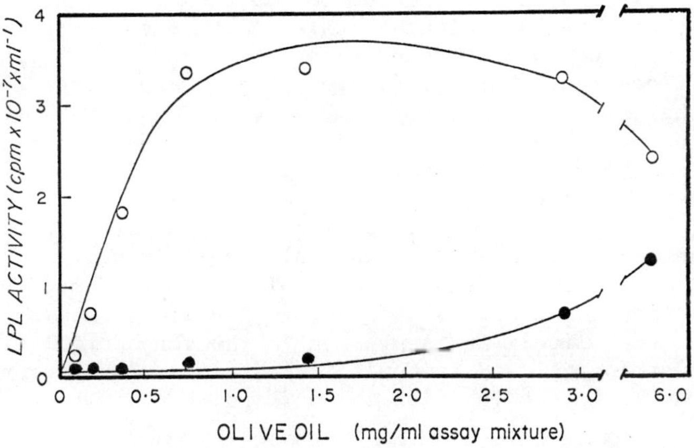

Atherosclerosis

Figure 6. The effect of various concentrations of olive oil on LPL inhibition. (○—○) control; (●—●) 625 μg inhibitor per mL assay. For details see Ref. Wagh and Olivecrona (1978).

weight, electrophoretic mobility and isoelectric point. Although the structure of the carbohydrate moiety for the porcine molecule has been predicted (Roberts and Wagh, 1976) the precise sequence of sugars remains to be confirmed and compared with that of the human molecule.

The second question is - what is the mechanism by which lipolipin inhibits LPL activity *in vitro*? We have postulated from kinetic studies that the inhibition of LPL may be at least partly due to covering of the lipid-water interphase by the inhibitor thereby denying the enzyme access to the lipid substrate (Wagh and Olivecrona, 1978). Although this has been suggested as a mechanism for the inhibition of pancreatic lipase by bovine serum albumin (Brockerhoff, 1971), further studies are needed to confirm our hypothesis.

Third, which cells in the intima synthesize lipolipin? It has been demonstrated that porcine valvular tissue contains a glycoprotein similar to that from porcine intima (Baig and Ayoub, 1976). Further immunological studies on the cross-reaction between anti-lipolipin against intimal lipolipin and valvular glycoprotein indicated that the precipitin bands formed due to cross-reaction fused with each other (unpublished data). It seems therefore, that lipolipin is present in the mesenchymal tissues. Further histological and immunological studies should provide information as to the localization of lipolipin in the vascular tissue.

Finally, we come to the question of the function of LPL inhibitor *in vivo*. Large blood vessels such as bovine thoracic aorta contain LPL (Henson and Schotz, 1975; Dicorleto and Zilversmit, 1975). Although the concentration of LPL inhibitor in the normal and diseased states is not known, it is possible that increased synthesis of the inhibitor during atherogenesis may result in decreased LPL activity thereby affecting the normal lipolytic process. If the inhibitor is a major constituent of total glycoprotein content in atherosclerotic aorta, the higher concentration of glycoproteins in atherosclerotic plaques as compared to that in the uninvolved regions of the aorta (Wagh et al, 1973) may be a reflection of an increased concentration of the inhibitor in diseased state. This speculation remains to be explored.

Acknowledgements. This work was supported in part by the Veterans Administration Research Funds (Project No. 9166) and a grant from the National Institutes of Health HL 20549. The kinetic studies on LPL inhibition by human intimal glycoprotein were done in collaboration with Professor Thomas Olivecrona at the University of Umea, Sweden. I wish to thank Mrs. Betty Blaydes for excellent technical assistance and Ms. Diane Butler for secretarial help.

Abstract. This paper briefly reviews the current understanding of a glycoprotein isolated from porcine and human aortic intima. Because the glycoprotein from either source inhibits lipoprotein lipase activity, *in vitro* of post-heparin plasma and of bovine milk, it is named "lipolipin" (lipoprotein lipase inhibitor). The carbohydrate moiety of the porcine molecule (MW 72,000) contains 1 mole each of fucose and sialic acid, 2 moles of mannose, 3 moles each of glucose and galactose and 4 moles of N-acetylglucosamine. The structure of the carbohydrate is predicted by sequential analysis. Although the molecule from the human atherosclerotic intima is not completely characterized, it appears that the porcine and human lipolipins share several features in common: molecular weight, isoelectric point (pI 4.3), electrophoretic mobility and ratio of carbohydrate constituents. Therefore, porcine tissue may serve as a model system to study the human molecule. Kinetic studies revealed that human lipolipin decreased both the basal and serum-stimulated activity of lipoprotein lipase. The inhibition was non-competitive with respect to serum. However, high levels of triglyceride substrate appeared to relieve the inhibitory effect. It is postulated that lipolipin might be involved in an interaction with emulsified lipid denying lipoprotein lipase access to its substrate. If lipolipin is a major constituent of total glycoprotein content in atherosclerotic aorta, the observed higher concentration of glycoproteins in atherosclerotic plaques as compared to that in the uninvolved regions of the aorta may be a reflection of an increased concentration of lipolipin in atherosclerosis.

Literature Cited

Anderson, J.C. (1976), International Rev. Conn. Tiss. Res., 7, 278.

Baig, M.M., and Ayoub, E.M. (1976), Biochemistry 15, 2585.

Berenson, G.S., and Fishkin, A.F. (1962), Arch. Biochem. Biophys. 97, 18.

Berenson, G.S., Radhakrishnamurthy, B., Dalferes, E.R., and Srinivasan, S.R. (1971), Human Path. 2, 57.

Brockerhoff, H. (1971), J. Biol. Chem. 246, 5828.

Buddecke, E. (1960), J. Physiol. Chem. 318, 33.

Dicorleto, P.E., and Zilversmit, D.B. (1975), Proc. Soc. Exptl. Biol. Med. 148, 1101.

Eisenberg, S., and Levi, R.I. (1975), Adv. Lipid Res. 13, 1.

Egelrud, T., and Olivecrona, T. (1972), J. Biol. Chem. 247, 6212.

Henson, L.C., and Schotz, M.C. (1975), Biochim. Biophys. Acta 409, 360.

Hernell, O., Egelrud, T., and Olivecrona, T. (1975), Biochim. Biophys. Acta 381, 233.

Ishii, M. (1971), Jap. Heart J. 12, 22.

Klemer, A., and Nager, C. (1967), Z. Naturforsch 22, 456.

Maier, V., and Buddecke, E. (1971), Hoppe-Seyler Z. Physiol. Chem. 352, 1338.

Manley, G., and Mullinger, R.N. (1967), Brit. J. Path. 58, 529.

Muller-Spreer, H.C., Werber, U., and Voight, K.D. (1960), Klin. Wschr. 38, 28.

Radhakrishnamurthy, B., and Berenson, G.S. (1966), J. Biol. Chem. 241, 2106.
Radhakrishnamurthy, B., Fishkin, A.F., Hubbell, G.J., and Berenson, G.S. (1964), Arch. Biochem. Biophys. 104, 19.
Roberts, B.I., and Wagh, P.V. (1976), Biochim. Biophys. Acta 439, 26.
Roberts, B.I., White, H.J., Read, R.C., and Wagh, P.V. (1974), Atherosclerosis 20, 533.
Robinson, D.S. (1970), Compr. Biochem. 19, 51.
Wagh, P.V. (1975), Adv. Exptl. Med. Biol. 52, 81.
Wagh, P.V., and Olivecrona, T. (1978), Atherosclerosis 29, 195.
Wagh, P.V., and Roberts, B.I. (1972), Biochemistry 11, 4222.
Wagh, P.V., Roberts, B.I., White, H.J., and Read, R.C. (1973), Atherosclerosis 18, 83.

RECEIVED February 13, 1978.

12

Carbohydrate Moieties of the Collagens and Collagen-Like Proteins in Health and Disease

WILLIAM T. BUTLER

Institute of Dental Research, University of Alabama in Birmingham, Birmingham, AL 35294

The fact that interstital collagens and other collagen-like proteins contain covalently bound carbohydrate has been known for many years. In 1935 Grassmann and Schleich (1) found that hide collagen had equal quantities of firmly bound glucose and galactose, and later studies from this group (2) suggested that the hexoses were bound through 0-glycosidic linkages. The stability of the binding of hexoses to collagen was demonstrated by Kühn, et al (3) who found that repeated reprecipitation of citrate-soluble collagen removed all the hexosamine, but only half the hexose. After three reprecipitations, the hexose level was 0.48%, and it remained constant throughout seven additional reprecipitations. This study revealed that hexose, but not hexosamine was an integral part of interstitial collagen.

Gross, et al (4) examined the amino acid and sugar contents of collagens from a variety of tissues and species. They found that for vertebrate collagens, purification or gelatinization drastically reduced the hexosamine content, but was less effective in removing hexose. The carbohydrate was identified as glucose and galactose with traces of other hexoses, pentoses and amino sugars. The hexose content was usually less than 1% for vetebrate and from 3 to 11% in invertebrate collagens. Similarly Blumenfeld, et al (5) found that ichthyocol (carp swim bladder collagen) contained glucose and galactose but no hexosamine or other sugars.

In contrast to the low content of sugars generally found in interstitial collagens of vertebrates, the collagen-like proteins of basement membranes contain much higher levels of carbohydrate (6-8). As in collagen, one type of carbohydrate consists of glucose and galactose units, but heteropolysaccarides with hexoses and hexosamines are also present.

0-8412-0452-7/78/47-080-213$05.00/0

THE NATURE OF THE LINKAGE

Prior to 1965 several studies on the nature of the carbohydrate of collagen were reported, but the results left many uncertainties. The studies of Butler and Cunningham (9,10) clarified the nature of the hexose attached to collagen. Starting with soluble guinea pig skin collagen and digesting to shorter units with collagenase and trypsin, glycopeptides were isolated. The compositions of the peptides suggested that glucose and galactose might be attached to the δ-hydroxyl group of hydroxylysine. The resistance of the hydroxylysine to periodate oxidation supported this conclusion (9). The nature of the linkage was conclusively shown by the isolation of a glycopeptide with stoichiometric amounts of hydroxylysine, glucose, and galactose recovered after cleavage of the peptide bonds of the collagen glycopeptide with 2N NaOH at 90° for 10 hr. The resistance to alkali showed that the attachment was O-glycosidic; this conclusion was also supported by the observation that mild acid hydrolysis (2N HCl, 110°, 30 min.) released 2 mol of reducing sugar. A linkage to the ε-amino group of hydroxylysine was ruled out by the reactivity of this group to fluorodinitrobenzene and by the electrophoretic mobility of the glycopeptide. These experiments thus showed that collagen contained a disaccharide consisting of glucose and galactose attached O-glycosidically to hydroxylysine (Glc-Gal-Hyl).

STRUCTURE OF THE CARBOHYDRATE MOIETY

The structure of the carbohydrate was elucidated by Spiro (11) using collagenous components from glomerular basement membranes. After isolation of a mixture of glycopeptides containing hydroxylysine-linked glucose and galactose, he showed that glucose, but not galactose, was readily liberated with 0.1 N H_2SO_4 at 100° for 2-20 hr. Thus glucose was in an external position, with galactose linked to hydroxylysine. Next a glucose and galactose-containing disaccharide was obtained after N-acetylation and mild acid hydrolysis of the glycopeptides. Glucose was shown to be attached to C-2 of galactose by studies employing periodate and galactose oxidase. This linkage was cleaved by α-glucosidase but not β-glucosidase. Next Glc-Gal-Hyl was isolated after alkaline hydrolysis of glycopeptides; the anomeric configuration of the galactose linkage to hydroxylysine was shown to be β by its susceptibility to β-galactosidase. These experiments thus indicated that Glc-Gal-Hyl had the structure 2-O-α-D-glucopyranosyl-O-β-D-galactopyranosylhydroxylysine (Figure 1). A similar structure was proposed by Kefalides (12).

The structure of Glc-Gal-Hyl in collagens from vertebrate and invertebrate sources has been shown to be the same as that found in basement membranes. Thus Cunningham and Ford (13) observed that glucose was split from the disaccharide of guinea pig skin collagen by mild acid hydrolysis, while galactose remained attached. They also reported finding galactosylhydroxylysine (Gal-Hyl) after alkaline hydrolysis of skin collagen. Using the amino acid analyzer, Spiro (14) introduced a separation technique for quantitation of Glc-Gal-Hyl and Gal-Hyl. Chromatography of alkaline hydrolysates of several collagen samples showed that they contained the same mono- and disaccharide units linked to hydroxylysine as basement membranes. Katzman, et al (15) utilized Glc-Gal-Hyl isolated from the sponge and showed that the structure was that proposed for basement membranes by Spiro (11). The hydroxylysine-linked disaccharide isolated from sea anenome, sea cucumber, and bovine cornea was shown to have an identical structure by a variety of techniques (15). Since these early studies, Glc-Gal-Hyl and Gal-Hyl have been shown to occur in a variety of collagen and collagen-like proteins.

OCCURRENCE IN THE DIFFERENT COLLAGEN TYPES

The discovery that cartilage contained a type of collagen distinct from that of skin and bone (16) ushered in a new era in collagen biochemistry. We now understand that at least three collagens with unique properties exist in the extracellular matrix of connective tissues (Table I). For more details concerning these collagens, the reader is referred to the review by Miller (17). Scientists have long recognized that the basement membranes contain collagenous components (6,18,19). A final resolution of the structures of the molecular species present in basement membranes has not been accomplished. At least one of these appears to have a triple-helical conformation referred to as $[\alpha 1(IV)]_3$ (20). However it is apparent that several other structures are present (21-23).

All three of the interstitial collagens contain hexose attached to hydroxylysine, but type II collagen has a significantly greater quantity of hydroxylysine and of hydroxylysine-linked carbohydrate (Table I). In general the collagenous portion of basement membranes contain much higher levels of hydroxylysine and of the associated glucose and galactose moieties.

It should be noted that other hydroxyproline-containing proteins with collagen-like (i.e. Gly-X-Y repeating) sequences exist. For example, the complement protein Cl_q contains several glycosylated hydroxylysines in a collagen-like sequence (24).

The exact locations of several of the carbohydrate moieties of the collagens has been documented during the studies on the primary structures of the various α chains. One site of glycosylation common to the four chains of the three interstitial collagens is Hyl-87 (Figure 2). The amino acid sequences around this site are similar in these chains (25-29). The occurrence of the disaccharide at this site suggests that it performs some

TABLE I: Structures and Characteristics of Interstitial Collagens and Collagen-Like Proteins of Basement Membranes

Type	Molecular Composition	Tissue Distribution	Hydroxylysine Content (Residues/1000)	Hexose[a] Content (Residues/1000)
I	$[\alpha 1(I)]_2\alpha 2$	Ubiquitous (Skin, Bone, Tendon Dentin, etc.)	5–15	2.5
II	$[\alpha 1(II)]_3$	Cartilage, Intervetebral Disc, Notocord	20–25	15–30
III	$[\alpha 1(III)]_3$	Ubiquitous (Especially Prominent in Fetal Skin, Arteries, Cirrhotic Liver)	5–8	2
IV	Several	Basement Membranes (Glomerulus, etc.)	40–50[b]	55[b]

[a] Glucose and galactose

[b] Data for the collagen-like portion of glomerular basement membranes. Taken from reference 19.

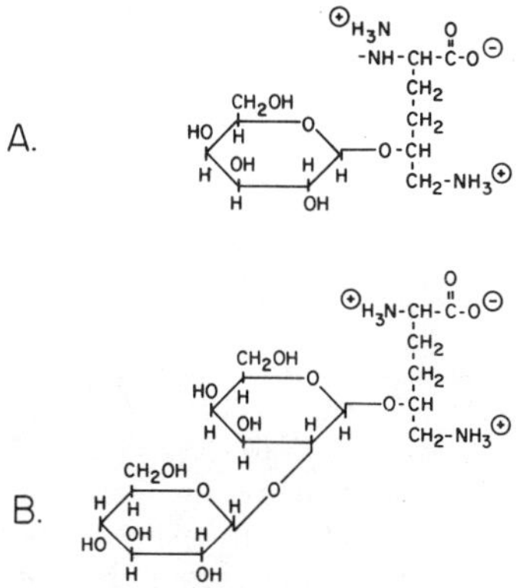

Figure 1. The structures of the glucose- and galactose-containing units isolated from the collagens and collagen-like proteins. (A) β-D-galactopyranosylhydroxylsine (Gal-Hyl), (B) 2-O-α-D-glucopyranosyl-O-β-D-galactopyranosylhydroxylysine (Glc-Gal-Hyl).

						87			
α1(I)	- GLY -	LEU -	HYP -	GLY -	MET -	HYL -	GLY -	HIS -	ARG -
						Glc - Gal			
α2	- GLY -	LEU -	HYP -	GLY -	PHE -	HYL -	GLY -	ILE -	ARG -
						Glc - Gal			
α1(II)	- GLY -	LEU -	HYP -	GLY -	VAL -	HYL -	GLY -	HIS -	ARG -
						Glc - Gal			
α1(III)	- GLY -	PHE -	HYP -	GLY -	MET -	HYL -	GLY -	HIS -	ARG -
						Glc - Gal			

Figure 2. Amino acid sequences around the disaccharide-bearing hydroxylysines common to types I, II, and III collagens

						174			
α1(I)	- GLY -	ALA -	ALA -	GLY -	ALA -	LYS -	GLY -	GLU -	ALA -
α1(III)	- GLY -	SER -	HYP -	GLY -	ALA -	LYS -	GLY -	GLU -	VAL -
α1(II)	- GLY -	ALA -	HYP -	GLY -	ALA -	HYL -	GLY -	GLU -	ALA -
						Gal			
α2	- GLY -	ALA -	HYP -	GLY -	PRO -	HYL -	GLY -	GLU -	LEU -
						Gal			

Figure 3. The amino acid sequences of α chains around the monosaccharide-bearing hydroxylysine of α2

functional role vital to the three collagens (e.g. in cross-linking). The α2 chain of type I collagen has a Gal-Hyl at position 174 (26,27) not found in α1(I) or α1(III) but present in α1(II) (Figure 3). Although these appear to be the *major* sites of carbohydrate in α1(I), α2 and α1(III), other positions may be partially glycosylated. For instance subintegral levels of Gal-Hyl are found near the COOH-terminus of α1(I) in calfskin collagen (26).

As stated before, the α1(II) chain of cartilage collagen has many more sites of carbohydrate attachment than the above. Studies on the covalent structure of bovine and chick α1(II) chains (28,30,31) from this laboratory have already located 13 glycosylated hydroxylysines (Figure 4), although only about 60% of the sequence is known. These hydroxylysines are not confined to any one area but occur throughout the chain. Comparison of the positions of these glycosylated hydroxylysines with homologous sequences in α1(I) and α2 shows that they are frequently occupied by lysine in the latter two chains. Thus the three chains have similar amino acid sequences with potential sites for carbohydrate attachment, but only for α1(II) are the post-translational modifications (i.e. hydroxylation and glycosylation) made. I will return to possible explanations for the high level of Glc-Gal-Hyl and Gal-Hyl in α1(II) in a later section.

BIOSYNTHETIC ATTACHMENT OF HEXOSE TO PROCOLLAGEN α CHAINS

Collagen is synthesized within connective tissue cells by the usual protein synthetic mechanism (32) but has several features which are unique (Figure 5). The initial form of the three α chains are elongated, compared to that found in fibrillar collagen (33-36). These "extensions" which occur on the NH_2-and COOH-terminal ends of the chains, are cleaved from the molecules during or after exit from the cell. During and after the translation of these procollagen chains on ribosomal complexes, several post-translation modifications take place. Lysyl and prolyl residues are hydroxylated to form hydroxylysyl and hydroxyprolyl residues, and galactose and glucose moieties are attached to certain of the hydroxylysines. Next the three chains of a procollagen unit associate, interchain disulfide bonds in the COOH-terminal extensions are formed, and folding into a triple-chain collagen helix takes place.

It is obvious that the attachment of galactose and glucose to collagen must be preceded by hydroxylation of lysyl residues. This reaction is catalyzed by lysyl hydroxylase (32,29,40), an enxyme which is similar in many respects to prolyl hydroxylase. Each of these mixed function oxidases requires molecular oxygen, α-ketoglutarate, ascorbate and ferrous iron (Figure 6) and each catalyzes hydroxylation of residues in the Y position of the repeating Gly-X-Y collagen sequence (37), though lysyl hydroxylase will apparently also act upon lysines in the X position (40).

87
α1(I) Bovine GLY - LEU - HYP - GLY - MET - HYL - GLY - HIS - ARG - GLY - PHE - SER
α1(II) Bovine GLY - LEU - HYP - GLY - VAL - HYL - GLY - HIS - ARG - GLY - THR - HYP
α2 Bovine GLY - LEU - HYP - GLY - PHE - HYL - GLY - ILE - ARG - GLY - HIS - ASN

99
α1(I) Bovine GLY - LEU - ASP - GLY - ALA - LYS - GLY - ASP - ALA - GLY - PRO - ALA
α1(II) Bovine GLY - LEU - ASP - GLY - ALA - HYL - GLY - GLU - ALA - GLY - ALA - HYP
α2 Bovine GLY - LEU - ASP - GLY - LEU - THR - GLY - GLN - HYP - GLY - ALA - HYP

108
α1(I) Bovine GLY - PRO - ALA - GLY - PRO - LYS - GLY - GLU - HYP - GLY - SER - HYP
α1(II) Bovine GLY - ALA - HYP - GLY - VAL - HYL - GLY - GLU - SER - GLY - SER - HYP
α2 Bovine GLY - ALA - HYP - GLY - VAL - HYL - GLY - GLU - HYP - GLY - ALA - HYP

174
α1(I) Rat GLY - ALA - ALA - GLY - ALA - LYS - GLY - GLU - ALA - GLY - PRO - GLN
α1(II) Bovine GLY - ALA - HYP - GLY - ALA - HYL - GLY - GLU - ALA - GLY - PRO -
α2 Bovine GLY - ALA - HYP - GLY - PRO - HYL - GLY - GLU - LEU - GLY - PRO - VAL

219
α1(I) Rat GLY - GLN - HYP - GLY - ALA - LYS - GLY - ALA - ASN - GLY - ALA - HYP
α1(II) Bovine GLY - ILE - HYP - GLY - ALA - HYL - GLY - SER - ALA - GLY - ALA - HYP
α2 Bovine GLY - LEU - HYP - GLY - ALA - HYL - GLY - ALA - ALA - GLY - LEU - HYP

252
α1(I) Rat GLY - ALA - HYP - GLY - PRO - LYS - GLY - ASN - SER - GLY - GLU - HYP
α1(II) Bovine GLY - PRO - LEU - GLY - PRO - HYL - GLY -
α2 Bovine GLY - ALA - THR - GLY - ALA - ARG - GLY - LEU - VAL - GLY - GLU - HYP

408
α1(I) Bovine GLY - PHE - HYP - GLY - PRO - LYS - GLY - ALA - ALA - GLY - GLU - HYP
α1(II) Bovine GLY - PHE - HYP - GLY - PRO - HYL - GLY - ALA - ASN - GLY - GLU - HYP
α2 Chick GLY - PHE - HYP - GLY - PRO - LYS - GLY - PRO - THR - GLY - GLU - HYP

420
α1(I) Bovine GLY - LYS - ALA - GLY - GLU - ARG - GLY - VAL - HYP - GLY - PRO - HYP
α1(II) Bovine GLY - LYS - ALA - GLY - GLU - HYL - GLY - LEU - HYP - GLY - ALA - HYP
α2 Chick GLY - LYS - HYP - GLY - GLU - LYS - GLY - ASN - VAL - GLY - LEU - ALA

531
α1(I) Bovine GLY - ASN - ASP - GLY - ALA - LYS - GLY - ASP - ALA - GLY - ALA - HYP
α1(II) Bovine GLY - THR - ASP - GLY - PRO - HYL - GLY - ALA - ALA - GLY - PRO - ALA
α2 Chick GLY - PRO - ASP - GLY - ASN - LYS - GLY - GLU - HYP - GLY - ASN - VAL

564
α1(I) Bovine GLY - LEU - HYP - GLY - PRO - LYS - GLY - ASP - ARG - GLY - ASP - ALA
α1(II) Bovine GLY - ILE - ALA - GLY - PRO - HYL - GLY - ASP - ARG - GLY - ASP - VAL
α2 Chick GLY - VAL - HYP - GLY - GLY - LYS - GLY - GLU - LYS - GLU - ALA - HYP

573
α1(I) Bovine GLY - ASP - ALA - GLY - PRO - LYS - GLY - ALA - ASP - GLY - ALA - PRO
α1(II) Bovine GLY - ASP - VAL - GLY - GLU - LYS - GLY - PRO - GLU - GLY - ALA - PRO
α2 Chick GLY - ALA - HYP - GLY - LEU - ARG - GLY - ASP - THR - GLY - ALA - THR

603
α1(I) Bovine GLY - ALA - HYP - GLY - ASP - LYS - GLY - GLU - ALA - GLY - PRO - SER
α1(II) Bovine GLY - ASP - VAL - GLY - GLU - HYL - GLY - GLU - VAL - GLY - PRO - HYP
α2 Chick GLY - GLY - ALA - GLY - ASP - ARG - GLY - GLU - GLY - GLY - PRO - ALA

648
α1(I) Bovine GLY - GLN - HYP - GLY - ALA - LYS - GLY - GLU - HYP - GLY - ASP - ALA
α1(II) Bovine GLY - GLN - PRO - GLY - ALA - HYL - GLY - GLU - GLN - GLY - GLU - ALA
α2 Chick GLY - GLU - HYP - GLY - ALA - LYS - GLY - GLU - ARG - GLY - PRO - LYS

657
α1(I) Bovine GLY - ASP - ALA - GLY - ALA - LYS - GLY - ASP - ALA - GLY - PRO - HYP
α1(II) Bovine GLY - GLU - ALA - GLY - GLN - HYL - GLY - ASP - ALA - GLY - ALA -
α2 Chick GLY - PRO - LYS - GLY - PRO - LYS - GLY - GLU - THR - GLY - PRO - THR

Figure 4. The sequences around several hydroxylysines and lysines of α1(I), α1(II), and α2 chains (taken from Ref. 31)

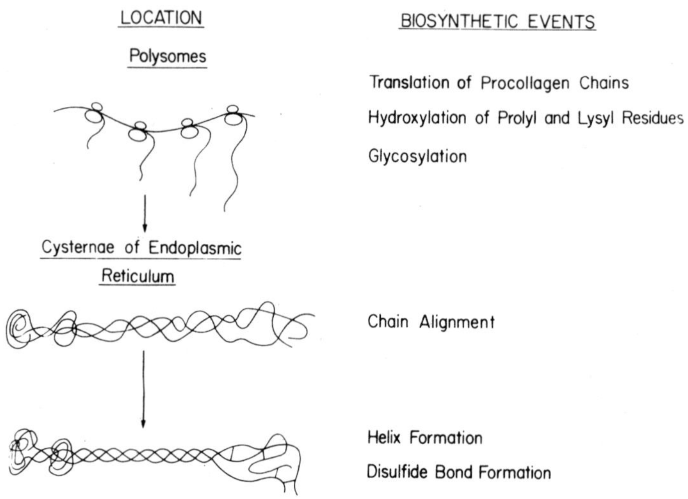

Figure 5. Intracellular steps in the biosynthesis of procollagen. Individual procollagen α chains are made by the usual protein biosynthetic mechanisms. The chains are then subjected to several post-translational modifications. After release from ribosomal complexes, three chains align, and triple-helices and interchain disulfide bonds at the COOH-terminal extremities are formed. Post-translational modifications cease upon formation of the triple-helix.

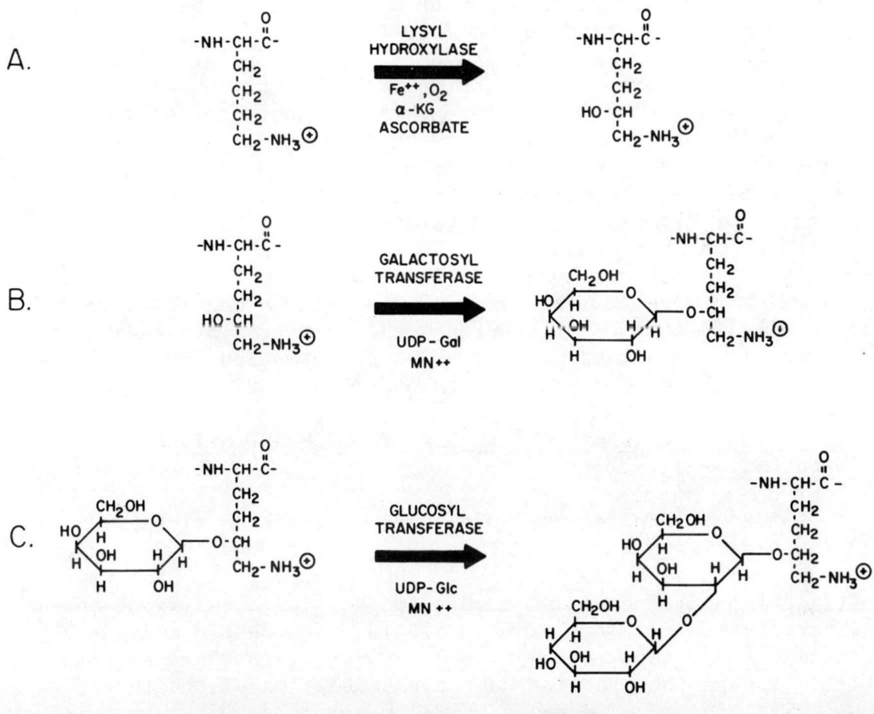

Figure 6. Three enzymatic steps in the biosynthetic attachment of hexose to collagen. (A) Formation of hydroxylysyl residues from lysines on the nascent procollagen α chains; (B) attachment of galactose to certain hydroxylysyl residues; and (C) attachment of glucose to certain Gal-Hyl moieties.

Results of experiments *in vitro* (37,38) and using cells in culture (41), indicate that the hydroxylation of lysyl residues does not occur after triple-helical conformation of the procollagen molecules is formed (Figure 5).

After hydroxylation, galactosyltransferase catalyzes the transfer of galactose from UDP-Gal to certain hydroxylysines (Figure 6); then the transfer of glucose from UDP-Glc to the C-2 of the galactose residues is catalyzed by glucosyltransferase. Both enzymes require divalent cations, perferably manganese and are probably located in the cisternae of the endoplasmic reticulum. Glucosyltransferase has been purified to homogeneity (42) and galactosyltransferase 1000-fold (43).

Similar to the hydroxylation reactions, the attachment of carbohydrate to procollagen chains ceases when they fold into a native, triple-helical structure (44). Using chick embryo tendon and cartilage cells in culture, Oikarinen, *et al* (45) have shown that the glycosylation of hydroxylysines was dramatically affected by the rate of triple-helix formation of the procollagen chains. For example, when the rate of helix formation was inhibited by incubation in the presence of 0.6 mM dithiothreitol, the glycosylation was more than doubled. Conversely when cartilage cells recovering from anoxia were used (a condition known to accelerate triple-helix formation), there was a marked decrease in the extent of hydroxylysine formation and an even greater decrease in glycosylation of hydroxylysine residues.

FACTORS CONTROLLING HYDROXYLYSINE FORMATION AND SUBSEQUENT GLYCOSYLATION

The foregoing discussion may help explain some observations on the variations in the content and occurrence of hydroxylysines and glycosylated hydroxylysines. The hydroxylysine content of α1(I) chains differs considerably in the type I collagens of different tissues. For example α1(I) of rat dentin collagen contains about three times as much hydroxylysine as does skin α1(I), though these chains have the same primary structure (46-48). This relative increase could be due to a higher activity of lysyl hydroxylase in odontoblasts; alternatively a slower rate of folding of dentin procollagen α chains into the triple-helical structure could occur, allowing a more complete hydroxylation of lysyl residues. The problem with the latter hypothesis is that one might expect an increased glycosylation of the hydroxylysines over that in skin α1(I) chains, if the rate of helix formation is the major factor controlling hydroxylation and glycosylation of nascent procollagen α chains. However the additional hydroxylysines do not appear to have increased levels of hexose (48). Kivirikko and Ristelli (32) speculate that the differences in hydroxylysine and glycosylated hydroxylysine contents of collagens in different tissues could be due in part to differences in the rates of triple-helix formation during synthesis in the various connective tissue cells.

The extensive comparison of the amino acid sequences near glycosylated hydroxylysines in the α1(II) chain with related sequences in α1(I) and α2 (Figure 4), suggests that the relatively high level of carbohydrate in type II collagen is not due to differences in amino acid sequences which favor the enzymatic reactions (i.e. hydroxylation and glycosylation). It appears that almost every lysine in procollagen α1(II) chains which is in the Y position of the Gly-X-Y repeating sequences, is hydroxylated and glycosylated. On the other hand, most of these lysines in α1(I) are either hydroxylated to a minimal extent or not at all and, are thus not glycosylated. It is curious that the extent of hydroxylation of α2 chains is greater than that of α1(I), but for the most part, this phenomenon is confined to the NH_2-terminal third of the chain.

The more extensive hydroxylation of lysines and glycosylation of the resultant hydroxylysines in type II procollagen could be due to a slower rate of triple-helix formation. In studies on the biosynthesis of type II procollagen by chick embryo cartilage cells, Uitto and Prockop (49) found that chain association and triple-helix formation of procollagen α1(II) chains required almost twice as long as that for type I collagen synthesized by tendon cells. These data therefore support the above hypothesis. Alternatively cartilage cells may simply have a higher activity for lysyl hydroxylase and the glycosyltransferases.

DISEASES AFFECTING THE LEVELS OF COLLAGEN-ASSOCIATED CARBOHYDRATE

One of the symptoms of diabetes is a thickening of basement membranes. Beisswenger and Spiro (50) have reported that diabetic glomerular basement membranes contain increased levels of hydroxylysine and of hydroxylysine-linked disaccharide units, compared to control specimens. Spiro (51) also found an increased activity of collagen glucosyltransferase activity in alloxan-diabetic rat kidneys, which was partially reversed by insulin administration. Cohen and Khalifa (52) studied the effect of diabetes on prolyl and lysyl hydroxylase activities in the rat glomerulus. Data from rats made diabetic by streptozotocin injections was compared to that from age-matched controls and from insulin-treated, streptoxotocin-diabetic rats. They found that lysyl hydroxylase activity was significantly higher in diabetic rat kidneys than in control animals, but that prolyl hydroxylase was unaffected. Administration of insulin reduced the lysyl hydroxylase activity to normal. These observations suggest that alterations of hydroxylation and glycosylation of lysines in basement membrane collagens might relate to the nephropathology of diabetes.

It should be pointed out that the differences in hydroxylysine and glycosylated hydroxylysine in diabetic kidneys seen by Beisswenger and Spiro (50) have not been noted by other investigators (53,54). This point is therefore one of controversy.

Osteogenesis imperfecta congenita (OI) is a genetic disease resulting in bones of severe fragility. Since collagen is the matrix or framework onto which apatite crystals are laid during bone formation, it has long been assumed that the basic defect in OI is some defect in collagen structure (55). Direct chemical analysis of tissues from OI patients have shown that bone and dentin collagens have increased levels of hydroxylysine (56). Recently Trelstad, et al (57) reported a more complete biochemical analysis of collagen from several tissues of an OI infant, and compared the results to these of tissues from age-matched controls. Hydroxylysine was doubled in OI bone collagen and increased by 55% in cartilage. In addition the galactose and glucose levels of OI collagen in both tissues were sharply increased. The data suggest that at least one form of OI is associated with increases in glycosylated hydroxylysines in type I collagen of bone and type II collagen of cartilage.

The actual effect which increased levels of hydroxylysine and the associated glycosides might have on the structure and function of collagens is unknown at this time. But it seems apparent from the foregoing disscussion that the control of post-translational modifications is critical in order to insure that the extracellular products (i.e. collagen monomers, fibrils, and fibers) are able to perform the intended functions.

ACKNOWLEDGEMENT

The research from this laboratory referred to in this publication was supported by United States Public Health Service Grant DE-02670.

LITERATURE CITED

1. Grassmann, W. and Schleich, J. (1935). Biochem. Z. 27, 230.
2. Grassmann, W., Hörmann, H., and Hafter, R. (1957). Hoppe-Seyler's Z. Physiol. Chem. 307, 87.
3. Kühn, K. Grassmann, W., and Hoffmann, U. (1959). Z Naturforsch. Teil B. 14, 436.
4. Gross, J., Dumsha, B., and Glazer, N. (1958) Biochim. Biophys. Acta 30, 293.
5. Blumenfeld, O.O., Paz, M.A., Gallop. P.M., and Seifter, S. (1963). J. Biol. Chem. 238, 3835.
6. Kefalides, N.A. and Winzler, R.J. (1966). Biochemistry 5, 702.
7. Spiro, R.G. (1967). J. Biol. Chem. 242, 1915.

8. Dische, Z. (1964). In Involvement of Small Blood Vessels in Diabetes, Ed. by Siperstein, E., Washington: Amer. Inst. Biol. Sci., p. 201.
9. Butler, W.T., and Cunningham, L.W. (1965). J. Biol. Chem. 240, 3449.
10. Butler, W.T., and Cunningham, L.W. (1966). J. Biol. Chem. 241, 3882.
11. Spiro, R.G. (1967). J. Biol. Chem. 242, 4813.
12. Kefalides, N.A. (1967). in "Proceedings of the Sixth Congress of the International Diabetes Foundation", Ed. by J. Ostman, Excerpta Med. Found., Amsterdam, P. 307.
13. Cunningham, L.W., and Ford (1968). J. Biol. Chem. 243, 2390.
14. Spiro, R.G. (1969). J. Biol. Chem 244, 602.
15. Katzman, R.L., Halford, M.H., Reinhold, V.N., and Jeanloz, R.W. (1972). Biochemistry 11, 1161.
16. Miller, E.J., and Matukas, V.J. (1969) Proc. Natl. Acad. Sci., USA 64, 1264.
17. Miller, E.J. (1976). Mol. Cell. Biochem. 13, 165.
18. Goodman, M., Greenspon, S.A., and Krakower, C.A. (1955). J. Immunol. 75, 96.
19. Kefalides, N.A. (1973). Intl. Rev. Connect. Tiss. Res. 6, 63.
20. Kefalides, N.A. (1971). Biochem. Biophys. Res. Commun. 45, 226.
21. Daniels, J.R. and Chu, G.H. (1975). J. Biol. Chem. 250, 3531.
22. Chung, E., Rhodes, R.K., and Miller, E.J. (1976). Biochem. Biophys. Acta 71, 1167.
23. Hudson, B.G., and Spiro, R.G. (1972). J. Biol. Chem. 247, 4229.
24. Redi, K.B.M. (1974). Biochem. J. 141, 189.
25. Butler, W.T. (1970). Biochemistry 9, 44.
26. Fietzek, P.P. and Kühn, K. (1976). Intl. Rev. Connect. Tis. Res. 7, 1.
27. Aguilar, H., Jacobs, H.G., Butler, W.T., and Cunningham, L.W. (1973). J. Biol. Chem. 248, 5106.
28. Butler, W.T., Miller, E.J. and Finch, J.E., Jr., (1976). Biochemistry 15, 3000.
29. Seyer, J.M., and Kang, A.H. (1977). Biochemistry 16, 1158.
30. Butler, W.T., Finch, J.E., Jr., and Miller E.J. (1977). Biochemistry 16, 4981.
31. Francis, G., Butler, W.T., and Finch, J.E., Jr., (1978). Biochem. J., in press.
32. Kivirikko, K.I. and Risteli, L. (1976). Med. Biol. 54, 154.
33. Bornstein, P. (1974). Annu. Rev. Biochem. 43, 567.
34. Fessler, L.I., Morris, N.P., and Fessler, J.H. (1975). Proc. Natl. Acad. Sci. USA 72, 4905.

35. Martin, G.R., Byers, P.H., and Piez, K.A. (1975). Adv. Enzymol. 42, 167.
36. Grant, M.E., Schofield, D.J., Kefalides, N.A., and Prockop, D.J. (1973). J. Biol. Chem. 248, 7432.
37. Kivirriko, K.I., Ryhänen, L., Anttinen, H., Bornstein, P., and Prockop, D.J. (1973). Biochemistry 12, 4966.
38. Ryhänen, L. and Kivirikko, K.I. (1974). Biochim. Biophys. Acta 343, 129.
39. Ryhänen, L. (1975). Biochim. Biophys. Acta 397, 50.
40. Ryhänen, L. (1976). Biochim. Biophys. Acta 438, 71.
41. Oikarinen, A., Anttinen, H., and Kivirikko, K.I. (1976). Biochem. J. 156, 545.
42. Myllylä, R., Anttinen, H., Ristelli, L. and Kivirikko, K.I. (1977). Biochim. Biophys. Acta 480, 113.
43. Risteli, L., Myllä, R., Kivirikko, K.I. (1976). Europ. J. Biochem. 67, 197.
44. Myllylä, R., Ristelli, L., and Kivirikko, K.I. (1975). Europ. J. Biochem. 58, 517.
45. Oikarinen, A., Anttinen, H., and Kivirikko, K.I. (1977). Biochem. J. 166, 357.
46. Butler, W.T. (1973). Ala. J. Med. Sci. 10, 103.
47. Butler, W.T., Finch, J.E., Jr. and Desteno. C.V. (1972). Biochim. Biophys. Acta 257, 157.
48. Butler, W.T. (1972). Biochem. Biophys. Res. Commun. 48, 1540.
49. Uitto, J. and Prockop, D.J. (1974). Biochemistry 13, 4586.
50. Beisswenger, P., and Spiro, R.G. (1970). Science 168, 596.
51. Spiro, R.G. (1973). New Engl. J. Med. 288, 1337.
52. Cohen. M.P. and Khalifa, A. (1977). Biochim. Biophys. Acta. 496, 88.
53. Kefalides, N.A. (1973). Adv. Metab. Disorders Suppl. 2, 167.
54. Westberg. N.A., and Michael, A.F. (1973). Acta Med. Scand. 194, 39.
55. McKusick, V.A. (1972). Hereditable Disorders of Connective Tissue, St. Louis: Mosby, p. 230.
56. Eastoe, J.E., Martens, P., and Thomas, N.R. (1973). Calc. Tiss. Res. 12, 91.
57. Trelstad, R.L., Rubin, D., and Gross, J. (1977). Lab. Invest. 36, 501.

RECEIVED March 13, 1978.

13

Molecular Pathology of Vascular Elastic Fiber—The Importance of the Glycoprotein Coat

M. M. LONG and D. W. URRY

Laboratory of Molecular Biophysics and the Cardiovascular Research and Training Center, University of Alabama Medical Center, Birmingham, AL 35294

J. BAKER and B. CATERSON

Department of Rheumatology, University of Alabama Medical Center, Birmingham, AL 35294

The tissues of ligaments, skin, lung and the arterial wall maintain their resilency, in large part due to the elastic fiber. Concentrating on the arterial wall, one sees that in the vascular tree, three major regions comprise the arterial wall; they are from lumen to vessel wall: the intima, the media and the adventia. Each are separated from one another by two dense bands of connective tissue; the internal elastic lamina and the external elastic lamina. Collagen fibers and elastic fibers are predominantly, but not exclusively, in these lamina. Further subdivision finds that the elastic fiber consists of two distinct molecular species, a single protein, elastin, which is surrounded by a glycoprotein sheath (1). Elastin, as a 70,000 m.w. protein, is insoluble because of covalent cross-links achieved primarily by combination and condensation of four lysine side chains to form desmosines and isodesmosines (2). These cross-links result in large bundles with a diameter of about 5-6 μm (3). The molecular details which ultimately render elastin elastomeric also render it susceptible to disease, for arterial wall elastic fiber is a site of calcium phosphate deposition and lipid binding (4,5). This calcification is a progressive process; in the first decade of life calcium salts comprise 0.6% of aorta dry weight, while in the eighth decade the percentage is from 6% to 12% (6). It is also rather specific to the elastic fiber. Martin and colleagues (7) showed that in the internal elastic lamina of rat aorta, zones of heavy mineralization occurred on the elastic fiber. Only when calcification was extensive did it spill over onto collagen fibers.

How this elastin calcifies has been a long standing problem of not inconsiderable interest, especially in view of the fact that the molecule is largely hydrophobic with few negatively charged groups. 90% of the amino acids of elastin have nonfunctional side chains; 33% are glycine; and 58% are hydrophobic residues. Of the remaining 10%, glutamyl and aspartyl residues constitute 2-3%, but 90% of these are blocked by being in their amide form, i.e. asparagine and glutamine (8). In 1971 (9), the

0-8412-0452-7/78/47-080-227$07.25/0

idea of neutral site binding/charge neutralization was developed to explain elastin's calcification, an idea consonant with elastin's amino acid composition and primary structure (Figure 1). The thought was that elastin bound calcium at the formally neutral yet polarizable carbonyl oxygens of the peptide moieties oriented in a configuration that offered selective binding. Calcium cations would be continuously bound until positive charge-charge repulsion between calcium ions balanced out the binding sites' calcium affinity. This initial calcium chelation then set the stage for the next step in calcification, phosphate deposition, because once calcium interacted with the elastin matrix, it became positively charged and sequestered polyvalent anions such as phosphate. Phosphate neutralized the positive charge of the calcium, allowed more calcium binding which in turn brought down more phosphate anions, forming the foundation for hydroxyapatite growth.

Experimental data support both aspects of the neutral site/ charge neutralization theory, that is (1) specific calcium binding to neutral sites - the peptide carbonyls, and (2) subsequent calcification of neutral elastin molecules. Figure 2 presents the cation titrations of blocked α-elastin with increasing concentration of $CaCl_2$ and NaCl. Binding was followed by monitoring the circular dichroism signal at 220 nm at each specific ion increment and plotting the difference between the initial ellipticity and the ellipicity at each ion addition. In TFE, calcium bound blocked α-elastin readily, sodium did not. These data are supportive of the neutral site carbonyl binding concept because the sample of α-elastin used was totally blocked. All the free aminos were blocked with formyl and all the free carboxyls with methyl, making the molecule uncharged. By elimination, one would predict then that calcium was binding at neutral sites, the carbonyl. Infrared spectroscopy provided data which directly implicated the peptide backbone as the calcium binding site. IR was used because it can follow different vibrational modes of the peptide moiety in molecules too large to be studied with nuclear magnetic resonance. The amide I C-O stretch of the carbonyl of the peptide moiety of blocked α-elastin coacervate was monitored as a function of calcium addition. Figure 3 shows that calcium caused the carbonyl frequency of 1662 cm^{-1} to shift to lower wave numbers and to become biphasic with minima at 1655 cm^{-1} and 1630 cm^{-1} (10). This is the same pattern exhibited by a synthetic elastin peptide which bound calcium cations via its peptide carbonyls as shown by PMR and CMR (11). These data led to the conclusion that the calcium ion specific site is the peptide carbonyl oxygen in blocked α-elastin coacervate. Since the coacervate approximates an *in vivo* conformation of elastin (12), these studies support the idea of neutral site binding to elastin when it is in a state having direct bearing on its biological state.

Not only does blocked α-elastin bind calcium and at neutral

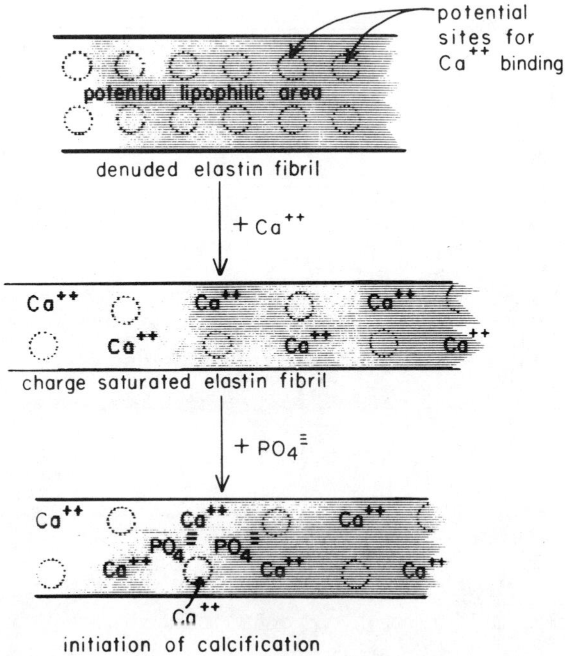

Technical Publishing Company

Figure 1. Schematics of neutral-site calcium ion binding to the elastin matrix, showing calcium sequestration followed by deposition of polyvalent anions, in this case, phosphate anions. Adapted from Ref. 46.

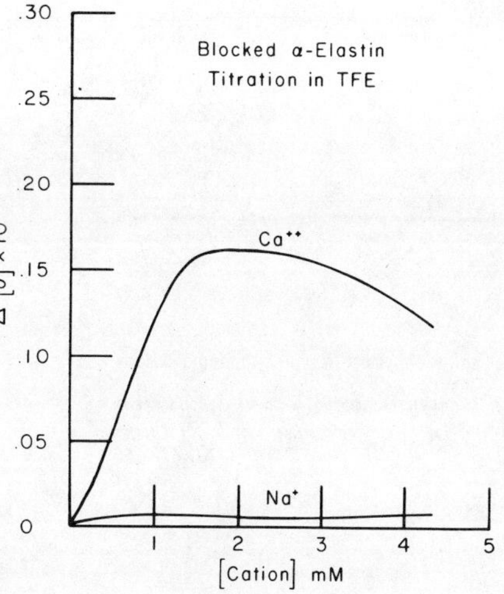

Figure 2. Ion titration of N*-formyl-*O*-methyl-ester α-elastin from aorta in trifluoroethanol (TFE) with aqueous increments of 0.1*M *$CaCl_2$ and 0.1*M *NaCl followed by circular dichroism. $\Delta[\theta]$ is the difference between the ellipticity of the free state and each state with incrementing amounts of cation.*

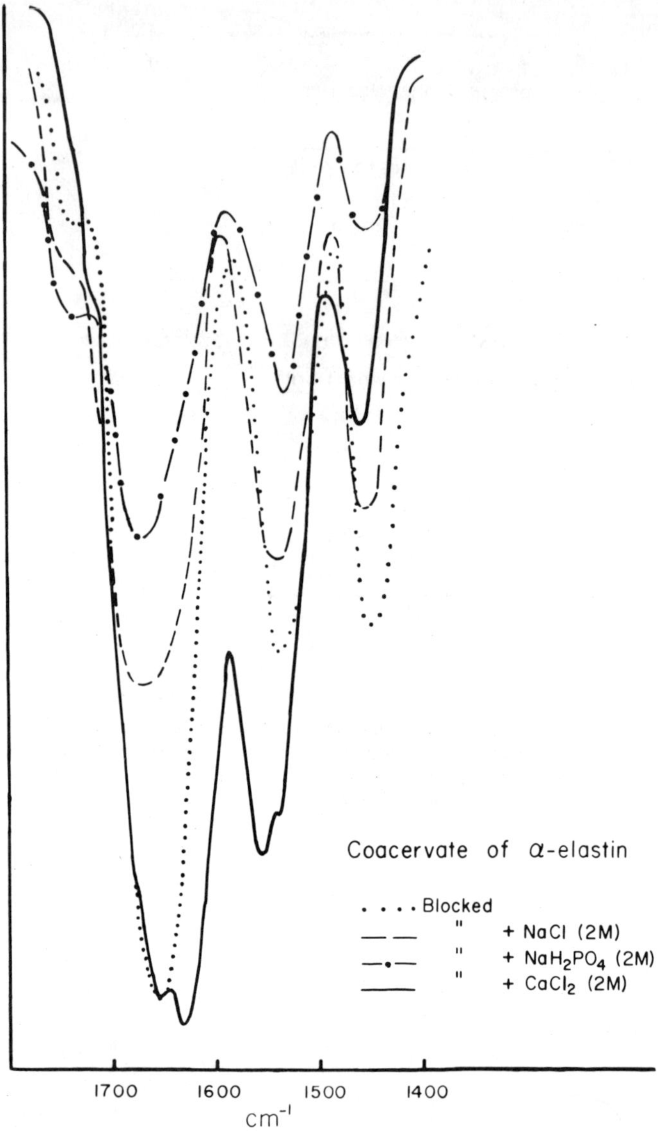

Biochemical and Biophysical Research Communications

Figure 3. IR spectra with percent transmission plotted vs. wave number of blocked α-elastin coacervate in the presence and absence of calcium, (——) 2M $CaCl_2$, (– – –) 2M NaCl, (– · –) 2M NaCl + KH_2PO_4 (10).

sites, but it also initiates calcification. Elastin was incubated at 37°C for 20 hours in a serum calcifying medium with (Figure 5) and without (Figure 4) calcium and phosphate (12). The secondary electron image of the control is seen in Figure 4b. The elastin matrix had cracked along the top and left hand corner, revealing the plexiglas support beneath. The surface appears smooth, the X-ray spectrum (Figure 4a) contains a peak for the aluminum coat and a small one for sulfur, and the X-ray map (Figure 4c) indicates no concentration of calcium ions on the sample when the EDAX was programmed to detect only calcium X-rays. The other sample, having been exposed to calcium and phosphate, now has a rough textured topography (Figure 5b), an X-ray spectrum indicating calcium (Figure 5a) and phosphate and a calcium map (Figure 5c) which reveals a uniform distribution of calcium, localized only on α-elastin. When Epon embedded calcified material is cross sectioned perpendicular to its surface (Figure 6), its secondary image, X-ray spectrum and X-ray map all demonstrate that calcification occurred throughout the bulk of the material (14). In summary, the CD studies and the SEM-EPM data demonstrate the validity of the neutral site/charge neutralization theory with respect to elastin calcium binding and calcification.

α-Elastin, being of 70,000 m.w., is much too large to study in molecular structural detail with nuclear magnetic resonance, both PMR and CMR which would be techniques of choice to delineate the calcium binding sites. Gray and Sandberg have sequenced about one half of tropoelastin, the precursor protein of elastin (15,16) and found repeating sequences, a tetramer, a pentamer, and a hexamer occurring approximately, four, six and five times in single sequences, respectively. They are as follows:

Val-Pro-Gly-Gly
Val-Pro-Gly-Val-Gly
Ala-Pro-Gly-Val-Gly-Val

Our laboratory synthesized these repeat peptides as monomers, dimers, trimers and high polymers. Because of their small molecular weight, relative to elastin and their defined number of different amino acid residues, these peptides are amenable to study with spectroscopic techniques, especially NMR to elucidate further the idea of neutral site binding/charge neutralization. Figure 7a is the ion titration of the pentamer N-formyl-Val-Pro-Gly-Val-Gly-O-Methyl followed with circular dichroism as in Figure 2. Of the three divalent cations, strontium chelated the best followed by calcium and then magnesium. Interaction with sodium and potassium was negligible—within instrumental error (17). Calcium and magnesium had the same initial slope, but after an [ion]/[peptide] ratio of about 0.25, the magnesium curve became biphasic, as if the pentamer peptide were losing the ion. This is undoubtedly the case because, while the peptide is

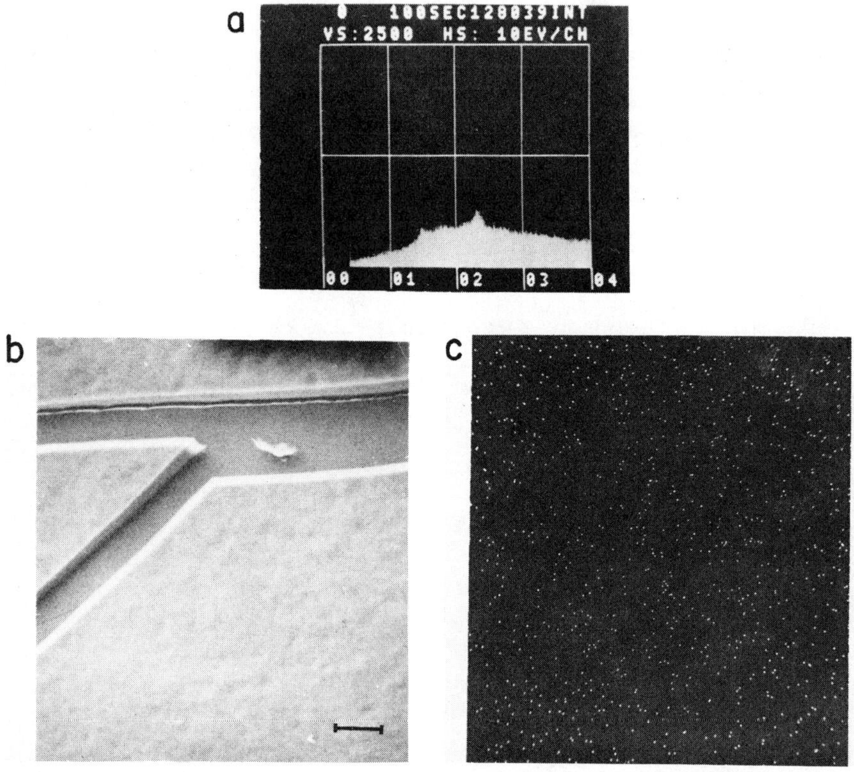

Calcified Tissue Research

Figure 4. Scanning electron microscopy and electron probe microanalysis data for blocked α-elastin incubated in a calcifying medium without calcium and phosphate (a) x-ray spectrum, (b) secondary electron image, and (c) calcium elemental map. The identification of the peaks in the x-ray spectrum (a) is as follows: 1.48 KeV peak is a scatter peak originating from the aluminum sample holder; 2.3 KeV peak is the $K_{\alpha 1}$ sulfur peak. Calcium and phosphorus x-rays were not detected. The vertical scale is from 0 to 250 cps. The secondary electron image (b) is at 850× magnification and demonstrates that the noncalcified elastin coacervate is a smooth film below which is visible the Plexiglas support in the cracked areas. Figure (c) is the calcium map of the area shown in Figure (b) with the EDAX set at 3,690 KeV energy (13).

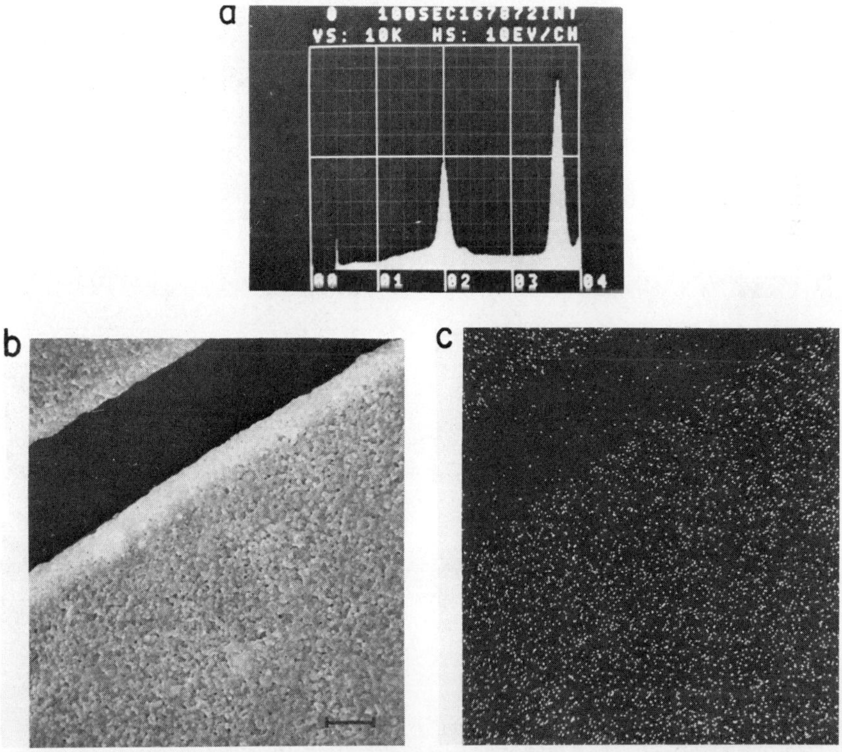

Calcified Tissue Research

Figure 5. This figure is arranged identically to Figure 4 and presents data obtained from blocked α-elastin coacervate which has been incubated in a calcifying medium with $CaCl_2$ and KH_2PO_4. The x-ray spectrum (a) contains a peak at 2.013 KeV corresponding to phosphorus $K_{\alpha 1}$ x-rays, a peak at 3.690 KeV corresponding to calcium K_α x-rays, and a small peak at 2.307 for sulfur. The vertical scale is from 0 to 10,000 cps. The secondary image is at 850× magnification showing that the calcified coacervate is very rough in appearance. Figure (c) is the calcium x-ray map which shows a dense population over areas of the field covered by coacervate. At 850× the calcium distribution appears uniform over the elastin matrix (13).

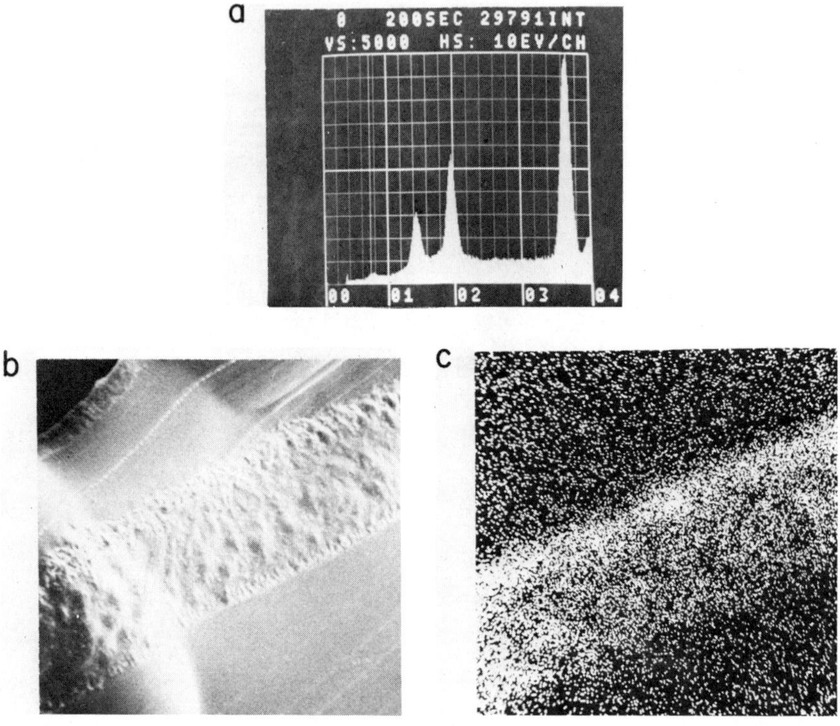

Calcified Tissue Research

Figure 6. Again the same arrangement as in Figures 4 and 5. In the x-ray spectrum (a), the peaks at 2.013 KeV and 3.690 KeV are from the phosphorus $K_{\alpha 1}$ *x-rays and calcium* K_{α} *x-rays respectively. The peak at 1.48 KeV is from the aluminum coating of the material. The secondary electron image is of cross-sectioned calcified elastin coacervate at 0.5μ thickness taken at 780× magnification. The depth of the coacervate calculates to be 10μ. The calcium x-ray map (c) shows that calcification extends the whole depth of the coacervate, although concentrated on the serum side (14).*

dissolved in trifluoroethanol, the ion is added in 1 μℓ aqueous aliquots. With water addition, a competition for the cation is set up between the peptide carbonyl oxygen and the water oxygen, each trying to attract the positively charged species. This competition can be used to an advantage to compare binding affinities among ions and peptides which have different stoichiometries and different water sensitivities. Figure 7b is a back-titration of both the calcium and magnesium peptide complexes with water to assess the strength with which the peptide carbonyl can secure the two cations. The midpoint of each complex's titration is indicated by an arrow. This point is interpreted to be the concentration of water where one half of the cation is bound to water and one half to the peptide carbonyl. Once the ratio of water concentration to peptide carbonyl concentration at this point is calculated it can be used to compare the binding affinities of the pentamer monomer relative to its high polymer or relative to the hexamer monomer and its high polymer. Table I summarizes these values:

TABLE I*

Peptide	Calcium Complex $\frac{[H_2O]}{[C=O]}$	Magnesium Complex $\frac{[H_2O]}{[C=O]}$
HCO-Val-Pro-Gly-Val-Gly-OMe	215	53
HCO-(Val-Pro-Gly-Val-Gly)$_n$-Val-OMe	350	97
HCO-Val-Ala-Pro-Gly-Val-Gly-OMe	270	66
HCO-(Val-Ala-Pro-Gly-Val-Gly)$_n$ Val-OMe	535	167

*Expanded from Reference 17

Figures 8a and 8b are comparable data for the hexamer repeat elastin peptide (18). In both its monomeric and polymeric form it chelated calcium and magnesium better than the pentamer repeat elastin peptide. Polymerization of both repeats also enhanced their binding potential. In numerical terms the calcium affinity constant for the hexamer monomer was about 0.5×10^6 at limiting peptide concentration (18). Since all of these synthetic peptides were neutral, the fact that they bound calcium tightly and selectively provides further support for neutral site binding.

NMR titrations of the synthetic elastin repeat peptide both prove that the calcium binding site is composed of peptide carbonyls and delineate which carbonyls actually are involved (11,18).

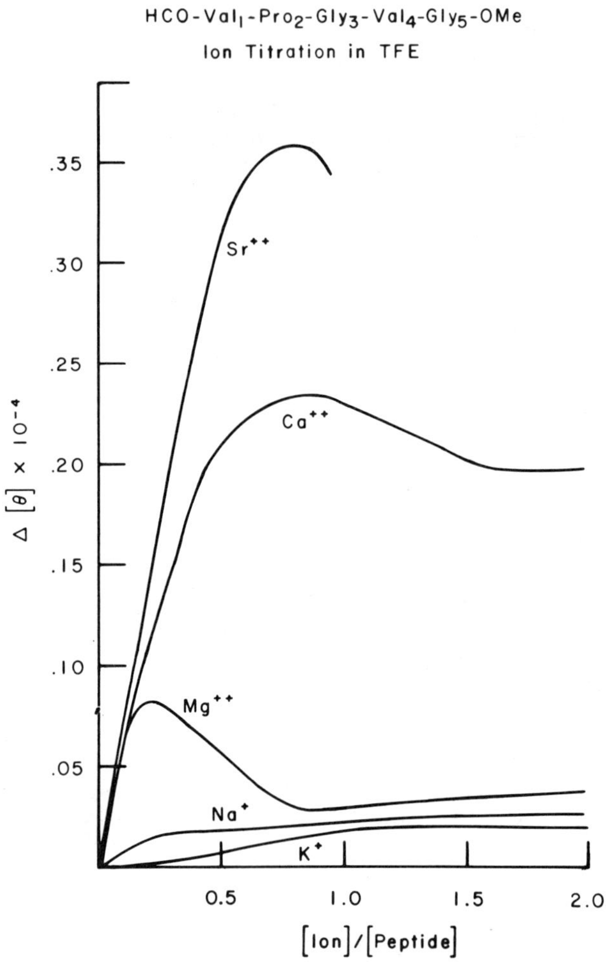

Calcium Binding Proteins and Calcium Function

Figure 7a. Ion titration of the repeat pentapeptide of elastin; cf. Figure 2 and text for discussion (17).

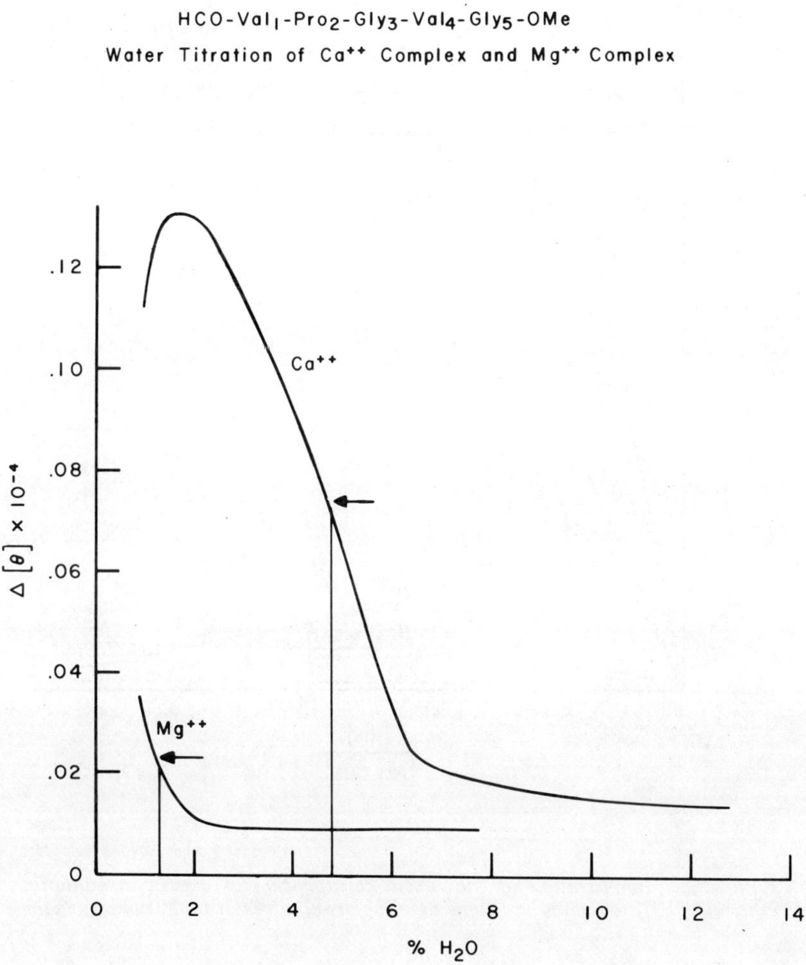

Figure 7b. Water titration of the calcium and magnesium complexes of the pentapeptide. See text for discussion (17).

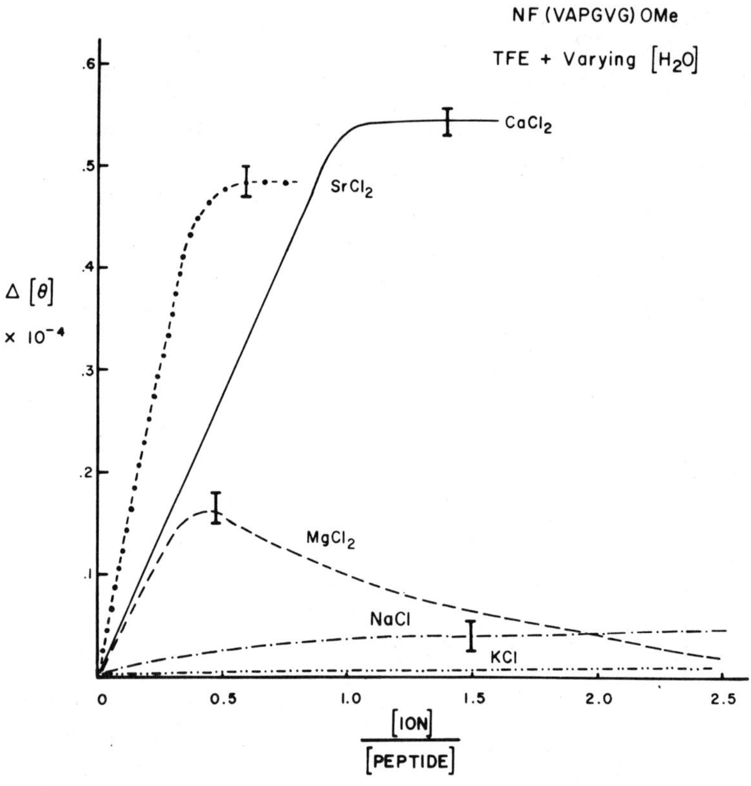

Archives of Biochemistry and Biophysics

Figure 8a. Ion titration of the repeat hexapeptide of elastin in trifluoroethanol (TFE) with ion additions as aqueous aliquots of 0.1M chloride salts (18).

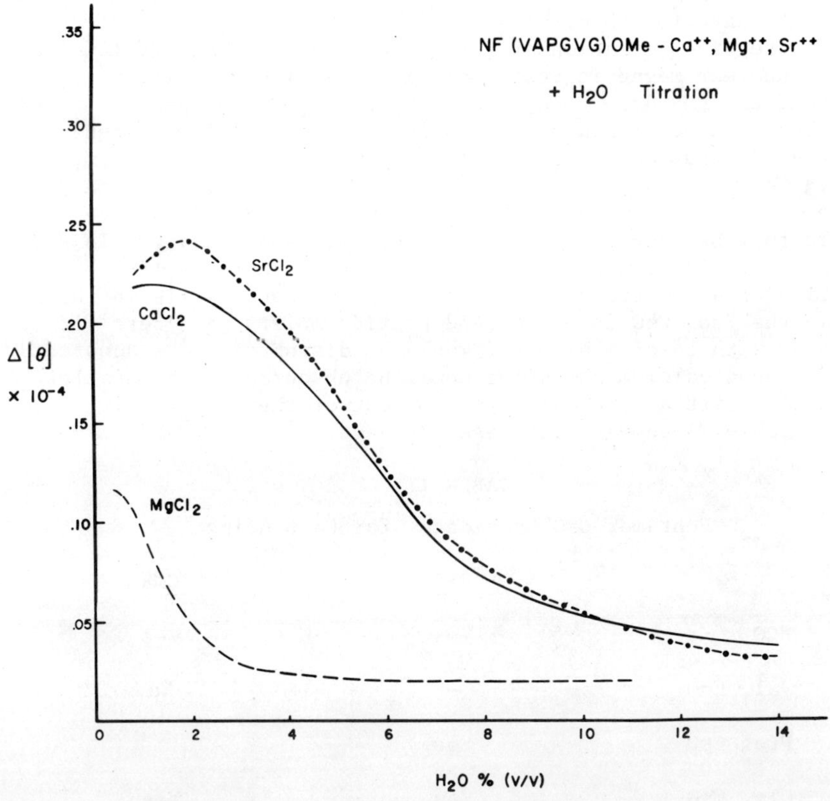

Archives of Biochemistry and Biophysics

Figure 8b. Water titration of the strontium, calcium, and magnesium ion complexes of the hexapeptide repeat (18).

PMR followed the calcium titration of the pentamer in trifluoroethanol with 3% water (Figure 9). The PMR signal, the chemical shift, is plotted as a function of increasing calcium concentration, with the peptide concentration held constant. Plotted are the protons of the formyl protecting group, Val_1 N<u>H</u>, Gly_3 N<u>H</u>, Val_4 N<u>H</u> and Gly_5 N<u>H</u> (<u>17</u>). Because of experimental conditions the valine and glycine NH were not delineated. The carbonyl of each respective peptide moiety interacted because all the resonances shifted downfield. These are the C=O of the Pro_2, the Gly_3 and the Val_4 residues. Since the Val_1 C=O precedes the proline and since the Gly_5 C=O is part of the methyl ester there are no peptide NH data for them.

Stepwise titration of the pentamer's carbonyls followed by C-13 nuclear magnetic resonance filled in the gaps in and confirmed the PMR data. In Figure 10 is plotted the peptide moiety carbonyl carbon resonances as a function of calcium concentration. There are four carbonyls which shifted downfield; the formyl, the Gly_3 C=O, the Val_4 C=O, and the Gly_5 C=O. The first three were predicted by the previous PMR data, while no PMR data were available for the last. The Pro_2 C=O curve is anomalous; while PMR indicated that the Pro_2 C=O did bind, here there is little indication of a downfield shift. The explanation may lie in the fact that the Pro_2 C=O is in the end peptide moiety of a beta turn which, with calcium binding, would be disrupted. The downfield shift upon calcium chelation could be obscurred by a parallel upfield shift as the carbonyl came out of the beta turn.

Table II summarizes these NMR data.

TABLE II

Pentamer C=O Candidates for Ca Binding

	PMR	CMR
HCO	Yes	Yes
Val_1 C=O	-	No
Pro_2 C=O	Yes	?
Gly_3 C=O	Yes	Yes
Val_4 C=O	Yes	Yes
Gly_5 C=O	-	Yes

All the carbonyls, but the Val_1, were involved in the calcium binding site of the pentamer. PMR and CMR titrations of the hexamer (<u>11</u>) repeat peptide indicate that again all the carbonyls of the HCO-Val_1-Ala_2-Pro_3-Gly_4-Val_5-Gly_6-OMe, but the Val_1, bound

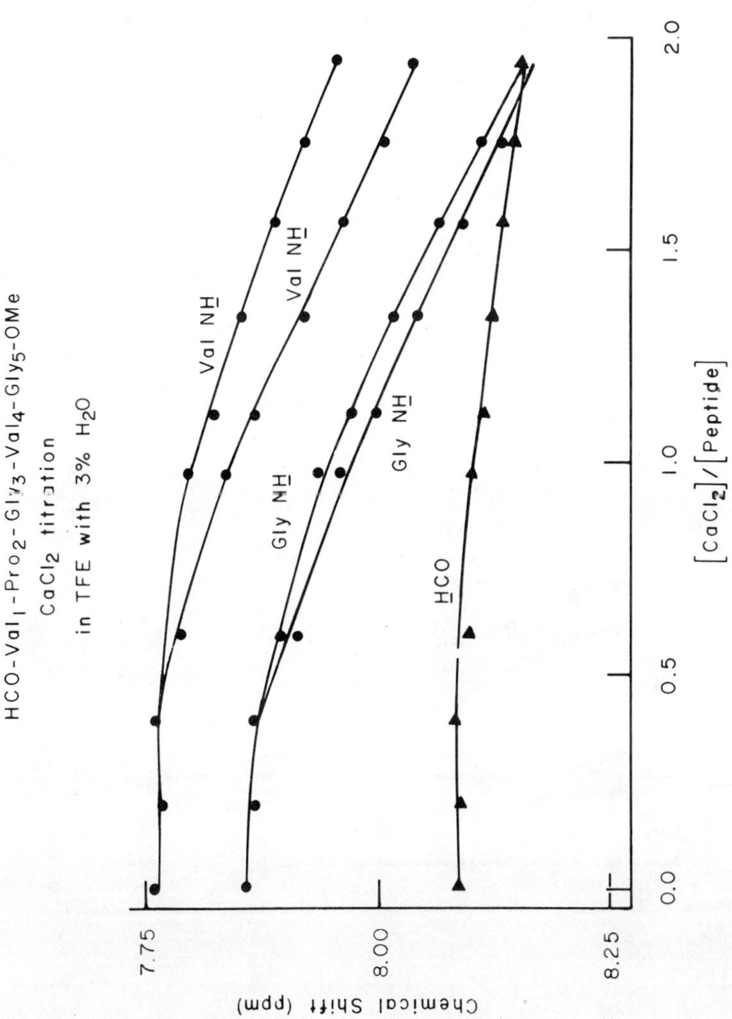

Figure 9. PMR titration data of $CaCl_2$ and the pentapeptide. The PMR signal, the chemical shift of each peptide NH, is plotted as a function of $[CaCl_2]$ which was added as a powder to the peptide in TFE (17).

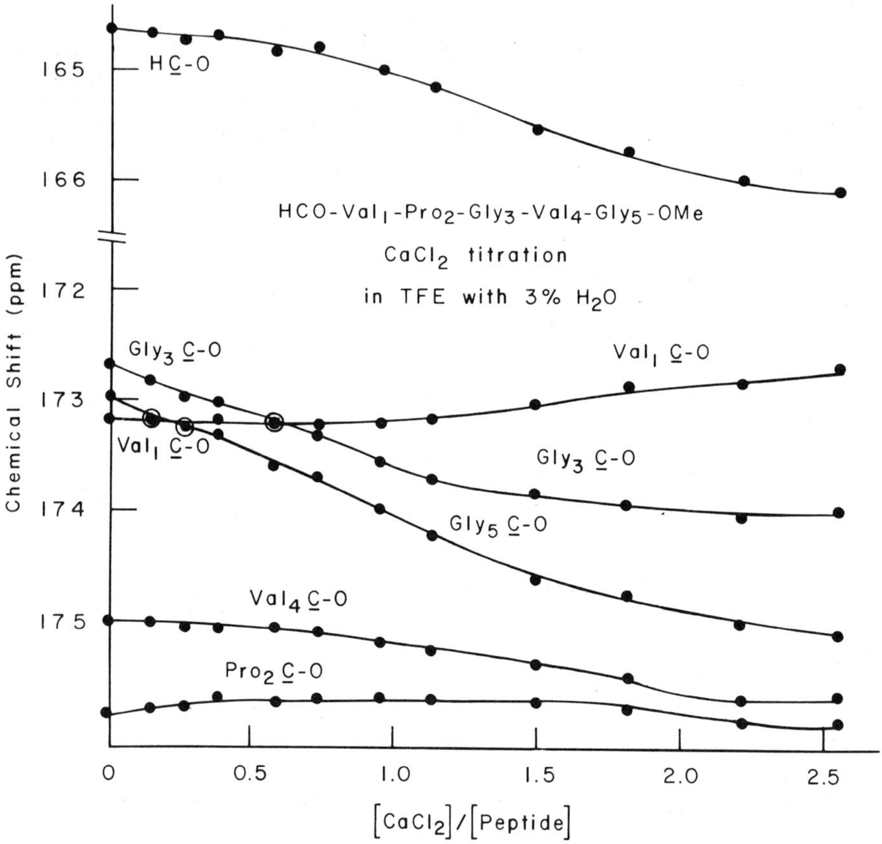

Calcium Binding Proteins and Calcium Function

Figure 10. CMR titration data as in Figure 9. Here the carbonyl chemical shift is plotted as a function of [$CaCl_2$] (17).

calcium. The major point from these studies is that the neutral elastin repeat peptides bind calcium via their carbonyls.

As with native α-elastin, the repeat elastin synthetic peptides also initiated calcification for when cross-linked, they formed insoluble matrices which served as substrates for hydroxyapatite growth (19). Figure 11 provides the scanning electron micrographs, Ca X-ray maps, and X-ray spectra of both the polypentapeptide (a,b,c) and polyhexapeptide (d,e,f) (20). Each was incubated with serum augmented with 3.0 mM exogenous $CaCl_2$ and KH_2PO_4 at 37°C for 72 hours. Figure 11b, which is the calcium X-ray map corresponding to the secondary electron image of Figure 11a, shows that the polypentapeptide calcified over its entire surface. In addition once cross-sectioned this calcified material has calcium and phosphate evenly distributed throughout its bulk (21). In contrast, the hexamer calcified only in discrete patches.

The information presented so far argue in favor of the neutral site/charge neutralization theory of initiation of elastin calcification. The evidence is so convincing that one could ask, since elastin and the repeat elastin peptides bind calcium and calcify readily _in vitro_, why does not elastin calcify just as spontaneously and rapidly _in vivo_? The fact that it does not points to the necessity of control mechanisms and/or protective barriers. Other _in vivo_ observations suggest the same. The extent of elastic fiber mineralization varies from tissue to tissue; the aorta and large arteries calcify well, while normal skin (with the notable exception of the skin of patients with Pseudoxanthoma elasticum) and lung calcify poorly (4). As mentioned before, elastic fiber calcification is a progressive disease, as if with time some intrinsic defense is breached.

Microscopic examination of the elastic fiber shows that it is composed of two structurally different components, a central area which is elastin protein and a peripheral sheath of microfibrils which in cross section appear tubular and of 10-12 nm in diameter (22-29). During the embryological development of elastic tissue, the microfibrils appear first in the extracellur matrix (22). In tissue culture, for example, after seven days the microfilaments are observable followed by the elastin by 14 and 21 days (30). At this stage the elastic fiber has attained roughly its mature EM appearance of a circular structure with a clear center and electron dense border. Once mature, the elastic fiber, especially the human fiber, is 10% microfibrils and 90% elastin (26,31). The proportion of microfibril to elastin varies with age though, from prenatal to mature, to senescent fibers (26,32,33). Aged elastic fibers appear to be deficient in peripheral microfibrils (26).

Ross and Bornstein (22), followed by Robert et al. (25) provided the chemical differentiation and characterization of the microfibrils vis à vis the elastin protein core. Using chemical isolation techniques and monitoring the reactions with electron

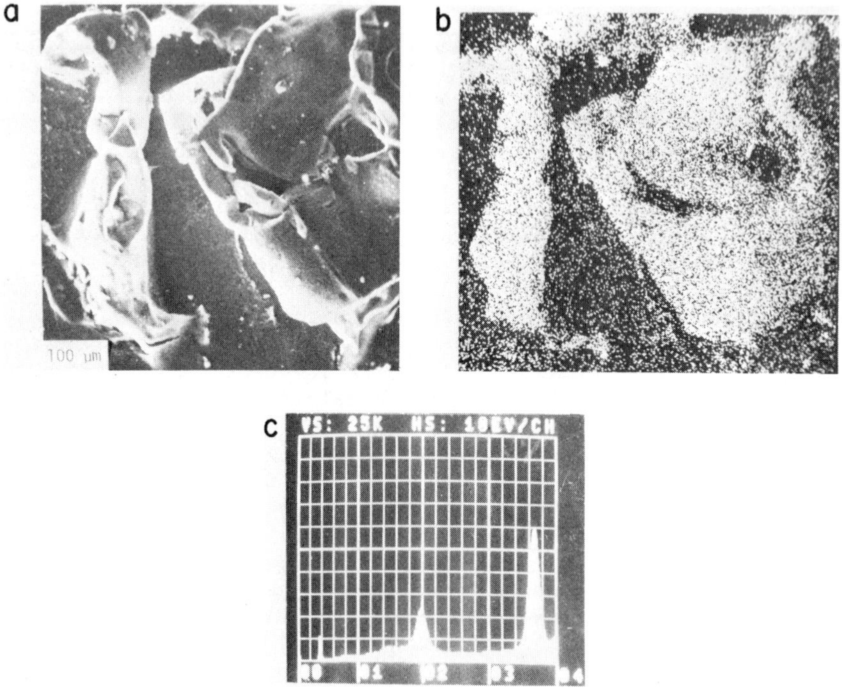

Fifth American Peptide Symposium

Figure 11. Figure 11a is an SEM photomicrograph of the cross-linked polypentapeptide which has been calcified in serum augmented with exogenous 3.0mM $CaCl_2$ and KH_2PO_4. Figure 11c is the x-ray elemental spectrum indicating the presence of calcium at 3.7 KeV, phosphorus at 2.0 KeV, and aluminum (from the coating) at 1.4 KeV in the sample. Figure 11b is the calcium x-ray map demonstrating that calcium phosphate clearly deposited on the synthetic elastin framework. Figures 11d, e, and f (right) *are the same as a, b and c except for the cross-linked polyhexapeptide* (20).

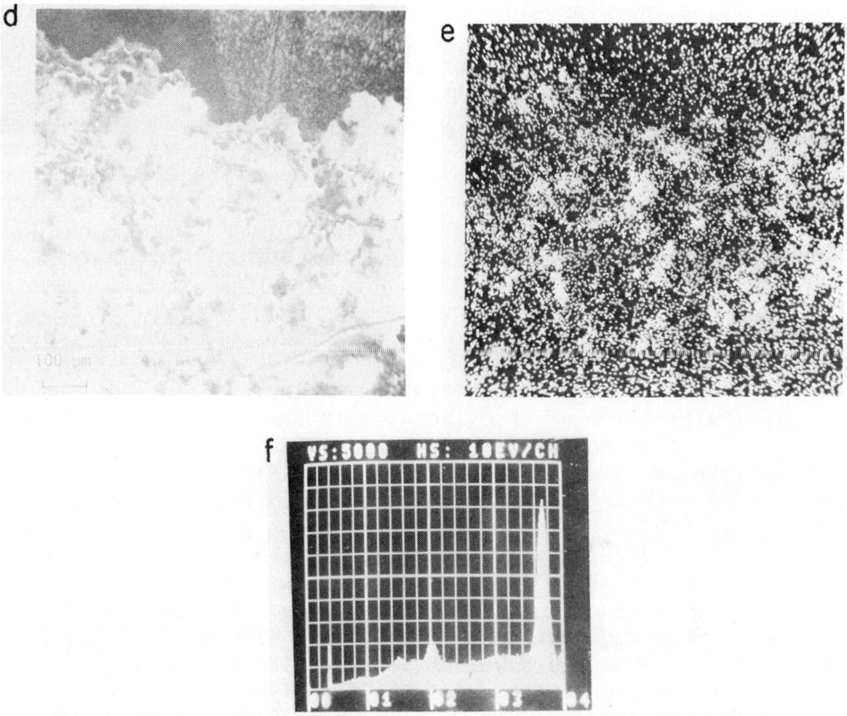
d
100 µm
e
f
VS:5000 HS: 10EV/CH

microscopy, Ross and Bornstein demonstrated that the elastic fiber microfilaments were digested by proteolytic enzymes including trypsin, chymotrypsin and pepsin which did not touch elastin. In contrast, hyaluronidase and β-glucuronidase were without effect. Analysis of the amino acid composition of the microfibrils solubilized by proteolytic enzymes, revealed a strikingly different distribution from that seen in elastin. They contained more polar and acidic amino acids, more sulfur presumably as cystine, and less neutral residues such as glycine, valine, alanine and proline. They lacked hydroxyproline, hydroxylysine and desmosines, clearly distinguishing them from both elastin and collagen. The high cystine content, 70-80 residues/1000 residues, was thought to represent numerous disulfide linkages which rendered the microfibrils relatively insoluble (26), and also provided a non enzymatic way to solubilize the sheath. The investigators found that dithiothreitol, a disulfide bond reducing agent, in 5M guanidine quantitatively released the microfibrils from the elastic fiber, and so provided another sample whose amino acid composition was that of the enzymatically cleaved material. Numerous sugar moieties were also found indicating the material was made up of one or more glycoproteins. The fibrils were 4.7% by weight hexose as measured with the anthrone reagent and 0.7% hexosamine as determined by ion exchange chromatography (22). The 1976 paper by Muir, Bornstein and Ross (34) describes the isolation of one of the presumptive proteins of the glycoprotein sheath and shows that it is of large molecular weight, i.e. 270,000. With this size and the amino acid and carbohydrate composition in mind, one could suggest that the glycoprotein sheath inhibits premature calcification by forming a shield around the protein core of the elastic fiber, screening it from calcium ions.

To test this idea, ideally one would wish to investigate the *in vitro* interaction of elastin and its glycoprotein(s) and in a calcifying medium, to see if calcification is inhibited. To date, these experiments have not been carried out, due in great part, to the difficulty in obtaining the glycoprotein. Several studies have been done, though, following the interaction of elastin (35) and tropoelastin (36) with more readily obtainable bovine nasal cartilage proteoglycan. Significant interaction was found by Radhakrishnamurthy, Ruiz and Berenson (37) with heparan sulfate and dermatan sulfate. In contrast, Wusteman and Gillard (38) demonstrated that mature fibers of elastic tissue from elastic cartilage had little specific association with hyaluronic acid.

Our laboratory studied these interactions with α-elastin and then extended this work to begin to define the sites on the elastin macromolecule which could be specific for interaction. This was done by investigating the interactive properties of each of the repeat elastin peptide high polymers synthesized here. Bovine nasal cartilage, isolated by Baker with a slightly

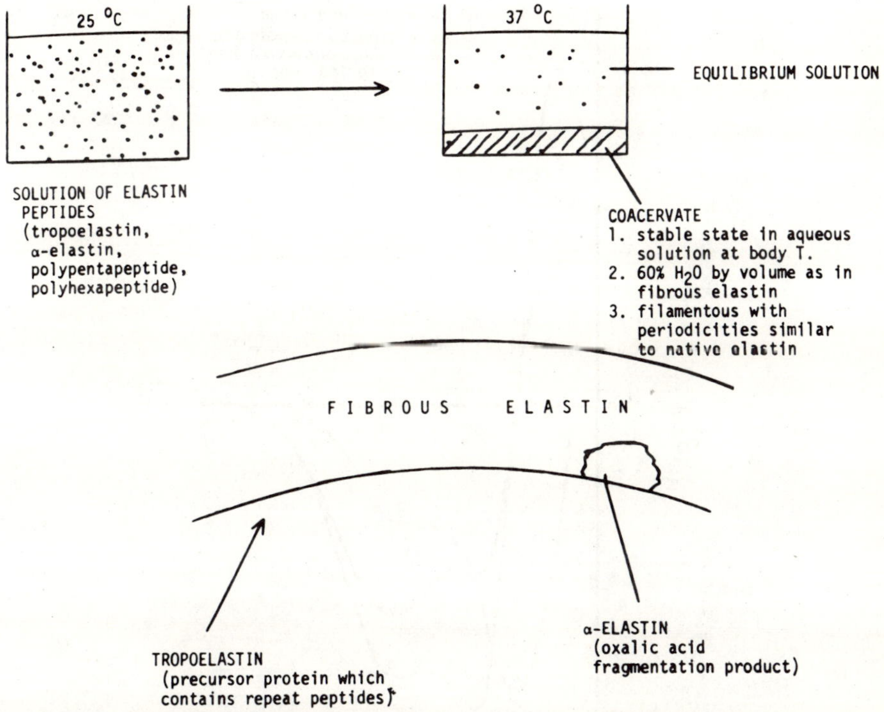

Figure 12. Diagram of the process of coacervation. Blocked α-elastin, tropoelastin, and polymers of the repeat peptides of elastin are soluble in aqueous solutions at low temperature. As the temperature increases, the solution becomes cloudy and with time a phase separation occurs with a yellow, viscous glue-like precipitate forming below an equilibrium solution. The temperature profile of coacervation follows this process by measuring turbidity as a function of temperature as depicted in Figures 15 and 16 (47).

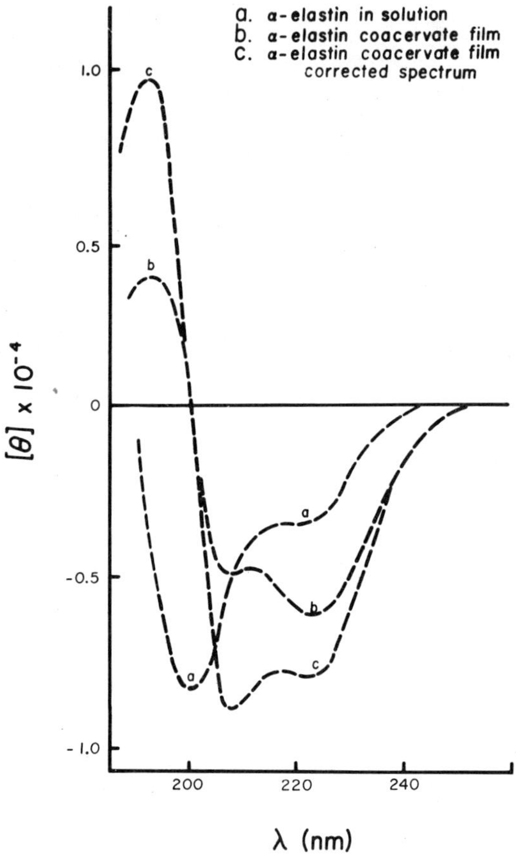

International Journal of Quantum Chemistry

Figure 13. CD spectra of α-elastin in solution (a) in 0.01M sodium acetate, pH=5, and in its coacervated state (b). Since the coacervate formed a light scattering film, curve b was corrected, giving curve c (41).

Biochimica et Biophysica Acta

Figure 14. (a) High resolution electron micrographs of tropoelastin coacervate which has been negatively stained, showing parallel aligned 5-nm filaments. As temperature increases, tropoelastin molecules go from a random distribution to the parallel array seen here, showing that elevated temperature enhances fiber formation (42).

Perspective in Biology and Medicine

(b) Electron micrograph of polypentapeptide coacervate negatively stained to bring out the fibrillar nature of the material. Twisted rope-like structures are observable (cf. left-handed structure at the arrow) (3).

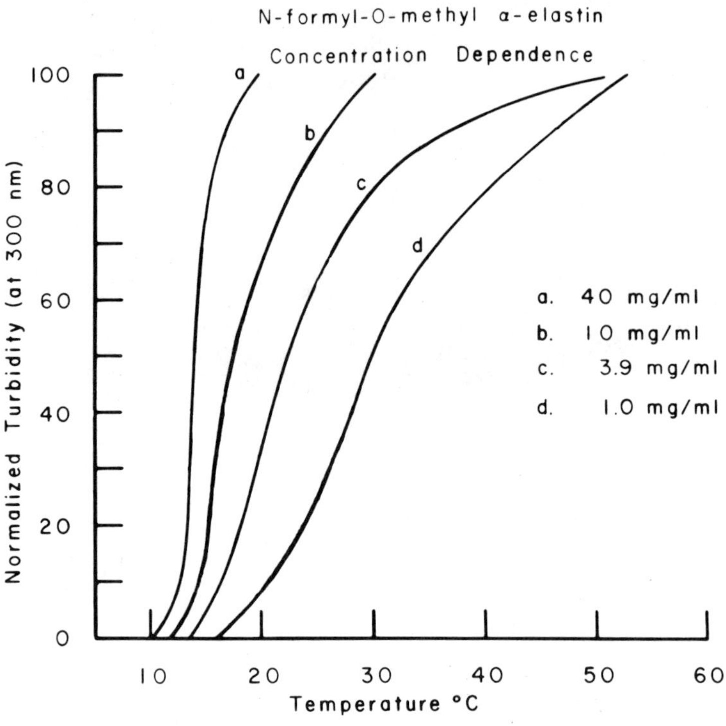

Perspective in Biology and Medicine

Figure 15. Blocked α-elastin temperature profiles of coacervation (cf. Figure 12) with varying concentrations of protein. That intermolecular interactions are responsible for coacervation is seen by its concentration dependency. That the process is cooperative is indicated by the increased steepness of the curve with increasing concentration, that is when two α-elastin molecules associate, the resultant conformation facilitates further association (3).

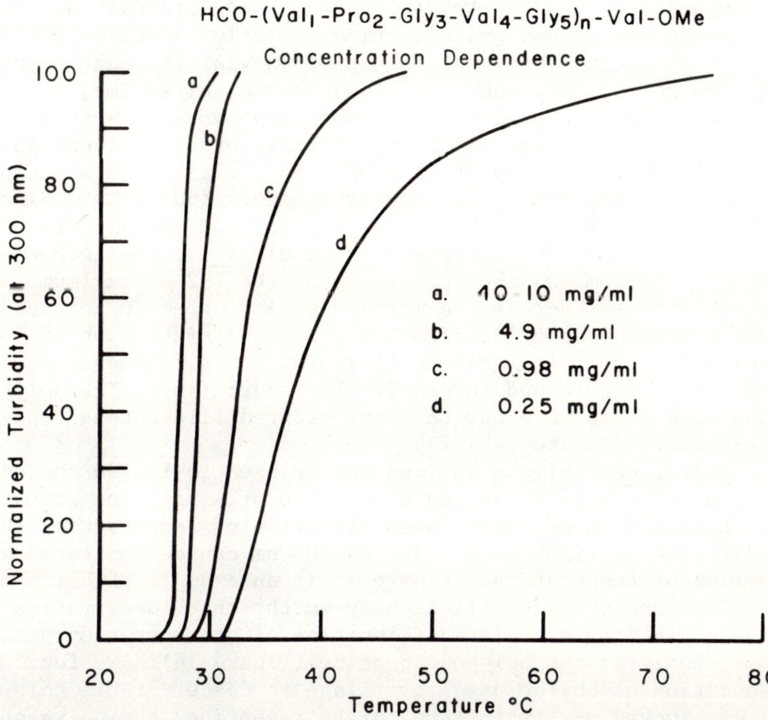

Perspective in Biology and Medicine

Figure 16. Polypentapeptide temperature profiles of coacervation with varying concentration of peptide. As with blocked α-elastin in Figure 15, the increased steepness with increased concentration indicates cooperativity (3).

modified technique of Hascall and Sajdera (39), and blocked α-elastin, the polytetrapeptide, the polypentapeptide, and the polyhexapeptide were mixed together and their temperature profile of coacervation recorded.

At this point, a brief description of elastin coacervation and a method to follow it is warranted. Figure 12 outlines the process of coacervation (40-41). Both tropoelastin, which is the precursor protein of fibrous elastin, and α-elastin which is an oxalic acid fragmentation product derived from the fiber, coacervate as do the synthetic repeat elastin peptide polymers. When aqueous solutions of these proteins and polypeptides go from room temperature to elevated temperature, for example from 25°C to body temperature at 37°C, they undergo a phase transition. The solutions become cloudy and with time, a yellow gluelike precipitate settles to the bottom of the reaction vial for each species except the polyhexapeptides. Its coacervate, with time, settles to a white, flocular precipitate. The coacervate is a state of increased intramolecular and intermolecular order. Figure 13 is the circular dichroism spectra of blocked α-elastin in solution at 25°C (curve a), and in its coacervate form (curve b). Since the coacervate forms a light scattering film on the spectroscopic cell, its spectrum was corrected (curve c) (42). The increase in the 222 nm minimum and the appearance of the 192 nm maximum correlate with an increase in intramolecular order to between 20% and 25% α-helix (43). The negatively stained micrograph of the tropoelastin coacervate (Figure 14a) proves its filamentous nature, one of increased intermolecular order (44). The polypentapeptide coacervate has the same ordered filamentous rope-like structure (Figure 14b) (3).

A rapid, sensitive technique can be used to follow the coacervation process (45). Since a solution of elastin or its repeat peptides turns cloudy with elevation of temperature, the turbidity due to light scattering at 300 nm can be monitored as a function of temperature. Figure 15 is an example of the nature of the data one obtains. It is note-worthy that coacervation of blocked α-elastin is a distinct function of concentration. This is true, too, for the polypentapeptide (Figure 16). We found that addition of bovine nasal cartilage to coacervatable solutions of blocked α-elastin, the polytetrapeptide, the polypentapeptide, or two different preparations of the polyhexapeptide resulted in changes in their temperature profile of coacervate mirroring changes in concentration observed with the free solutions. These observations are summarized in Table III.

TABLE III

Proteoglycan Interaction with Elastin and Elastin Repeat Peptides*

Blocked α-Elastin	+
$H(VPGG)_nVOMe$	-
$NF(VPGVG)_nVOMe$	-
$H(APGVGV)_nVOMe$	+
$H(APG\phi GV)_nVOMe^1$	+

*Reaction conditions were as follows: [proteoglycan] = 5 mg/ml, [Blocked α-elastin] = 5 mg/ml, [$H(VPGG)_nVOMe$] = 1.0 mg/ml, [$NF(VPGVG)_nVOMe$] = 0.5 mg/ml and 0.25 mg/ml. [$H(APGVGV)_nVOMe$] = 1 mg/ml and 0.5 mg/ml and [$H(APG\phi GV)_nVOMe$] = 1 mg/ml and 4 mg/ml in aqueous solutions. The proteoglycan was added in weighed amounts of dried material directly to the elastin solution in the spectrophotometer cell which was continuously vibrated to insure proper mixing as the complex coacervation proceeded. 1: ϕ is Val or Lys.

ACKNOWLEDGMENT: This work was supported in part by National Institutes of Health, Grant No. HL-11310. The authors thank Dr. W. D. Thompson for assistance with the proteoglycan-peptide temperature profiles of coacervation.

LITERATURE CITED:

1. Ross, R., Bornstein, P., J. Cell. Biol. 40, 366, 1969.
2. Partridge, S. M., Gerontologia, 15, 85, 1969.
3. Urry, D. W., Persp. Biol. Med., 21, 265, 1978.
4. Yu, S. Y., Blumenthal, H. T., in The Connective Tissue, ed. by B. M. Wagner and D. E. Smith, Williams and Williams Co., Baltimore, 1967, pp. 17-49.
5. Kramsch, D. M., Franzblau, C., Hollander, W., Adv. Exp. Med. Biol., 43, 193, 1974.
6. Yu, S. V., Blumenthal, H. T., J. Gerontol., 18, 119, 1963.
7. Martin, G. R., Schiffman, E., Bladen, H. A., Nylen, M., J. Cell. Biol., 16, 243, 1963.
8. Kagan, H. M., Lerch, R. M., Biochim. Biophys. Acta, 436, 223, 1976.
9. Urry, D. W., Proc. Natl. Acad. Sci., U.S.A., 68, 810, 1971.
10. Long, M. M., Urry, D. W., Starcher, B. C., Biochem. Biophys. Res. Commun., 56, 14, 1974.
11. Urry, D. W., Ohnishi, T., Bioinorg. Chem., 3, 305, 1974.
12. Cox, B. A., Starcher, B. C., Urry, D. W., Biochim. Biophys. Acta, 317, 209, 1973.
13. Cox, B. A., Starcher, B. C., Urry, D. W., Calcif. Tissue Res., 17, 219, 1975.

14. Urry, D. W., Hendrix, C. F., Long, M. M., Calcif. Tissue Res., 21, 57, 1976.
15. Foster, L. A., Bruenger, E., Gray, W. R., Sandberg, L. B., J. Biol. Chem., 248, 2876, 1973.
16. Gray, W. R., Sandberg, L. B., Foster, J. A., Nature (London), 246, 461, 1973.
17. Long, M. M., Urry, D. W., Thompson, W. D., in Calcium Binding Proteins and Calcium Function, ed. Wasserman, R. H. et al., Elsevier North-Holland, N.Y., 1977, p. 419.
18. Long, M. M., Ohnishi, T., Urry, D. W., Arch. Biochem. Biophys., 166, 187, 1975.
19. Urry, D. W., Long, M. M., Okamoto, K., Volpin, D., Roveri, N., Ripamonte, A., Ala. J. Med. Sci., 14, 255, 1977.
20. Rapaka, R. S., Okamoto, K., Long, M. M., Urry, D. W., Fifth American Peptide Symposium, University of California, San Diego, La Jolla, California, 1977.
21. Urry, D. W., Long, M. M., Hendrix, C. F., Okamoto, K., Biochemistry, 15, 4089, 1976.
22. Ross, R., Bornstein, P., J. Cell. Biol., 40, 366, 1969.
23. Ross, R., Bornstein, P., in Chemistry and Molecular Biology of the Intercellular Matrix, ed. Balazs, E. A., Academic Press, London and New York, 1970, p. 641.
24. Ross, R., J. Cell. Biol., 50, 172, 1971.
25. Robert, B., Szigetti, M., Derouette, J. C., Robert, L., Bouissou, H., Fabre, M. T., Eur. J. Biochem., 21, 507, 1971.
26. Ross, R., J. Histochem. and Cytochem., 21, 199, 1973.
27. Robert, L., Kadar, A., Robert, B., Adv. Exp. Med. Biol., 43, 85, 1974.
28. Robert, L., Biol. Med. (Paris), 4, 1, 1975.
29. Ross, R., Philos. Trans. R. Soc. Lond. (Biol), 271, 247, 1975.
30. Daoud, A. S., Fritz, K. E., Jarmolych, J., Augustyn, J., Mawhinney, T. P., Adv. Exp. Med. Biol., 82, 928, 1977.
31. McCullagh, K. G., Derouette, S., Robert, L., Exp. Mol. Pathol., 18, 202, 1974.
32. DeBacker, M., Pizieux, O., Moschetto, Y., Paroi Arterielle, 3, 71, 1975.
33. Giro, M. G., Castellani, I., Volpin, D., Connect. Tissue Res., 2, 231, 1974.
34. Muir, L. W., Bornstein, P., Ross, R., Eur. J. Biochem., 64, 105, 1976.
35. Podrazky, V., Adam, M., Experientia, 31, 523, 1975.
36. Adam, M., Podrazky, V., Experientia, 32, 430, 1976.
37. Radhakrishnamurthy, B., Ruizitt, and Berenson, G. S., Adv. Exp. Med. Biol., 82, 160, 1977.
38. Wusteman, F. S., Gillard, G. C., Experientia, 33, 721, 1977.
39. Hascall, V. C., Sajdera, S. W., J. Biol. Chem., 244, 2384, 1969.

40. Urry, D. W., Faraday Disc. Chem. Soc., No. 61, 205, 1976.
41. Urry, D. W., Long, M. M. and Mitchell, L. W., Int. J. Quantum Chem.: Quantum Biology Symp. No. 3, 107, 1976.
42. Urry, D. W., Biochim. Biophys. Acta, 265, 115, 1972.
43. Starcher, B. C., Saccomani, G., Urry, D. W., Biochim. Biophys. Acta, 310, 481, 1973.
44. Cox, B. A., Starcher, B. C., Urry, D. W., J. Biol. Chem., 249, 997, 1974.
45. Urry, D. W., Long, M. M., Cox, B. A., Ohnishi, T., Mitchell, L. W., Jacobs, M., Biochim. Biophys. Acta, 371, 597, 1974.
46. Urry, D. W., Res./Develop., 24, 30, 1973.
47. Urry, D. W., Long, M. M., Mitchell, L. W., Int. J. Quantum Chem.: Quantum Biol. Symp., No. 3, p. 107, 1976.

RECEIVED December 29, 1977.

14

Membrane Glycoproteins—Dynamics and Affinity Isolation

REUBEN LOTAN and GARTH L. NICOLSON

Department of Developmental and Cell Biology, University of California, Irvine, CA 92717

Although the exact structure and organization of biological membranes has not been determined in detail, a fair amount of information has been assembled indicating that cellular membranes generally conform to several basic principles. The first of these is that the bulk membrane lipids such as the phospholipids are arranged in a planar bilayer configuration which is in a predominantly fluid state under physiological conditions (1-4). The lipid bilayer is interrupted at certain sites by tightly bound integral (or intrinsic) membrane proteins and glycoproteins which are inserted to varying degrees into the bilayer, and these proteins and glycoproteins are capable of rapid lateral movement within the fluid planar lipid matrix. Cellular membranes are asymmetric with respect to the distribution of specific lipids in each half of the bilayer and proteins and glycoproteins at either surface, and plasma membranes often have non-uniform distributions of proteins, glycoproteins, lipids and glycolipids in the membrane plane (5). In addition, most plasma membranes are not autonomous structures; they are linked to other cellular organelles by a cytoskeletal system composed of microfilaments, microtubules and perhaps intermediate filaments (Fig. 1). However, in its simplest form a biological membrane can be viewed as a two-dimensional solution of a mosaic of integral membrane proteins and glycoproteins embedded in a fluid lipid bilayer with peripheral (loosely bound) proteins and glycoproteins attached at the inner and outer membrane surfaces, respectively. The striking asymmetry of this structure allows receptors for hormones, antibodies, viruses, lectins and other agents to be present exclusively on the outer surface where they are exposed to the extracellular environment. Asymmetric arrangement of these components is also well suited to allow the vectorial flow of information across the membrane. The other most important feature of this type of molecular arrangement is that components can diffuse laterally within the membrane plane permitting rapid and reversible changes in membrane topography. Measurements of the lateral mobility of membrane glycoproteins indicate that some diffuse freely in the membrane while others are

0-8412-0452-7/78/47-080-256$05.00/0

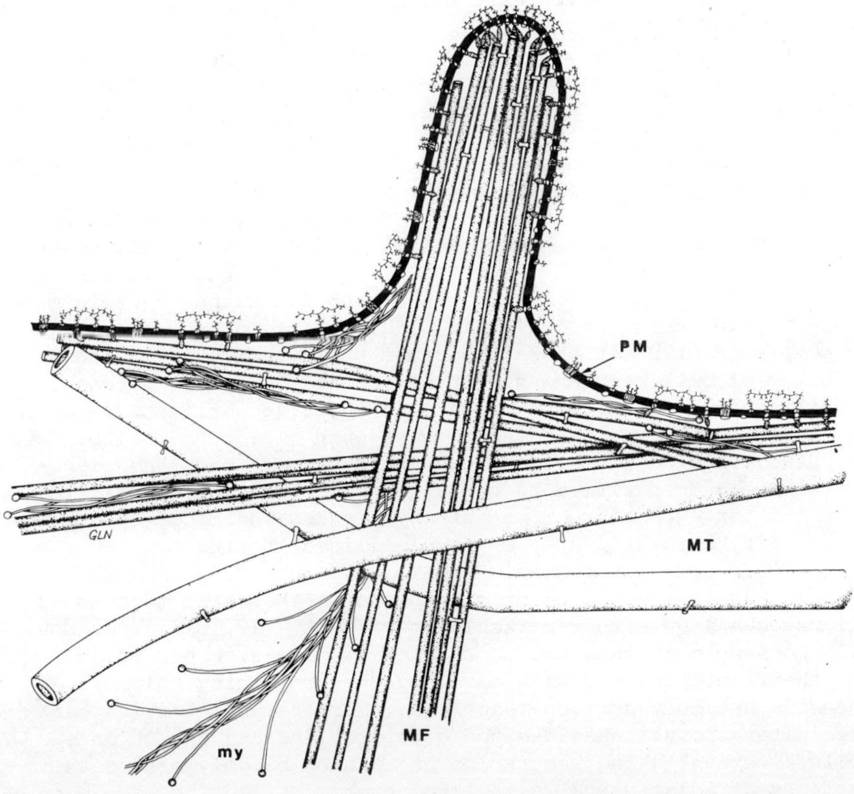

Cell Surface Reviews

Figure 1. Hypothetical structure of a plasma membrane (PM) including possible interactions between membrane-associated microtubule (MT) and microfilament (MF) systems involved in trans-membrane control over cell surface receptor mobility and distribution (5)

restrained and are unable to undergo rapid lateral movements (4,5). This graded hierarchy of mobilities for different components in the plasma membrane and even within certain specific regions of the plasma membrane suggests that a cell can control the mobilities of certain membrane components in order to maintain certain specific topographic arrays or "patterns" of cell surface receptors that are probably important in identification and a variety of other cellular phenomena requiring trans-membrane communication. Since these topographic arrangements are dynamic and not static, rapid and reversible changes in membrane topography could occur in response to both intra- and extracellular stimuli.

MEMBRANE GLYCOPROTEINS

The most important class of cell surface recpetors involved in trans-membrane receptor processes are the plasma membrane glycoproteins. Most of these appear to be integral components which are intercalated into the hydrophobic plasma membrane core to depths dependent on the amounts of hydrophobic surface revealed by three-dimensional folding of the molecules into their "native" structures. Some of these integral membrane glycoproteins span the plasma membrane and have hydrophilic regions of their structures protruding at both the extracellular and intracellular membrane surfaces (6-10). Also, it has been found that certain membrane glycoproteins can exist in oligomeric complexes (11-13) or in complexes with peripheral membrane proteins (14-16). The majority of cell membrane glycoproteins are integral components stabilized by hydrophobic forces, and their solubilization requires the use of chaotropic agents or detergents (1). Once solubilized and stabilized in buffered detergent solutions, membrane glycoproteins can be purified by conventional methods of gel filtration and ion exchange chromatography (in non-ionic detergents) or by affinity chromatography on insolubilized lectins.

LECTIN AFFINITY CHROMATOGRAPHY

Lectins are proteins or glycoproteins that bind to mono- and oligosaccharides with remarkable specificity (17,18). They usually possess more than one saccharide binding site per molecule, and their interactions with carbohydrate-containing polymers resemble antibody-antigen reactions. For example, lectins can precipitate polysaccharides and glycoproteins and agglutinate cells by crosslinking saccharide-containing biopolymers on the surfaces of adjacent cells. A large number of lectins have been isolated and purified, and in Figure 2 we have listed some common lectins available that bind certain oligosaccharides found in soluble glycoproteins (19-23) and cell membrane glycoproteins (24-28). Binding of lectins to glycoproteins or even cells can be inhibited and in many cases reversed by use of specific simple sugars or glycosides, and this information can be used to elucidate possible carbohydrate determinants on cell surface components (Figure 2). However, lectin binding to macromolecules carrying hetero-oligosaccharide side chains can be influenced by a variety of factors which cannot be adequately reproduced in experiments where simple monosaccharide-lectin interactions are measured. Nevertheless, the binding of a lectin in a specific and reversible manner to a macromolecule is generally accepted as an indication that the molecule in question contains certain carbohydrate sequences, but the unequivocal characterization of carbohydrate sequence in a lectin receptor requires its isolation in a pure form from the cell membrane followed by chemical elucidation of its structure.

Since most membrane glycoproteins are almost insoluble in neutral aqueous solutions, proteolytic enzymes have been used to cleave and remove soluble surface glycopeptides from cells. Such glycopeptides were then tested for their ability to inhibit cell agglutination, lymphocyte stimulation or binding of lectins to cells (29-32). The chemical structure of a few lectin-reactive glycopeptides has been determined (31-32). However, since the behavior of membrane glycoproteins depends not only on their oligosaccharide side chains but also on interaction of the protein moiety with other membrane components, studies on intact molecules rather than investigations of fragmented portions of the membrane components are required. For this purpose techniques have been developed to isolate and purify membrane glycoproteins in buffered detergent solutions by lectin affinity chromatography. Using a wide variety of insolubilized lectins, affinity chromatography has proven to be very useful in purifying an increasing number of membrane glycoproteins (Table 1).

Lectin Affinity Purification of RBC Glycoproteins

The erythrocyte membrane has been intensively investigated in an attempt to understand the structure and function of its components (33). As a result of these efforts the human red blood cell (HuRBC) membrane has proven to be a useful model for the development of methods for isolation of lectin receptors. Indeed, several laboratories have succeeded in purifying HuRBC membrane components to homogeneity using lectin affinity chromatography techniques (Table 1). Findlay (34) applied solubilized RBC ghost membrane glycoproteins onto Con A-Sepharose or LCA-Sepharose, and after the unbound material was removed by washing, the addition of methyl α-mannoside to Con A-Sepharose detached a glycoprotein which was identified as the major integral glycoprotein of HuRBC -- the component migrating on gel electrophoresis as Band III, a glycoprotein having a mol. wt. of ∿100,000 that contains 5-8% carbohydrate. Use of the LCA-Sepharose column yielded the major sialoglycoprotein, glycophorin, which contains nearly 60% carbohydrate (24). Band III has also been purified on Con A-Sepharose by Ross and McConnell (35) who combined it with egg lecithin, erythrocyte lipids, cholesterol and glycophorin to obtain vesicles which were capable of sulfate transport. Two distinct classes of HuRBC membrane glycoproteins with differing carbohydrate compositions and lectin reactivities have been separated by Adair and Kornfeld (36) using affinity chromatography on WGA- and RCA_I-Sepharose. The WGA-Sepharose column bound and specifically released a single

Table 1: ISOLATION AND CHARACTERIZATION OF MEMBRANE GLYCOPROTEINS USING IMMOBILIZED LECTINS

Material Purified	Source of Membrane	Membrane Solubilizing Agent	Affinity Adsorbent	Ref.
Band III	HuRBC	TX-100	Con A-Seph	34
Band III + glycophorin	HuRBC	TX-100	LCA-Seph	34
Band III	HuRBC	DTAB	Con A-Seph	35
Band III + 2 other glycoproteins	HuRBC	TX-100	RCA_I-Seph	36
Glycophorin	HuRBC	TX-100	WGA-Seph	36
Glycophorin A	HuRBC	SDS	WGA-Seph	37
Asialoglycophorin (T-Ag)	HuRBC + NANase	DDG (Empigen BB)	PNA-Seph	38
Major glycoprotein	Bovine RBC	Emulphogene	RCA_I-Seph	40
Glycoproteins	Pig lymphocytes	NaDOC	Con A-Seph	41
Glycoproteins	Pig lymphocytes	NaDOC	LCA-Seph	42
HLA antigens	Hu lymphocytes	NaDOC	LCA-Seph	43
Component 5.1	Ra thymocytes	TX-100	Con A-Seph	44
T-25, T-200	Mu thymocytes	NP-40, NaDOC	Con A, PA-Seph	45
Major antigen	Rat thymocytes	NaDOC	LCA-Seph	46
H-2K, H-2D, Ia	Mu thymocytes	NP-40	Con A-Seph	47
IgM, IgD	Mu B lymphocytes	NP-40	Con A-Seph	47
Glycoproteins:				
MW 20,000-80,000	Mu lymphocytes	TX-100	PA-, LCA, RCA_I-Seph	48
MW >100,000	Mu lymphocytes	TX-100	WGA, BPA-Seph	48
IgM, IgD	Mu B lymphocytes	TX-100	PA-Seph	48
H-2K, H-2D, Ia	Mu lymphocytes	TX-100	PA-Seph	48
Synaptic glycoproteins	Rat brain	NaDOC	LCA-, WGA-Seph	49
Synaptic glycoproteins	Rat brain	SDS	Con A-, UEA-Seph	50,51
Nicotinic acetylcholine receptor	Rat brain or electric eel	TX-100	Con A, WGA-Seph, RCA_I-glass beads	52

Table 1 (con't.): ISOLATION AND CHARACTERIZATION OF MEMBRANE GLYCOPROTEINS USING IMMOBILIZED LECTINS

Material Purified	Source of Membrane	Membrane Solubilizing Agent	Affinity Adsorbent	Ref.
Tissue factor	Bovine brain	NaDOC	Con A-Seph	53
Rhodopsin	Bovine retina	CTAB	Con A-Seph	54
Major glycoprotein	Hu platelets	LIS	Con A-Seph	55
WGA-receptors	Mu L1210 lymphoma	LIS	WGA-Seph	56
Con A-receptor	Mu L-929 cells	LIS	Con A-Seph	57
WGA-receptors	Hu HeLa cells	SDS	WGA-Seph	58
Galactoproteins _a_ and _b_	Ham embryo fibroblasts (NIL)	DDG	RCA_I-Seph	59
Glycoproteins	Mu Ehrlich ascites carcinoma	NaDOC	RCA_I-, Con A-, WGA, SBA-Seph	60
Con A-receptors	Rat spermatozoa	TX-100	Con A-Seph	61
Viral glycoprotein	Enveloped viruses (influenza, Sendai)	NaDOC	LCA-Seph	62
Major glycopeptides	Mu mammary tumor virus	Serdox NNP10	Con A-Seph	63
Membrane-bound enzymes (cf, 5'-nucleotidase, alkaline phosphatase)	_Dictyostelium discoidum_	NaDOC	Con A-Seph	64

Abbreviations: BPA, _Bauhinia purpurea_ agglutinin; CTAB, cetyltrimethylammonium bromide; DDG, dimethyldodecylglycine (Empigen BB); DTAB, dodecyltrimethylammonium bromide; Emulphogene, alkoxypoly(ethyleneoxy)ethanol; Ham, hamster; Hu, human; HuRBC, human red blood cells; Mu, murine; NaDOC, sodium deoxycholate; NP-40, Nonidet P-40, polyoxyethyleneglycol(6-7)-p-t-octylphenol; Ra, rabbit; SDS, sodium dodecyl sulfate; Seph, Sepharose; Serdox NNP10, a nonionic detergent, Servo, Delden, The Netherlands; TX-100, Triton X-100, polyoxyethyleneglycol(9-10)-p-t-octylphenol.

glycoprotein (glycophorin) inhibitor of WGA, *Agaricus bisporus* agglutinin and PHA. The RCA_I-Sepharose column adsorbed several glycproteins containing binding sites for RCA_I and *Abrus precatorius* agglutinin, while most of the glycoprotein receptors for *A. bisporus* agglutinin, PHA, LCA and WGA were not retained by the column. Purification of glycophorin in a one-step preparative procedure was obtained by Kahane et al. (37) after SDS solubilization of HuRBC ghosts in the presence of relatively high salt and adsorption to WGA-Sepharose. The receptor for peanut agglutinin was isolated from neuraminidase-treated, dimethyl dodecylglycine-solubilized HuRBC by chromatography on PNA-polyacrylhydrazido-Sepharose (38). In this last procedure cell surface galactose residues on intact neuraminidase-treated HuRBC were radiolabeled by oxidation with galactose oxidase followed by reduction with NaB^3H_4 prior to their purification. The use of polyacrylhydrazido-Sepharose afforded a high capacity, charge-free and non-leaching supporting matrix which exhibited very low nonspecific adsorption and high recoveries (38,39). The material eluted from PNA-polyacrylhydrazido-Sepharose with lactose consisted of two glycoproteins (MW ∿82,000 and ∿46,000, respectively) which resembled asialoglycophorin in electrophoretic migration and amino acid composition. Like asialoglycophorin, the PNA receptor formed precipitin bands with SBA and RCA_I but failed to precipitate with Con A, *Dolichos biflorus* agglutinin and WGA (38).

In an effort to determine the degree to which the carbohydrate side chains of HuRBC membrane glycoproteins vary between species, Emerson and Kornfeld (40) used non-ionic detergent solubilized bovine erythrocyte membranes as a source of glycoproteins. RCA_I-Sepharose adsorbed the 230,000 mol. wt. major glycoprotein of bovine erythrocyte membrane which contains ∿80% carbohydrate, mostly linked o-glycosidically. Virtually all the carbohydrate of the membrane (excluding glycolipids) was borne by this glycoprotein.

Lectin Affinity Purification of Lymphoid Cell Membrane Components

Lymphoid cell membranes contain a wide variety of glycoprotein molecules compared to RBC; therefore, lectin affinity columns are generally useful for partial purification or enrichment of glycoproteins. For example, almost all the glycoproteins of pig lymphocyte plasma membrane solubilized in sodium deoxycholate (NaDOC) were adsorbed on columns of Con A-Sepharose (41) or LCA-Sepharose (42). The latter immobilized lectin was much more efficient in purifying the solubilized membrane glycoproteins. Con A-Sepharose bound 20% of the applied material in a nonspecific manner, and the yield of glycoproteins off the Con A-Sepharose column was low (5%), whereas LCA-Sepharose yielded more than twice the amount obtained with Con A-Sepharose, and the recovery was 95%. SDS-polyacrylamide gel electrophoresis (SDS-PAGE) of the saccharide-eluted LCA-Sepharose fractions revealed at least ten protein bands that stained for carbohydrate with periodate-Schiff reagent. Purified LCA-receptors inhibited LCA-induced lymphocyte transformation 10 times more effectively compared to the unfractionated membrane glycoproteins, and at least one of the LCA-receptor glycoproteins was also capable of inhibiting PHA-induced lymphocyte stimulation. In the above studies further characterization was not performed to elucidate the nature and function of the lectin reactive materials; however, in other reports defined components have been isolated by lectin affinity chromatography. For example, the histocompatibility antigens of human lymphocytes (HLA) were solubilized in sodium deoxycholate and then purified by chromatography on LCA-Sepharose (43).

Improvements have been made in lectin affinity chromatography techniques to increase specificity and yield as well as reduce nonspecific adsorption. In an attempt to increase the selectivity of lectin-Sepharose chromatography, Schmidt-Ullrich et al. (44) adsorbed Triton X-100-solubilized rabbit thymocyte membrane glycoproteins onto Con A-Sepharose as in other procedures. However, the elution was performed in a stepwise fashion using increasing concentrations of the hapten methyl α-glucoside. This procedure allowed preferential elution of receptors according to their relative binding affinities to Con A. Using these procedures a sialoglycoprotein that has been designated component 5.1 because of its electrophoretic mobility in SDS-PAGE was eluted in a homogeneous form. This Con A receptor has an apparent MW of 55,000 daltons but tends to aggregate in solution. When this preparation was examined further, two molecules were found that differ in their carbohydrate side chains, and these could be recognized by cross immune electrophoresis. Trowbridge et al. (45) reported that specific ^{125}I-labeled surface antigens (T25 and T200) on mouse thymocytes could be purified from NaDOC-solubilized membranes by adsorption to either Con A- or PA-Sepharose. Upon analysis by SDS-PAGE these components separated into several species, possibly due to heterogeneity of their carbohydrate side chains. Fabre and Williams (46) isolated a major antigen from the surfaces of rat thymocyte membranes by specific adsorption on LCA-Sepharose, and this behavior demonstrated that the antigen is a glycoprotein. Nilsson and Waxdal (47) also labeled proteins on the surface of murine lymphocytes by lactoperoxidase-catalyzed ^{125}I-iodination or by culturing the cells in media containing ^{3}H-leucine or ^{3}H-fucose. Cell membranes were solubilized in NP-40 before passing the solubilized glycoproteins through Con A-Sepharose. The material specifically eluted from Con A-Sepharose was characterized by immunoprecipitation with specific antisera, and it was found that Con A-receptors included antigens coded by the histocompatibility-2 complex such as H-2K, H-2D and Ia. In addition, the immobilized Con A also adsorbed immunoglobulins M and D from solubilized B lymphocyte membranes. Essentially the same results were obtained recently by Iwata et al. (48) who extensively fractionated murine lymphoid cell surface components on a battery of lectin-Sepharose columns (see Table 1). This last study demonstrated that the use of several lectin-Sepharose columns for fractionation of cell membrane components narrows the range of isolated components. For example, PLA-, LCA- and RCA_I-Sepharose specifically bound glycoproteins of MW in the range 20,000-80,000 daltons, whereas glycoproteins in the higher MW (>100,000) range that were not adsorbed on the latter columns were preferentially bound by WGA- and BPA-Sepharose (48).

Lectin Affinity Chromatography of Nerve Cell Components

Lectin affinity chromatography has been especially useful in the purification of carbohydrate-containing macromolecules from nervous tissue. Glycoproteins present in NaDOC-solubilized synaptic plasma membranes prepared from rat brains that were injected with ^{3}H-fucose 16 hr earlier have been fractionated by chromatography on LCA- or WGA-Sepharose (49). In these studies LCA-Sepharose adsorbed 40-45% of the applied radioactivity and bound most of the 7-8 glycoproteins known to be present in the synaptic membranes; whereas WGA-Sepharose bound only 25-30% of the radioactivity suggesting that some of the glycoproteins retained on the LCA-Sepharose were not bound by the WGA column. Therefore, Gurd and Mahler (49) decided to use both of these affinity columns in tandem such that the sample applied onto LCA-Sepharose was allowed to pass into the WGA-Sepharose column. After washing both columns and sequentially eluting with the appropriate monosaccharide inhibitors for each lectin, four fractions were obtained: LCA-negative, WGA-negative; LCA-negative, WGA-positive; LCA-positive, WGA-negative; and LCA-positive, WGA-positive. When analyzed by SDS-PAGE, each of the fractions had distinct patterns with some overlap of molecular weight classes. Zanetta et al. (50,51) have fractionated rat brain synaptic membranes dissolved in SDS on Con A- or UEA-Sepharose. Elution from the former affinity column was with methyl α-glucoside; however, L-fucose was unable to release bound material but could inhibit binding to the UEA-Sepharose. Increasing the SDS concentration in the elution buffer (without saccharides) released the adsorbed material, and subsequent PAGE analysis revealed a large number of carbohydrate containing components. Part of the heterogeneity in glycoproteins eluted by these procedures was attributed to the fact that the synaptic membranes were derived from a heterogeneous population of neurons.

Specific, functional glycoproteins have been isolated from solubilized brain tissue by lectin-Sepharose chromatography. Nicotinic acetylcholine receptor, Triton X-100 solubilized from a crude, post-nuclear membrane fraction of rat cerebral cortex or from membranes of *Torpedo californica* electroplax, was retained on immobilized Con A, WGA or RCA_I (52). In another study partial purification of coagulant arylamidase and alkaline phosphatase enzymes solubilized in NaDOC from bovine brain was achieved by adsorption on Con A-Sepharose (53).

Lectin Affinity Chromatography of Components from Other Cell Types

In contrast to the multitude of lectin receptors found in synaptic membranes, several systems have been described in which a single receptor was obtained. For example, solubilization of bovine retinal rod outer segments with cetyltrimethylammonium bromide followed by chromatography on Con A-Sepharose yielded a homogeneous rhodopsin (54). The major human platelet membrane glycoprotein (MW ∿100,000) was solubilized by lithium diiodosalicylate--phenol extraction of ^{125}I-labeled membranes and purified on Con A-Sepharose (55). Using similar extraction methods, four WGA-receptors of MW 40,000-60,000 were isolated from L1210 cells (560, and a single Con A receptor having an apparent MW of 100,000 and at least five carbohydrate side chains was isolated from mouse L929 cells (57). Multiple WGA receptors of MW in the range 40,000-300,000 were isolated from SDS-solubilized membranes of ^{125}I-labeled HeLa cells (58). Extraction of hamster embryo fibroblasts (NIL) labeled with galactose-oxidase and NaB^3H_4 with 8M urea or solubilization with diemthyldodecylglycine followed by chromatography on RCASepharose allowed purification of galactoprotein a (LETS) and galactoprotein b (59). Galactoprotein a was accompanied by an actin-like protein, and galactoprotein b appeared compositionally heterogeneous. While most of the above investigators did not report problems in using lectin-Sepharose columns, an extensive study by Nachbar et al. (60) demonstrated that in cell membranes of Ehrlich ascites carcinoma cells solubilized in NaDOC there are several carbohydrate containing components that adsorb to Con A-, WGA-, SBA- and RCA-Sepharose columns. Using ^{14}C-glucosamine-labeled membrane components the recoveries from the various columns were determined, and it was found that specific elution of the WGA- and the Con A-Sepharose columns resulted in release of only 55-70% of the bound material. The remaining tightly bound material may have been irreversibly held to the column by hydrophobic interactions. The receptors for the different lectins appeared similar on SDS-PAGE, and some immobilized lectins purified the same glycoprotein receptors. Purification of glycoproteins from solubilized enveloped viruses (62,63) and of membrane-associated enzymes on lectin affinity columns has also been described (Table 1).

Although one-step purification to homogeneity of a membrane glycoprotein has only been reported for HuRBC glycophorin (37), most studies indicate that affinity chromatography on immobilized lectin columns is the method of choice, at least in preliminary stages of membrane glycoprotein purification. In earlier studies (34,36,41,42) no attempt was made to radioactively label cell membrane components prior to solubilization and chromatography, but more recent investigations have demonstrated that prelabeling membrane components either by lactoperoxidase-catalyzed ^{125}I-iodination (47,55) or neuraminidase followed by galactose oxidase and NaB^3H_4 (38,57) facilitates the fractionation procedure and insures that the isolated components originate from the plasma membrane surface. In addition, the labeled components may be detected by autoradiography or fluorography after separation by SDS-PAGE, even if they are obtained in small quantities.

Many of the saccharides found in membrane glycoproteins are

ubiquitous; therefore, only partial purification should be expected when a single lectin-Sepharose column is used. The use of a hapten gradient in the elution step (44) or combinations of immobilized lectins with different sugar specificities (48,49) has proven to be very powerful in the separation of membrane glycoproteins. The choice of detergent for the solubilization of membranes and stabilization of membrane glycoproteins prior to application onto lectin-Sepharose columns seems in many reports to have been random, or in the best examples, empirical. This problem has been eliminated in part by a systematic analysis of the effects of several zwitterionic, cationic, anionic and non-ionic detergents on glycoprotein binding to and specific elution from five lectin-Sepharose columns (39). Our results (Figure 3) indicate that for most lectins, NaDOC caused marked decreases in binding affinity. The zwitterionic detergent dimethyl dodecylglycine decreased the efficiency of the Con A- and SBA-Sepharose, while the non-ionic detergents NP-40 and Triton X-100 were suitable for all the immobilized lectins tested (39). Future use of specific lectin affinity chromatography techniques will be greatly facilitated by optimizing various lectins, detergents, solution conditions, columns and techniques for specific elution of bound glycoproteins.

DYNAMICS OF CELL SURFACE LECTIN RECEPTORS

One of the most useful approaches for studying cell surface glycoprotein dynamics has been the utilization of lectins as probes for fluorescent or electron microscopy (18,65). Observing the movement and redistribution of cell surface-bound lectins has allowed estimates to be made on the rates of lateral diffusion and, in certain cases, internalization of the lectin-receptor complexes (5). These experiments have been generally conducted with fluorescein-labeled lectins and UV-fluorescent microscopy or ferritin-, hemocyanin- or peroxidase-labeled lectins and electron microscopy (65). By labeling prefixed or unfixed cell surfaces with these probes and then observing the alterations in surface distributions of the probes under a variety of experimental conditions, a dynamic picture of lectin receptor movements has been obtained in several biological systems. These observations have led investigators to conclude the following: First, the inherent topographic distributions of most lectin receptors (as found on prefixed cells or cells labeled at 0°C) are uniform and random across the entire cell surface. Some exceptions have been noted, however, in specialized cells of high asymmetry such as sperm where WGA-receptors were found to be localized only in certain regions of the sperm head (66) and nerve cells where Con A receptors were found to be relatively immobile and predominately in the synaptic cleft (67). Most lectin receptor redistribution studies have revealed that ligand-receptor complexes undergo rapid change from random, uniform distributions to form clusters and patches (68-72). In some cell types the clustered ligand-receptor complexes coalesce to form a single polar "cap" or aggregate of cross-linked ligand-receptor complexes. After redistribution the clus-

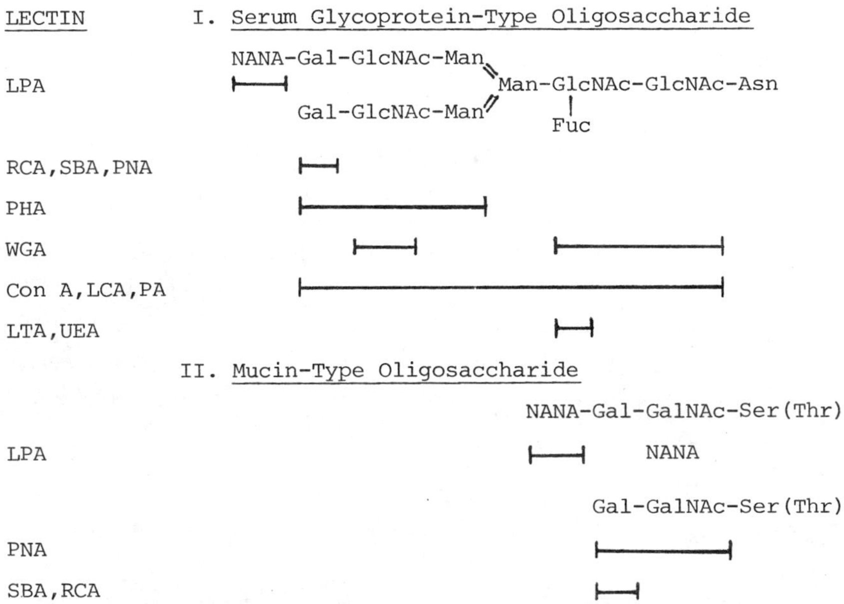

Figure 2. Possible binding region for lectins on oligosaccharide side chains of a hypothetical cell membrane glycoprotein. LPA, Limulus polyphemus *agglutinin; RCA,* Ricinus communis *agglutinin; SBA, soybean agglutinin; PNA, peanut agglutinin; PHA, phytohemagglutinin from* Phaseolus vulgaris; *WGA, wheat germ agglutinin; Con A, concanavalin A; LCA,* Lens culinaris *agglutinin; PA, pea agglutinin; LTA,* Lotus tetragonolobus *agglutinin; UEA,* Ulex europeus *agglutinin.*

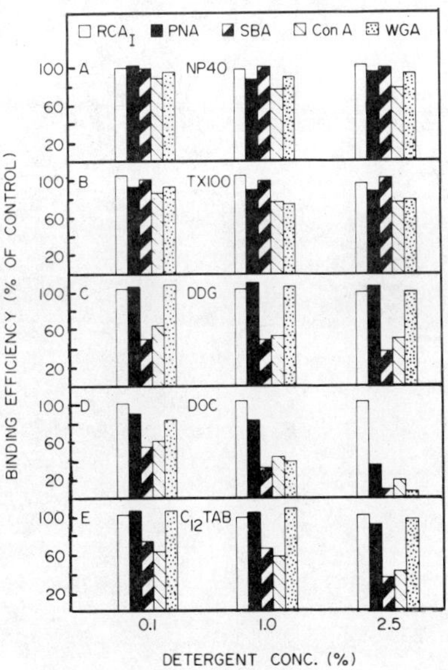

Biochemistry

Figure 3. Effect of increasing concentrations of detergents on the glycoprotein binding efficiencies of immobilized lectins (for experimental details see *Ref.* 39)

tered or capped ligand-receptor complexes may be internalized by endocytosis or shed from the cell. The extents and rates of redistribution, internalization and shedding appear to be controlled by the cell type and the biochemical nature of the ligand and its receptors (reviewed in 4,18,73). It has been noted that while a lectin at one concentration can induce redistribution of its receptors into polar caps, at other concentrations (usually much higher), it can inhibit the movement of different cell surface receptors (74,75).

Quantitation of the rotational and lateral mobilities of cell surface glycoproteins on cell surfaces has been achieved by use of fluorescein isothiocyanate-labeled Con A. Shinitzky et al. (76) and Inbar et al. (77) have studied the rotational mobilities of Con A receptor complexes on cell surfaces by fluorescence polarization. The recent introduction of the fluorescence photobleaching recovery technique allowed Jacobson et al. (78) and Schlessinger et al (79) to measure lateral diffusion coefficients for the Con A-receptor complexes within surface membranes of several cultured cells. Such studies and others indicated that the mobilities of cell membrane glycoproteins are determined to a large extent by their interactions with other membrane components and membrane-associated structures such as microfilaments and microtubules (4,5). It is hoped that isolation and characterization of specific membrane glycoproteins will increase the understanding of the exact molecular structure and organization of cell membranes.

Acknowledgments: Supported by PHS NCI Contract CB-74153, PHS NCI Grant CA-15122 and American Cancer Society Grant BC-211A to G.L. Nicolson.

LITERATURE CITED

1. Singer, S.J. & Nicolson, G.L. (1972) Science 175, 720.
2. Edidin, M. (1974) Ann. Rev. Biophys. Bioeng. 3, 179.
3. Singer, S.J. (1974) Ann. Rev. Biochem. 43, 805.
4. Nicolson, G.L. (1976) Biochim. Biophys. Acta 457, 57.
5. Nicolson, G.L., Poste, G. & Ji, T.H. (1977) In "Dynamic Aspects of Cell Surface Organization", Vol. 3 of "Cell Surface Reviews" (G. Poste & G.L. Nicolson, eds.), p. 1, North-Holland, Amsterdam.
6. Bretscher, M.S. (1971) J. Mol. Biol. 59, 351.
7. Segrest, J.P., Kahne, I., Jackson, R.L. et al. (1973) Arch. Biochem. Biophys. 155, 167.
8. Morrison, M., Mueller, T.J. & Huber, C.T. (1974) J. Biol. Chem. 249, 2658.
9. Hunt, R.C. & Brown, J.C. (1975) J. Mol. Biol. 97, 413.
10. Gahmberg, C.G. (1977) In "Dynamic Aspects of Cell Surface Organization", Vol. 3 of "Cell Surface Reviews" (G. Poste & G.L. Nicolson, eds.), p. 371, North-Holland, Amsterdam.
11. Pinto da Silva, P. & Nicolson, G.L. (1974) Biochim. Biophys. Acta 363, 311.
12. Wang, K. & Richards, F.M. (1974) J. Biol. Chem. 249, 8005.
13. Hynes, R.O. & Pearlstein, E.S. (1976) J. Supramol. Struct. 4, 1.
14. Nicolson, G.L. & Painter, R.G. (1973) J. Cell Biol. 59, 395.
15. Ji, T.H. & Nicolson, G.L. (1974) Proc. Natl. Acad. Sci. U.S.A. 71, 2212.
16. Elgsaeter, A., Shotton, D.M. & Branton, D. (1976) Biochim. Biophys. Acta 426, 101.
17. Lis, H. & Sharon, N. (1973) Ann. Rev. Biochem. 43, 541.
18. Nicolson, G.L. (1974) Int. Rev. Cytol. 39, 89.
19. Spiro, R.G. (1973) Adv. Protein Chem. 27, 349.
20. Spiro, R.G. & Bohyroo, V.D. (1974) J. Biol. Chem. 249, 5704.
21. Toyoshima, S., Fukada, M. & Osawa, T. (1974) Biochemistry 11, 4000.
22. Thomas, D.B. & Winzler, R.J. (1969) J. Biol. Chem. 244, 5943.
23. Kornfeld, S. & Kornfeld, R. (1971) In "Glycoproteins in Blood Cells and Plasma" (G.A. Jamieson & T.J. Greenwalt, eds.), p. 50, J.B. Lippincott, Philadelphia.
24. Tomita, M. & Marchesi, V.T. (1975) Proc. Natl. Acad. Sci. U.S.A. 72, 2964.
25. Newman, R.A., Glöckner, W.M. & Uhlenbruck, G.G. (1976) Eur. J. Biochem. 64, 373.
26. Finne, J. (1975) Biochim. Biophys. Acta 412, 317.
27. Funakoshi, I., Nakada, H. & Yamashina, I. (1974) J. Biochem. (Tokyo) 76, 319.
28. Nakada, H., Funakoshi, I. & Yamashina, I. (1975) J. Biochem. (Tokyo) 78, 863.
29. Kornfeld, S. & Kornfeld, R. (1969) Proc. Natl. Acad. Sci. U.S.A. 63, 1439.
30. Kornfeld, R. & Kornfeld, S. (1970) J. Biol. Chem. 245, 2536.

31. Kornfeld, S., Rogers, J. & Gregory, W. (1971) J. Biol. Chem. 246, 6581.
32. Akiyama, Y. & Osawa, T. (1972) Hoppe-Seyler's Z. Physiol. Chem. 353, 323.
33. Marchesi, V.T., Furthmayr, H. & Tomita, M. (1976) Ann. Rev. Biochem. 46, 667.
34. Findlay, J.B.C. (1974) J. Biol. Chem. 249 4398.
35. Ross, A.H. & McConnell, H.M. (1977) Biochem. Biophys. Res. Commun. 74, 1318.
36. Adair, W.L. & Kornfeld, S. (1974) J. Biol. Chem. 249 4696.
37. Kahane, I., Furthmayr, H. & Marchesi, V.T. (1976) Biochim. Biophys. Acta 426, 464.
38. Carter, W.G. & Sharon, N. (1977) Arch. Biochem. Biophys. 180, 570.
39. Lotan, R., Beattie, G., Hubbell, W.L. et al. (1977) Biochemistry 16, 1787.
40. Emerson, W.A. & Kornfeld, S. (1976) Biochemistry 15, 1697.
41. Allan, D., Auger, J. & Crumpton, M.J. (1972) Nature New Biol. 236, 23.
42. Hayman, M.J. & Crumpton, M.J. (1972) Biochem. Biophys. Res. Commun. 47, 923.
43. Snary, D., Goodfellow, P., Hayman, M.J. et al. (1974) Nature New Biol. 247, 457.
44. Schmidt-Ullrich, R., Wallach, D.F.H. & Hendricks, J. (1975) Biochim. Biophys. Acta 382, 295.
45. Trowbridge, I.S., Nilsen-Hamilton, M., Hamilton, R.T. et al. (1977) Biochem. J. 163, 211.
46. Fabre, J.W. & Williams, A.F. (1977) Transplantation 23, 349.
47. Nilsson, S.F. & Waxdal, M.J. (1976) Biochemistry 15, 2698.
48. Iwata, M., Ide, H., Terao, T. et al. (1977) J. Biochem. (Tokyo) 82, 661.
49. Gurd, J.W. & Mahler, H.R. (1974) Biochemistry 13, 5193.
50. Zanetta, J.P., Morgan, I.G. & Gombos, G. (1975) Brain Res. 83, 337.
51. Zanetta, J.P., Reeber, A., Vincendon, G. et al. (1977) Brain Res. 138, 317.
52. Salvaterra, P.M., Gurd, J.M. & Mahler, H.R. (1977) J. Neurochem. 29, 345.
53. Pitlick, F.A. (1976) Biochim. Biophys. Acta 428, 27.
54. Steinemann, A. & Stryer, L. (1973) Biochemistry 12, 1499.
55. Nachman, R.L., Hubbard, A. & Ferris, B. (1973) J. Biol. Chem. 248, 2928.
56. Jansons, V.K. & Burger, M.M. (1973) Biochim. Biophys. Acta 291, 127.
57. Hunt, R.C., Bullis, C.M. & Brown, J.C. (1975) Biochemistry 14, 109.
58. Kramer, R.H. & Canellakis, E.S. (1977) Proc. Am. Assoc. Cancer Res. 18, 56.
59. Carter, W.G. & Hakomori, S.-I. (1977) Biochem. Biophys. Res. Commun. 76, 299.

60. Nachbar, S.M., Oppenheim, J.D. & Aull, F. (1976) Biochim. Biophys. Acta 419, 512.
61. Fournier-Delpech, S., Danzo, B.J. & Orgebincrist, M.C. (1977) Ann. Biol. Anim. Biochem. Biophys. 17, 207.
62. Hayman, M.J., Skehel, J.J. & Crumpton, M.J. (1973) FEBS Lett. 29, 185.
63. Westenbrink, F., Koonstra, W. & Bentvelzen, P. (1977) Eur. J. Biochem. 76, 85.
64. Crean, E.V. & Rossomando, E.F. (1977) Biochem. Biophys. Res. Commun. 75, 488.
65. Nicolson, G.L. (1978) In "Advanced Techniques in Biological Electron Microscopy" (J.K. Koehler, ed.), Vol. 2, Springer-Verlag, New York (in press).
66. Nicolson, G.L., Usui, N., Yanagimachi, R. et al. (1977) J. Cell Biol. 74, 950.
67. Kelly, P., Cotman, C.W., Gentry, C. et al. (1976) J. Cell Biol. 71, 487.
68. de Petris, S. & Rabb, M.C. (1974) Eur. J. Immunol. 4, 130.
69. Rosenblith, J.Z., Ukena, T.E., Yin, H.H. et al. (1973) Proc. Natl. Acad. Sci. U.S.A. 70, 1625.
70. Nicolson, G.L. (1973) Nature New Biol. 243, 218.
71. de Petris, S., Raff, M.C. & Mallucci, L. (1973) Nature New Biol. 244, 275.
72. Goldman, R., Sharon, N. & Lotan, R. (1976) Exp. Cell Res. 99, 408.
73. de Petris, S. (1977) In "Dynamic Aspects of Cell Surface Organization", Vol. 3 of "Cell Surface Reviews" (G. Poste & G.L. Nicolson, eds.), p. 643, North-Holland, Amsterdam.
74. Loor, F., Forni, L. & Pernis, B. (1972) Eur. J. Immunol. 2, 203.
75. Yahara, I. & Edelman, G.M. (1972) Proc. Natl. Acad. Sci. U.S.A. 69, 608.
76. Shinitzky, M., Inbar, M. & Sachs, L. (1973) FEBS Lett. 34, 247.
77. Inbar, M., Shinitzky, M. & Sachs, L. (1973) J. Mol. Biol. 81, 245.
78. Jacobson, K., Wu, E.-S. & Poste, G. (1976) Biochim. Biophys. Acta 433, 215.
79. Schlessinger, J., Koppel, D.E., Axelrod, D. et al. (1976) Proc. Natl. Acad. Sci. U.S.A. 73, 2409.

RECEIVED April 4, 1978.

Secreted Glycoproteins and Cell-Surface Antigens of Cancer Cells

Introduction

MARY CATHERINE GLICK

Department of Pediatrics, University of Pennsylvania Medical School, Children's Hospital of Philadelphia, Philadelphia, PA 19104

Valuable information has been gained from the study of glycoproteins from cells in culture and there is little doubt that these will continue to be productive systems (1,2). When studying oncogenesis, however, two factors should be considered: firstly, most cancers are of epithelial origin; and secondly, these cells may be subject to more restraints and expressions than cells in culture. Therefore, the final definition of tumor associated events must eventually come from studies *in vivo*.

An excellent example of an *in vitro* system which was shown subsequently to have a close corollary *in vivo* was the glycoprotein shed from fibroblasts in culture. Table 1 summarizes some of the studies which have been concerned with these glycoproteins associated with the matrix of fibroblasts, and gives a recent reference to each study. Fibronectin, the name proposed for all of these high molecular weight glycoproteins described by various names in many cell types, illustrates a macromolecule which can also be found in humans (3). This latter glycoprotein(s) has long been described as cold insoluble globulin (4) and is a major plasma glycoprotein.

The papers which follow have used *in vivo* systems to address the problems of cell shedding and surface antigens. It is shown, for example, that in some cases, shed tumor products mask the surface antigens, and that this material is in some way related to the intestinal flora of the experimental animal. Other papers show that certain tumor associated glycoproteins or glycolipids in one individual may be normal antigens or developmental glycoproteins in another individual. All of these findings are milestones on the road to cancer therapy and diagnosis.

Dr. Ward Pigman realized the importance of such studies and yet fully realized the drawbacks of working with whole animals. He and his colleagues thus devoted considerable effort in the past few years to defining a system which may turn out to be a successful combination of both approaches. This system possesses some of the advantages of cells in culture and yet retains many of the characteristics of an *in vivo* system (5). The isolated

TABLE 1

High molecular weight glycoproteins associated with the cell matrix

CIG	cold insoluble globulin	Mosher (J. Biol. Chem. 250, 6614, 1975)
gap a	galactoprotein a	Gahmberg and Hakomori (J. Biol. Chem. 250, 2447, 1975)
FSA	fibroblast surface antigen	Keski-Oja, et al (Biochem. Biophys. Res. Commun. 74, 699, 1977)
LETS	large external transformation sensitive	Hynes and Destree (Proc. Natl. Acad. Sci. USA 74, 2855, 1977)
Z	zeta protein	Wickus, et al (Cold Spring Harbor Symp. Quant. Biol. 39, 541, 1974)
CSP	cell surface protein	Yamada, et al (Biochemistry 16, 5552, 1977)
MFGP	major fucosylated glycoprotein	Sear, et al (Adv. Exp. Med. Biol. 79, 375, 1977)

epithelial cells from rat colonic mucosa represent a readily available system to study cell secretion. It is clear that much information will be forthcoming from the use of cell preparations such as this and others described here (6).

Literature Cited

1. Glick, M.C., and Flowers, H. (1978) In The Glycoconjugates, Vol. II, M.I. Horowitz and W. Pigman, eds. Academic Press, New York. In press.
2. Hughes, R.C. (1976) Membrane Glycoproteins. Butterworths, London.
3. Vuento, M., Wrann, M., and Ruoslahti, E. (1977) FEBS Lett. 82, 227-231.
4. Edsall, J.T., Gilbert, G.A., and Scheraga, H.A. (1955) J. Am. Chem. Soc. 77, 157-161.
5. Perret, V., Lev, R., and Pigman, W. (1977) Gut 18, 382-385.
6. Starling, J.J., Hixson, D.C., Capetillo, S.C., Davis, E.M., Neri, G., and Walborg, E.F., in these proceedings.

RECEIVED April 7, 1978.

15

Masking of Cell-Surface Antigens by Ectoglycoproteins

JOHN F. CODINGTON

Laboratory for Carbohydrate Research, Massachusetts General Hospital, Boston, MA 02114

The surfaces of mammalian cells possess a wide variety of antigenic activities, that may characterize the cell with regard to individual (1, 2), and tissue (3) of origin. Specific antigens at the cell surface may reflect the stage of embryonic development (4), and antigens may result from environmental factors, such as exposure to viruses (5), chemicals (6), or radiation (7). The impact of such factors can result in neoplasia, and the surfaces of neoplastic cells then possess antigens which have resulted from these factors (7, 8, 9) or they may possess derepressed embryonic or fetal antigens (10). These antigens, known as tumor associated or tumor specific antigens, cannot usually be detected at the surfaces of normal cells similar to those from which the neoplasm was derived.

The effectiveness of an immune surveillance system in the defense against cancer in humans has been debated widely (11, 12). An immune response to tumor specific antigens has often proven sufficient to cause rejection of some experimental tumors, although in established neoplasms the manifestation of a similarly successful defense mechanism is rare (12). A number of explanations have been proposed to explain the capacity of tumor cells to survive and grow progressively in immunologically hostile environments. Of particular concern is the survival mechanism(s) employed by metastatic or suspended tumor cells in the blood, lymph, or other body fluids. Masking of cell-surface antigens to protect them from immune defense factors of the host has been proposed by a number of investigators (13-22). Evidence to support the existence of at least four different types of masking has been presented. These include masking of cryptic antigenic sites, usually galactose terminated, by sialic acid (14, 15); nonspecific coulombic or charge repulsion due to cell-surface sialic acid residues (16); masking by adsorbed exogenous glycoprotein (17, 18) or proteoglycan (19) material, which may be specific for particular cell-surface receptors; and masking of nearby antigenic sites by large endogenous glycoprotein molecules (13, 20-22). The present discussion will be limited to evidence

0-8412-0452-7/78/47-080-277$05.00/0

in support of masking by endogenous glycoproteins and the possible relevance of this phenomenon to human cancer.

Occurrence

Evidence in support of antigen masking by large endogenous ectoglycoproteins has been presented, to my knowledge, for only two ascites tumor systems, the mammary adenocarcinoma, TA3, of the strain A mouse and the IgA-synthesizing plasmacytoma cell, 58-8, of the BALB/c mouse. In each case, evidence to support this mechanism has involved comparison with cells derived from the same tumor in which no evidence for masking was found.

In the TA3 tumor system several nonstrain-specific sublines were investigated, as well as six nonstrain-specific TA3-Ha/A.CA hybrid cloned lines, which had resulted from fusion of cells from the nonstrain-specific TA3-Ha subline and normal embryonic fibroblasts of the A.CA mouse (22). In these studies only masking of the major histocompatibility antigen (H-2^a) was investigated. Each nonstrain-specific cell line was found to possess at its cell surface an abundance of rod-like glycoprotein (epiglycanin) molecules of high molecular weight. These molecules are believed to be involved in masking. Fortuitously, a strain-specific ascites subline, TA3-St, which had been derived from the same spontaneous solid tumor as the TA3-Ha cell, was available for comparison. The TA3-St ascites cell was not found to possess any epiglycanin at its surface. As described below, immunological, chemical, immunochemical, and physical studies support the hypothesis of masking by epiglycanin molecules.

In the plasma cell tumor, 58-8, which has been less extensively studied than the TA3 tumor, the capacities to absorb anti-H-2^a antisera or rabbit anti-58-8 antisera were observed to decrease over five to six passages in the ascites form in the mouse (passage 8 to passage 13) (23). After passage 13 no absorption of either antibody could be observed. Removal of protein material by brief treatment with pronase, however, restored the capacity to absorb each antibody. Moreover, after incubation of the pronase-treated cells for several hours in culture at 37°, the cells again resisted absorption of antisera, suggesting resynthesis of a masking glycoprotein. These characteristics, which are closely related to those of nonstrain-specific TA3-Ha cells, suggest masking by endogenous molecules. To my knowledge, no chemical or immunochemical evidence for a masking glycoprotein has been reported.

Glycoproteins related in composition to epiglycanin have been isolated from hepatoma ascites cells by Funakoshi and co-workers (24) and from B-16 melanoma ascites cells by Bhavanandan and Davidson (25). To my knowledge, no firm evidence in support of a masking function for the glycoproteins has been published.

Early Chemical and Immunochemical Studies of the TA3 Ascites Sublines

In 1962 Gasic and Gasic reported, on the basis of histochemical studies, that the TA3-Ha cell possessed at its surface a high concentration of sialic acid-containing glycoprotein material (26). This finding was confirmed by our laboratory, which reported in 1970 that large amounts of carbohydrate and protein material could be removed by proteolysis from viable TA3-Ha ascites cells (27). Soon thereafter we described the isolation of a glycoprotein fraction of high molecular weight, which had been removed from viable ascites cells by proteolysis (28). This material possessed a high concentration of serine, threonine, and *N*-acetylgalactosamine, components associated with glycoproteins with a high proportion of *O*-glycosyl-linked carbohydrate chains. The cell-surface glycoprotein from which these fragments were derived was later termed epiglycanin (29). The significance of this finding was not recognized until comparison with a strain-specific subline of the same tumor, TA3-St, revealed that the TA3-St ascites cell did not possess any detectable amount of this fraction (20, 21).

At about this time several laboratories described some striking biological differences between the same two ascites sublines, the nonstrain-specific TA3-Ha cell and the strain-specific TA3-St subline (21, 30-35). Friberg (30) reported marked differences in karyotypes of the two cell lines, and Hauschka *et al.* (32) confirmed a near diploid chromosome number for the TA3-Ha cell. The work of several groups (30-34) showed clearly that the TA3-Ha cell was significantly less immunogenic than the TA3-St cell and far more immunoresistant. Of particular significance was the finding of Friberg (31) that the TA3-St cell possessed the capacity *in vitro* to absorb many fold more anti-H-2^a antibody to the major histocompatibility antigen (H-2^a) than the TA3-Ha cell. Sanford *et al.* (21) later confirmed this observation and showed, as did Friberg and Lillihook (35), that disruption of the cell structures by lyophilization resulted in a marked increase in the number of H-2^a antigenic sites exposed on the surface of the TA3-Ha cell, but this treatment resulted in no increase in antigenic sites exposed on the TA3-St cell surface. These observations, coupled with the finding that the TA3-Ha, but not the TA3-St, cell possessed at its surface an abundance of epiglycanin molecules, led us to suggest at that time that allotransplantability and the reduced capacity to absorb anti-H-2^a antibody was due to masking of surface antigens by epiglycanin molecules (20, 21).

Springer *et al.* (36) demonstrated that epiglycanin was a potent inhibitor of hemagglutination by the lectin from *Vicia graminea* seeds, a lectin of high specificity for structures at the surfaces of human erythrocytes of N blood group specificity.

Later Cooper et al. (37) developed a method for the quantitation of epiglycanin by the use of this lectin. Although receptors for the Vicia graminea lectin, which are present in epiglycanin, are abundant at the TA3-Ha cell surface, no receptors for this lectin were detected at the surface of the TA3-St ascites cell (20, 38).

The Origin of Nonstrain Specificity in the TA3 Ascites Cells

The TA3-Ha ascites subline was initially strain specific (32). Loss of strain specificity in the TA3-Ha subline occurred during routine passage as an ascites cell in the syngeneic strain A mouse (32). Although this event occurred in ascites cells following passage 200, neither the exact passage(s) nor the circumstances of the occurrence were reported. Although epiglycanin may have appeared at the TA3-Ha cell surface at this time, its presence was not observed until after passage 520 by means of chemical methods (28).

Another opportunity to investigate this phenomenon, however, occurred recently in our laboratory (39, 40). Strain specific TA3-St cells were inoculated intraperitoneally into strain A mice suffering from incipient acute interstitial pneumonia. Later two micoorganisms, Pasteurella pneumotropica and Pasteurella multocida, were isolated from the sick mice. Cells were harvested routinely on day 7 and tested for the presence of receptors for the lectin from Vicia graminea seeds at their surfaces. The determination of lectin receptors was part of a project in progress at that time in collaboration with a graduate student, D. K. Miller, and with the laboratory of Dr. A. G. Cooper.

Whereas, TA3-St ascites cells normally demonstrate essentially no lectin receptors, the cells were now able to adsorb a high level of Vicia graminea lectin (39). It was shown that the new cell line, which was called TA3-MM, possessed at its surface a high concentration of epiglycanin molecules (40). Similar to the nonstrain specific TA3-Ha ascites cell, the TA3-MM cell line was able to grow progressively in mice of foreign strains. The origin and characteristics of the TA3-MM subline are described in two papers prepared for publication (39, 40).

The new allotransplantable cell line, TA3-MM, was approximately tetraploid with an average of about 80 chromosomes, which was 12 more than was possessed by the TA3-St cell (approximately 68) and twice the number of the TA3-Ha cell (41-42) (39). Although about 20% greater in diameter than the TA3-Ha ascites cell, it possessed about the same amount (2.3 mg/10^9 cells) of epiglycanin per cell (40). Comparison of the compositions of purified fractions of epiglycanin from TA3-Ha and TA3-MM cells are presented in Table I. Values are almost identical for the two cell lines.

One may speculate that nonstrain specificity in the TA3-Ha ascites cell line also occurred under the influence of pneumonia associated microorganisms, but no evidence is available to support

TABLE I

Carbohydrate and Amino Acid Compositions of Epiglycanin Isolated from the Ascites Fluid of Syngeneic Mice Bearing TA3 Ascites Cells

	TA3-Ha	TA3-MM[a]
Carbohydrate Components	(moles relative to N-acetylgalactosamine)	
N-Acetylgalactosamine	1.0	1.0
Galactose	1.4	1.4
N-Acetylglucosamine	0.6	0.3
N-Acetylneuraminic Acid	0.4	0.6
Amino Acid Components	(residues per 1000 residues)	
Aspartic Acid	24	31
Threonine	307	306
Serine	252	223
Glutamic Acid	32	36
Proline	93	107
Glycine	81	92
Alanine	131	131
Valine	26	12
Isoleucine	1	11
Leucine	33	49
Lysine	12	n.d.

[a] Values were obtained for the TA3-MM/1 subline (40).

this contention. Moreover, the two cell lines demonstrate many different characteristics at this time. It is, of course, conceivable that after many passages the TA3-MM cell will become similar to the TA3-Ha cell, but it is also plausible that the two original strain specific ascites sublines, TA3-Ha and TA3-St, were dissimilar in important respects and some of their dissimilarities have persisted after loss of strain specificity. It is clear, however, that in each subline allotransplantability is associated with the presence at the cell surface of a large number (approximately 4 x 10^6) of molecules of epiglycanin.

Topographical Features of the TA3-Ha and TA3-St Cells

Electron microscopy revealed major differences in the topography of the TA3-Ha and TA3-St ascites cells (41). Scanning electron microscopy showed the surface of the TA3-Ha cell to consist of long microvilli distributed evenly over the entire cell surface, as shown in Figure 1; whereas, the TA3-St ascites cell surface consisted of irregularly-spaced ridges or folds (Figure 2) (41). By dark field lightmicroscopy, microvilli of viable TA3-Ha cells in suspension appeared to be of greater length than the diameter of the cells and the microvilli appeared to have a cilia-like motion (42).

By transmission electron microscopy the two cell types were hardly distinguishable at a magnification of about 10,000, with the exception of the morphological differences in the micro-extensions, those of the TA3-Ha cell being more erect, more regularly spaced, and longer than those of the TA3-St cell (41). Each cell type appeared to contain two distinct types of virus particles (41). This observation was consistent with immunologic findings which demonstrated that the TA3-Ha, TA3-St, and TA3-MM ascites cells contained the same three types of virus particles, the murine mammary tumor virus, the Rauscher leukemia virus, and the pneumonia virus of mice (39).

After glutaraldehyde and osmium tetroxide fixing, high resolution electron microscopy (> 50,000 times magnification) revealed a network of fine filaments extending from the surface of the TA3-Ha ascites cell. On occasion these filaments appeared to aggregate into thicker, more rod-like structures, as shown in Figure 3. These extended 200-400 nm, sometimes 500 nm, from the plasma membrane. By contrast, the TA3-St cell, when fixed similarly, possessed no visible extracellular coat (41).

Evidence that these extracellular structures at the TA3-Ha cell surface represented sialic acid-containing glycoproteins, probably epiglycanin, was obtained by the use of polycationic ferritin labeling. Anionic groups, labeled with the electron-dense polycationic ferritin, appeared to exist several hundred nm from the plasma membrane of the TA3-Ha ascites cell. By contrast, this reagent appeared to be attached only in close proximity to the plasma membrane of TA3-St cells treated in a similar way. After prior treatment with neuraminidase to remove the cell-surface sialic acid relatively few ferritin particles were visualized at the surface of either cell (41). It has been shown that in the TA3-Ha cell epiglycanin contains approximately 60% of the surface sialic acid (43).

Isolation of Epiglycanin

Epiglycanin was originally identified as glycopeptide fragments cleaved from viable cells by proteolysis and fractionated by gel filtration (28). The molecular weight of this material (Fraction I), as calculated from sedimentation equilibrium and electron microscopy data, ranged from 80,000 to 460,000 (44). The yields of epiglycanin obtained by this method averaged about 2.3 mg/10^9 cells for either the TA3-Ha or TA3-MM ascites sublines (Table I). Further fractionation has revealed marked heterogeneity in the proteolysis products with regard to carbohydrate composition (44). This variability in the composition of carbohydrate is due to variations in the relative proportion of each of five different types of O-glycosyl-linked carbohydrate chains (29, 43, 45).

We have observed that epiglycanin is shed from the cell surfaces of each nonstrain-specific TA3 cell line which we have investigated. It may be detected in the growth medium at levels below 5 ng/ml by a hemagglutination inhibition method (37). By this method epiglycanin has been detected in the serum of strain A mice bearing TA3-Ha ascites cells on day 2 or 3 following intraperitoneal inoculation of the cells.

Isolation of epiglycanin from the ascites fluid followed precipitation of the bulk of other protein and glycoprotein material by brief exposure to 0.3 M perchloric acid at 0°. Fractionation by passage of the perchloric acid-free solution through a long column of Sepharose 4B gave nearly pure epiglycanin of composition similar to that of proteolysis derived material (43). Yields of epiglycanin by this method averaged about 0.2 to 0.3 mg/10^9 cells, and the molecular weight of the materials isolated from either TA3-Ha or TA3-MM ascites cells was approximately 500,000, as determined by sedimentation equilibrium or electron microscopy (40). By shadow-casting electron microscopy (H. S. Slayter, personal communication), approximately 60-70% of the material obtained in this way ranged from 450 to 500 nm in length.

Whether epiglycanin shed into the growth medium represents the entire native molecule or only the major portion cleaved from the cell surface by membrane-bound proteases has not been determined. This material demonstrated little tendency to aggregate, in contrast to epiglycanin isolated after solubilization with detergents or, alternatively, with lithium diiodosalicylate, followed by phenol extraction (43). By gel filtration chromatography small yields of non-aggregated

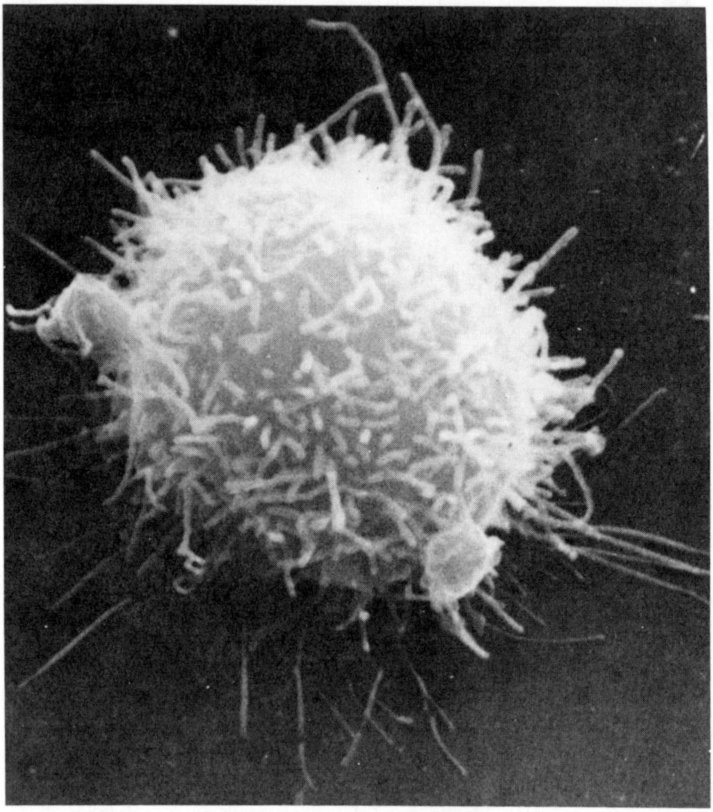

Figure 1. Scanning electron micrograph of the TA3-Ha ascites cell. Long, regularly-spaced microvilli cover the cell surface. ×4300. (Courtesy of S. C. Miller and E. D. Hay.)

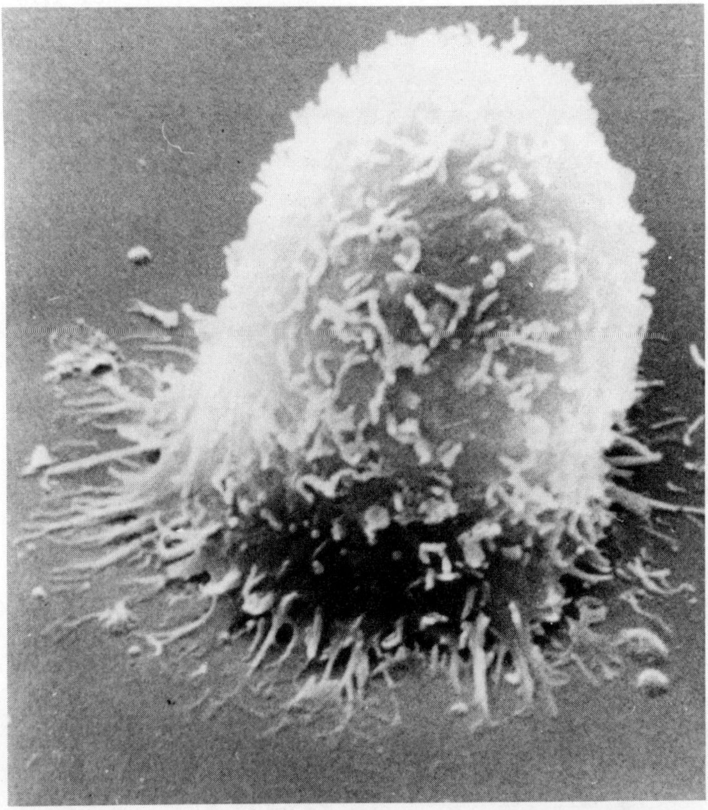

Figure 2. Scanning electron micrograph of the TA3-St ascite cell. The cell surface is characterized by numerous irregularly spaced folds or ridges. ×4300. (Courtesy of S. C. Miller and E. D. Hay.)

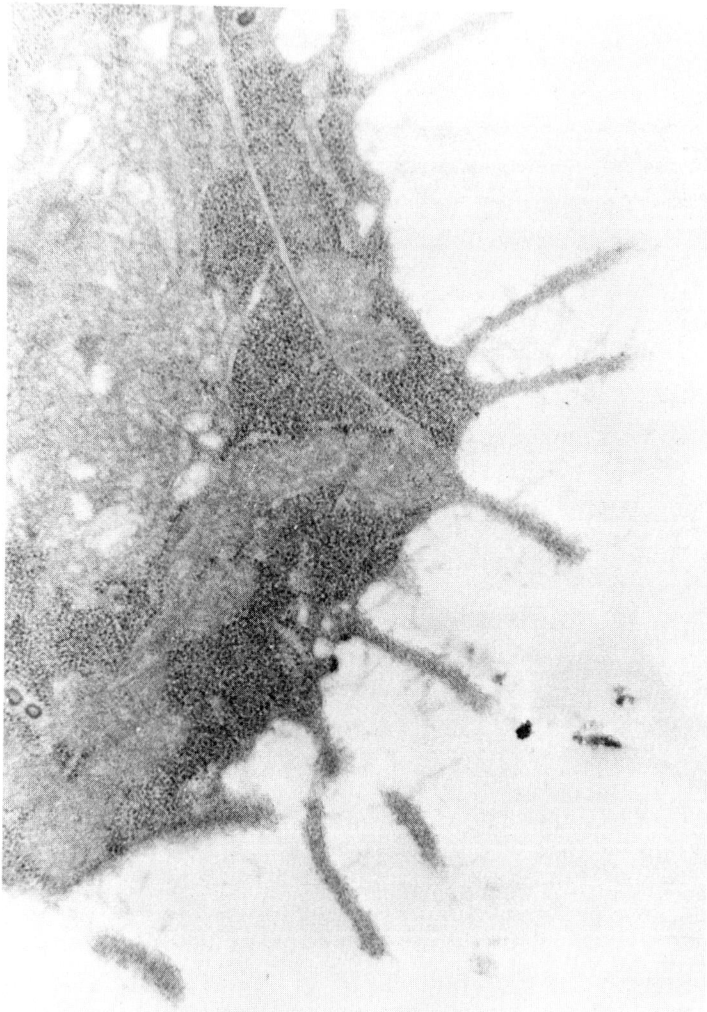

Figure 3. Transmission electron micrograph of the TA3-Ha ascites cell surface. Cells were fixed with glutaraldehyde and osmium tetroxide. Extramembranous structures 200–500 nm in length are observed on the cell surface. ×43,000. (Courtesy of S. C. Miller and E. D. Hay.)

epiglycanin were obtained by either of these two methods. From gel filtration effluent volumes and by measurement of the molecular lengths on electron micrographs (H. S. Slayter, personal communication) the molecular size of epiglycanin obtained in these ways appeared to be similar to that of ascites fluid-derived epiglycanin. It seems probable, however, that in the case of a molecule as large as epiglycanin (approximately 1300 amino acids in an extended chain), the addition of a peptide segment of 20 to 30 amino acids could not be readily detected.

Physicochemical Characterization of Epiglycanin

The carbohydrate and amino acid compositions of epiglycanin from the TA3-MM and TA3-Ha sublines are similar (Table I). The percentage of carbohydrate, however, appears to be consistently higher (80-90%) in TA3-MM material than in TA3-Ha epiglycanin (75-80%). Five types of carbohydrate chains have been isolated from epiglycanin as their N-acetylgalactosaminitol derivatives, and their sequences proposed (13, 43, 45). A pentasaccharide, containing two galactose, one N-acetylglucosamine, one N-acetylgalactosamine, and one sialic acid residue was suggested on the basis of compositional data obtained by gas-liquid chromatography (43). Recent analyses, however, (45) suggest that a hexasaccharide with two sialic acid residues, rather than one, as suggested for the pentasaccharide, may be present. In this chain it would be expected that the second sialic acid residue is attached by a 2→6 linkage to N-acetylgalactosamine. A sialic acid (2→3) galactose linkage has been previously proposed (29). The number of carbohydrate chains present in a molecule of 500,000 molecular weight, consisting of approximately 1300 amino acid residues, has been found to be 500-600 (29, 43).

All samples of epiglycanin which we have investigated have contained a small amount of mannose (approximately 0.3%). It was not known whether the presence of this residue indicated that a glycoprotein impurity were present or if the mannose were a component of an asparagine-linked carbohydrate chain in epiglycanin. By the use of affinity chromatography with a Concanavalin A - Sepharose 4B column it has recently been possible to obtain from proteolytically-cleaved fragments (Fraction I) a fraction of epiglycanin free of mannose. It was found, however, that epiglycanin isolated from the ascites fluid (molecular weight, 500,000) was completely adsorbed on a similar affinity column and could be eluted from the column with methyl-α-mannopyranoside. These results suggest that epiglycanin does contain N-glycosyl-linked carbohydrate material. On the basis of 0.3% mannose there could be no more than nine residues of mannose per molecule of epiglycanin, enough for one, possibly two, asparagine-linked carbohydrate chains.

It has not been possible to obtain amino acid sequence data for epiglycanin isolated from the ascites fluid, since the amino group at the N-terminal end of the molecule is blocked. This observation suggests that the N-terminal amino group is extracellular, and that the carboxy terminal is rooted in the cell, as it is in the erythrocyte glycoprotein which contains M and N blood group activity (46).

By shadow casting electron microscopy all samples of epiglycanin have appeared as similar rod-like particles. Widths were consistently 2.5 nm, and lengths varied from about 70 nm for the smallest proteolytically-cleaved fragment (44) to 450-500 nm for material isolated without proteolytic digestion (40).

Further Evidence for Masking in the TA3 Tumor

The hypothesis of masking was tested in a series of six hybrid cell lines which had resulted from fusion *in vitro* of TA3-Ha cells and normal embryonic fibroblasts of the A.CA mouse (47). These lines had been adapted to grow in the ascites form in strain A mice and possessed only the major histocompatibility complex of the TA3-Ha parent, H-2^a (48, 49). Each of the hybrid lines exhibited the capacity to grow progressively in mice of foreign strains; each showed a reduced capacity, compared to the TA3-St strain specific cell line, to absorb anti-H-2^a antibody; and each of the hybrid lines possessed epiglycanin at its cell surface (22).

The relative values for the three parameters, allotransplantability (% of foreign mice killed by the tumor), resistance to antibody absorption (absorptive capacity plotted in reverse order of magnitude), and amount (mg) of epiglycanin per 10^9 cells are plotted in Figure 4. The results strongly suggest that with increasing amounts of epiglycanin the cells increase their capacity to grow in mice of foreign strains and possess a reduced capacity to absorb antibody to H-2^a antigen. These results support the hypothesis that in the TA3 murine tumor system a function of epiglycanin is the masking of cell-surface antigens (22).

Relevance to Human Cancer of Masking by Endogenous Glycoproteins

The discovery of epiglycanin and the correlation of its presence at the cell surface with allotransplantability was unprecedented. Furthermore, since the initial report of its isolation (28) in 1972 and biological function (20, 23) in 1973 very little evidence for the occurrence of a similar phenomenon in other tumor systems has been reported. Three plausible explanations for this may be suggested: (a) It may be that masking of this type does not often occur in neoplasia, but is associated only with certain stages in the cancer process and occurs only under certain conditions. (b) Suitable experimental models for the study of this phenomenon have not been available. (c) Investigators have not always utilized experimental procedures

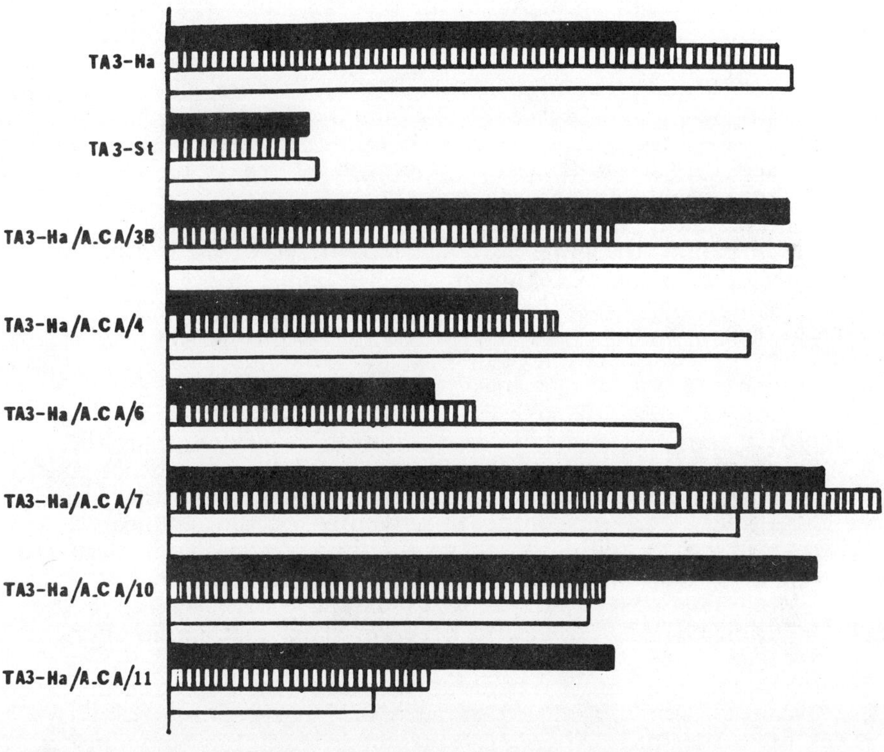

Figure 4. Comparison of three characteristics of six TA3-Ha/A.Ca hybrid, the TA3-Ha, and the TA3-St ascites cell lines. The relative amounts of epiglycanin on the cell surfaces (▬) are plotted with the capacities of the cells to resist absorption of anti-H-2^a antibody (▥) and their capacities to grow progressively in mice of foreign strains (▭). The units and scales used in plotting this graph are similar to those described in Ref. 22.

suitable for a study of this type. It is my contention that all of these explanations may apply to this situation. The appearance of large extended molecules, such as epiglycanin, on tumor cell surfaces probably occurs only under certain circumstances in established tumors - with probably disastrous consequences for the host. Furthermore, it may be difficult to establish "spontaneously-developed" masking-glycoprotein-synthesizing cells as experimental models for study, or in the past when cells of this type did develop in experimental systems they may have been overlooked.

Loss of strain specificity in the TA3 tumor system is known to have occurred only twice, and only on the second occurrence were the circumstances recorded (39, 40). As described in this report the capacity of strain-specific cells to synthesize epiglycanin occurred in the TA3-St subline *in vivo* in hosts infected by microorganisms. The mechanism by which this change took place in the TA3-Ha cell is not known, and it may by that the biochemical event which initiated the synthesis of epiglycanin could occur under the influence of other, seemingly unrelated factors. During the period of tumor development in humans, which is often of many years duration, the patient may be subjected to many types of environmental influences, among them invasions by viruses and bacteria. Under such circumstances changes in the character of tumors may possibly occur. Indeed, the occurrence of changes in the morphology and growth characteristics of human tumors is well documented (50). Of particular relevance to this discussion may be the appearance of invasiveness, metastasis, and a more rapid rate of cell proliferation. Indeed, evidence from the TA3 murine mammary carcinoma system suggest that the presence of epiglycanin-like glycoprotein molecules at the surfaces of human tumor cells could possibly explain an increase in virulence in some human cancers.

Comparison of the growth rates of TA3 ascites cells in the peritoneal cavity of syngeneic mice revealed that cells bearing epiglycanin at their surfaces, namely those of the TA3-Ha and TA3-MM lines, grew more rapidly than those (Ta3-St) without epiglycanin. Friberg (30) attributed differences in growth rates of TA3-Ha and TA3-St lines to an initial lag period in TA3-St growth. We have found that during a 7 day growth period the average doubling time for the TA3-MM and TA3-Ha cells is approximately 15 hours, compared to a doubling time of about 20 hours for the TA3-St cell. A plausible explanation for the increased growth rate in the epiglycanin-bearing cells may involve masking of tumor associated antigens, although to my knowledge no comparison of the properties of the tumor-associated antigens in the TA3 sublines has been reported.

Recent reports of Miller *et al.* (51) and Cooper and coworkers (52) describe properties in TA3 sublines which are consistent with a role in metastasis by epiglycanin. It was shown that the appearance of epiglycanin depends upon environmental factors and that, under certain circumstances epiglycanin may become lost, or become nondetectable, at the TA3-Ha cell surface. After several weeks of growth in culture epiglycanin could not be detected at the surface of the TA3-Ha cell (51). However, after passage in the mouse in ascites form, epiglycanin was again present in abundance at the cell surface.

It was recently reported by Hagmar and Ryd (53) that the TA3-Ha cell was not transplantable when passaged from the ascites form in the strain A mouse to a subcutaneous site in a mouse of a foreign strain. However, Cooper *et al.* (52) have demonstrated that with continuous passage of TA3-Ha ascites cells subcutaneously in syngeneic mice the concentration of epiglycanin increased, as evidenced by the appearance of an increased concentration of receptors for the *Vicia graminea* lectin (38), with the appearance of subcutaneous allotransplantability. These circumstances for tumor growth may more nearly simulate those of the more slowly-growing solid tumors in humans. These results demonstrate correlation between the presence of epiglycanin and allotransplantability in solid tumors and suggest that masking by endogenous glycoprotein molecules may occur in solid tumors.

A possible relationship between epiglycanin concentration and metastatic properties was also demonstrated by this group (52). Variants derived from TA3-Ha cells by prolonged subcutaneous growth in syngeneic mice were selected on the basis of their capacity to grow in sites distant from the tail vein injection site, *i.e.*, beyond the lung, in the liver, spleen, or kidney. It was found that those variants which possessed the greatest concentration of epiglycanin demonstrated the greatest metastatic property. The most "metastatic" of the cells investigated was a TA3-Ha variant which possessed 2-3 times as much epiglycanin at its surface as the established TA3-Ha ascites cell line. Thus, a correlation has been demonstrated between the presence of epiglycanin and the capacity of cells in the TA3 murine carcinoma to pass freely in the circulatory system to distant sites before establishing foci of tumor growth.

Any conclusions regarding the relationship between masking by large endogenous ectoglycoproteins and cancer in humans must be considered as purely speculative at this time. Nevertheless, the investigations of the TA3 mammary carcinoma in the mouse described here suggest, as was stated in 1975 (13), that "similar-type transformations, resulting in the appearance of epiglycanin-like macromolecules, could occur in other spontaneously derived tumors. Such phenomena may have relevance to human cancers, particularly to the occurrence of metastases or to the recurrence of cancer in seemingly cured patients."

Acknowledgements

This investigation was supported by grants from the National Institutes of Health, U.S. Public Health Service (CA-18600 to J. F. Codington and CA-08418 to R. W. Jeanloz).

Literature Cited

1. Kahan, B. D., and Reisfeld, R. A., Eds., "Transplantation Antigens", Academic Press, New York, 1972.
2. Nathenson, S. G., and Cullen, S. E., Biochim. Biophys. Acta (1974) 344, 1-25.
3. Moscona, A.A., Proc. Soc. Exptl. Biol. Med. (1956) 92, 410-416.
4. Moscona, A., Ed., "The Cell Surface in Development", J. Wiley and Sons, Inc., New York, 1974.
5. Old, L. J., and Boyse, E. A., Ann. Rev. Med. (1964) 15, 167
6. Prehn, R. T., Canad. Cancer Conf. (1963), 387.
7. Klein, G., in Davis, B. D., and Warren, L., Eds., "The Specificity of Cell Surfaces", pp. 165-180, Prentice-Hall, Inc., New Jersey, 1967.
8. McKhann, C. F., and Jagarlamoody, S. M., in Day, S. B., and Good, R. A., Eds., "Membranes and Viruses in Immunopathology", pp. 577-599, Academic Press, New York, 1972.
9. Prehn, R. T., in Nowotny, A., Ed., "Cellular Antigens", pp. 308-313, Springer-Verlag, New York, 1972.
10. Granatek, C. H., Hanna, M. G., Jr., Hersh, E. M., Gutterman, J. U., Mavligit, G. M., and Candler, E. L., Cancer Res. (1976) 36, 3464-3470.
11. Smith, R. T., and Landy, M., Eds., "Immune Surveillance", Academic Press, New York, 1970.
12. Klein, G., and Klein, E., Proc. Natl. Acad. Sci. U.S.A. (1977) 74, 2121-2125.
13. Codington, J. F., in "Cellular Membranes and Tumor Cell Behavior", pp. 399-419, Williams and Wilkins Co., Baltimore, 1975.
14. Jeanloz, R. W., and Codington, J.F., in Rosenberg, A., and Schengrund, C. -L., Eds., "Biological Roles of Sialic Acid", pp. 201-238, Plenum Press, New York, 1976.
15. Bekesi, J. G., Roboz, J. P., and Holland, J. F., Ann. N.Y. Acad. Sci. (1976) 277, 313-331.
16. Apffel, C. A., and Peters, J. H., J. Theor. Biol. (1970) 26, 47-59.
17. Whitehead, R. H., Roberts, G. P., Thatcher, J., Teasdale, C., and Hughes, L. E., J. Natl. Cancer Inst. (1977) 58, 1573-1576.
18. Dorval, G., Witz, I. P., Klein, E., and Wigzell, H., J. Natl. Cancer Inst. (1976) 56, 523-527.
19. Lippman, M., Nature (1968) 219, 33-36.
20. Codington, J. F., Sanford, B. H., and Jeanloz, R. W., J. Natl. Cancer Inst. (1973) 51, 585-591.

21. Sanford, B. H., Codington, J. F., Jeanloz, R. W., and Palmer, P. D., J. Immunol. (1973) 110, 1233-1237.
22. Codington, J. F., Klein, G., Cooper, A. G., Lee, N., Brown, M. C., and Jeanloz, R. W., J. Natl. Cancer Inst. (1978) in press.
23. Ohno, S., Natsu-ume, S., and Migita, S., J. Natl. Cancer Inst. (1975) 55, 569-578.
24. Funakoshi, I., Nakada, H., and Yamashina, I., J. Biochem. (1974) 76, 319-333.
25. Bhavanandan, V. P., and Davidson, E. A., Biochem. Biophys. Res. Commun. (1976) 70, 139-145.
26. Gasic, G., and Gasic, T., Nature (1962) 196, 170.
27. Codington, J. F., Sanford, B. H., and Jeanloz, R. W., J. Natl. Cancer Inst. (1970) 45, 637-647.
28. Codington, J. F., Sanford, B. H., and Jeanloz, R. W., Biochemistry (1972) 11, 2559-2564.
29. Codington, J. F., Linsley, K. B., Jeanloz, R. W., Irimura, T., and Osawa, T., Carbohydr. Res. (1975) 40, 171-182.
30. Friberg, S., Jr., J. Natl. Cancer Inst. (1972) 48, 1463-1476.
31. Friberg, S., Jr., J. Natl. Cancer Inst. (1972) 48, 1477-1489.
32. Hauschka, T. S., Weiss, L., Holdridge, B. A., Cudney, T. L., Zumpft, M., and Planinsek, J. A., J. Natl. Cancer Inst. (1971) 47, 343-359.
33. Grohsman, J., and Nowotny, A., J. Immunol. (1972) 109, 1090-1095.
34. Lippman, M. M., Venditti, J. M., Kline, I., and Elam, D. L., Cancer Res. (1973) 33, 679-684.
35. Friberg, S., Jr., and Lilliehook, B., Nature New Biol. (1973) 241, 112-113.
36. Springer, G. F., Codington, J. F., and Jeanloz, R. W., J. Natl. Cancer Inst. (1972) 49, 1469-1470.
37. Cooper, A. G., Codington, J. F., and Brown, M. G., Proc. Natl. Acad. Sci. U.S.A. (1974) 71, 1224-1228.
38. Codington, J. F., Cooper, A. G., Brown, M. C., and Jeanloz, R. W., Biochemistry (1975) 14, 855-859.
39. Cooper, A. G., Codington, J. F., Miller, D. K., and Brown, M. C., prepared for publication.
40. Codington, J. F., Cooper, A. G., Miller, D. K., Slayter, H. S., and Jeanloz, R.W., prepared for publication.
41. Miller, S. C., Hay, E. D., and Codington, J. F., J. Cell Biol. (1977) 72, 511-529.
42. Codington, J. F., and Bruns, R. R., unpublished results.
43. Codington, J. F., Van den Eijnden, D. H., and Jeanloz, R. W., in Harmon, R. E., Ed., "Cell Surface Carbohydrate Chemistry", pp. 49-66, Academic Press, New York, 1978.
44. Slayter, H. S., and Codinton, J. F., J. Biol. Chem. (1973) 248, 3405-3410.
45. Van den Eijnden, D. H., Evans, N., Codington, J. F., and Jeanloz, R. W., prepared for publication.

46. Marchesi, V. T., and Furthmayr, H., Ann. Rev. Biochem. (1976) 45, 667-698.
47. Wiener, G., Klein, G., and Harris, H., J. Cell Sci. (1971) 8, 681-692.
48. Klein, G., Friberg, S., Jr. Wiener, F., J. Natl. Cancer Inst. (1973) 50, 1259-1268.
49. Friberg, S., Jr., Klein, G., Wiener, F., J. Natl. Cancer Inst. (1973) 50, 1269-1286.
50. Berenblum, I., in Florey, H. W., "General Pathology", pp. 645-667, W. B. Saunders Co., Philadelphia, 1970.
51. Miller, D. K., Cooper, A. G., Brown, M. C., and Jeanloz, R. W., J. Natl. Cancer Inst. (1975) 55, 1249-1252.
52. Cooper, A., Morgello, S., Miller, D., and Brown, M., in "Cancer Invasion and Metastasis: Biologic Mechanisms and Therapy", pp. 49-64, Raven Press, New York, 1977.
53. Hagmar, B., and Ryd, W., Transplantation (1977) 23, 93-97.

RECEIVED February 27, 1978.

16

Glycoconjugate Alterations in Malignant and Inflammatory Disease of the Colon

YOUNG S. KIM, JAMES S. WHITEHEAD, BADER SIDDIQUI, and DEAN TSAO

Gastrointestinal Research Laboratory, Veterans Administration Hospital and University of California, San Francisco, CA 94121

In the United States large bowel cancer has the largest incidence among all malignant diseases in men and women except for easily detectable and curable skin cancer. As is the case with most tumors, the survival rate and prognosis of patients with colonic cancer depends upon the grade and the extent of malignancy. Low grade, more differentiated colonic tumors as well as disease confined to the mucosa and submucosa are associated with longer patient survival. The prognosis is dismal for those who have a high rate of tumor growth or nonlocalized disease (1). The final outcome rests with a balance between the biological properties of the tumor and the type and extent of the host response. This appears to be reflected in the wide variation in the clinical development of the disease and in the growth rate and changes in biochemical and immunochemical properties of the tumor.

In considering development of methods for early diagnosis of colonic cancer in those patients of high risk, such as individuals with inflammatory bowel diseases or intestinal polypoid lesions, and development of rational methods to control colonic neoplasia, it is important to identify and characterize organ specific tumor markers and to clarify the molecular mechanisms involved in the altered biochemical and biological properties that accompany malignant transformation in the colon.

The involvement of cell surface components in cell-cell interaction, in growth regulation of cells and in the process of malignant transformation has been reported by many investigators (2-5). Much of our current knowledge has been obtained from studies on virally transformed murine or avian fibroblasts. Although most human carcinomas, including colonic tumors, are epithelial in type and origin, relatively little is known about cell surface properties of normal and cancerous epithelial cells. Since glycoproteins and glycolipids are major components of surface membranes of mammalian cells, especially those of the intestinal mucosa, and glycoconjugates have been implicated in diverse cellular phenomena, our laboratory has been examining these

0-8412-0452-7/78/47-080-295$05.00/0

components in normal and cancerous colonic mucosal cells. Three experimental models are being used for our investigation: 1) human cancerous and normal colonic tissues obtained at surgery, 2) human colonic cancer cells maintained in monolayer culture, and 3) experimental colonic cancer in rodents induced by a chemical carcinogen, 1,2-dimethylhydrazine.

Some of the approaches that we are using in the study of membrane glycoconjugates in normal and cancerous colonic cells include: 1) the isolation and chemical and immunochemical characterization of membrane glycoconjugates, 2) the identification and characterization of both membrane and cytoplasmic glycoconjugates using fluorescein- or radio-labeled lectins, 3) external cell surface radio-labeling with galactose oxidase and iodium borotritide or with lactoperoxidase-catalyzed iodination, and 4) a study of the metabolism and turnover of membrane glycoconjugates using *in vivo* and *in vitro* incorporation of radio-labeled precursors. In the present paper, we will confine the discussion primarily of our data on the chemical and immunochemical properties of human colonic tumors and the isolation and characterization of surface membrane glycoproteins.

Studies in Human Colonic Adenocarcinomatous Tissues.

Many antigenic activities have been ascribed to glycoproteins and glycolipids. For example, gastrointestinal mucins and intestinal mucosal cell surface membrane glycoconjugates have been reported to possess blood group activities.

Immunofluorescence studies on tissues indicated that ABH blood group activity was low or absent in primary gastrointestinal tumors as well as in other epithelial tumors (6). The loss of these antigenic activities may be due to either decreased synthesis or increased degradation of these antigens in tumor tissues. To determine if deletion of blood group activity occurs in colonic cancer tissues and to clarify the metabolic basis for such alterations, the levels of blood group antigens were measured in cancerous and adjacent normal colonic mucosa using hemagglutination inhibition methods (7). The results are shown in Table I. The data reported here are for tissues taken from blood group A patients. The levels of these antigens were determined by measuring the ability of membrane glycopeptides to inhibit the agglutination of red blood cells by various agglutinins. These agglutinins bind the specific sugar structures indicated in the parentheses. A negative sign indicates no detectable antigen. The greater the number of pluses, the higher the level of antigen. In all patients examined, blood group A activity could not be detected in the cancer tissue but normal levels were present in adjacent tissues. Two patients who had high levels of Le^a activity in normal mucosa had no detectable Lewis A activity in the cancer tissue. The use of *Ricinus communis* lectin (RCA), which recognizes specific sugar structures including the immediate precursor of the H structure, detected higher levels of terminal galactose residues in the cancerous tissues. No consistent change in Con A-binding glycopeptides was observed.

Since alterations in blood group antigenicity and lectin reactivity suggested changes in carbohydrate composition, the membrane fraction of cancerous and adjacent tissues was subjected to chemical analysis.

We measured the amounts of 3 neutral sugars, 2 aminosugars and sialic acid in normal and cancerous tissues. The results indicated that in general the glycoprotein fraction of the cancer tissues contained less of each of the sugars than did the normal tissues (Figure 1). Fucose, glucosamine, galactosamine and sialic acid were reduced significantly in the cancerous tissues, while galactose was reduced to a lesser extent. The mannose content was only slightly reduced.

Not only was the total sugar content reduced, but each sugar was reduced to a different extent. This is evident when the molar ratios of each sugar normalized to mannose are compared (Figure 2). For example, N-acetylgalactosamine which had a molar ratio to mannose of 1.5 in normal tissue was reduced to 0.4 in cancer tissues. This reduction may be directly related to the loss in blood group A activity since N-acetylgalactosamine is a part of this blood group determinant. To establish whether the observed alterations in antigen levels and in carbohydrate content were due to decreased synthesis or to increased degradation of the carbohydrate moiety, we examined two groups of enzymes: glycosyltransferases, involved in their biosynthesis, and glycosidases, which degrade them.

The A, B or O blood group of an individual is determined by the presence or absence of two specific glycosyltransferases which are products of the A or B gene. The relationship between blood groups of an individual and the presence of these two glycosyltransferases is shown in Table II. Only individuals of blood type A or AB contain a specific N-acetylgalactosaminyltransferase in their intestinal mucosa, while only those of blood types B and AB have a specific galactosyltransferase. Subjects who are blood type O lack both of these enzymes. The immediate precursor for the A and B determinants is the O(H) structure, so that it can be assumed that all individuals who secrete A. B or O blood group substances have the glycosyltransferases required to synthesize the precursor oligosaccharide. Four other glycosyltransferases shown in Table II are not dependent on the individual's blood type.

Figure 3 shows some of the glycosyltransferase activities assayed in tumors and in tissues taken from the histologically normal adjacent mucosa. Also shown are the levels of these glycosyltransferases in normal colonic tissues taken from patients with diseases other than cancer (diverticulitis or volvulus of the colon). There were no significant differences between the two

Table I

Hemagglutination Inhibition by the Membrane Glycopeptides of the Normal and Cancerous Mucosa (7)

	Dolichos biflorus (A)		Anti-Le^a serum (Le^a)		Ricinus communis (Gal)		Con A (Man,GlcN)	
Patients	NL	CA	NL	CA	NL	CA	NL	CA
1	++*	-	-	-	-	+	+++	+++
2	++++	-	++++	-	-	++	++	++++
3	++++	-	++++	-	-	++	++++	++++
4	++++	-	-	-	-	-	++++	++++
5	++++	-	++++	++++	-	+++	++++	++++
6	+++	-	++++	++++	-	+	++++	++

*-, no inhibition up to 300 ug membrane protein; +, inhibited by 200-300 ug; ++, inhibited by 100-200 ug; +++, inhibited by 50-100 ug; and ++++, inhibited by 50 ug.
A, blood group A; Le^a, blood group Lewis A; Gal, galactose; Man, mannose; GlcN, N-acetylglucosamine.

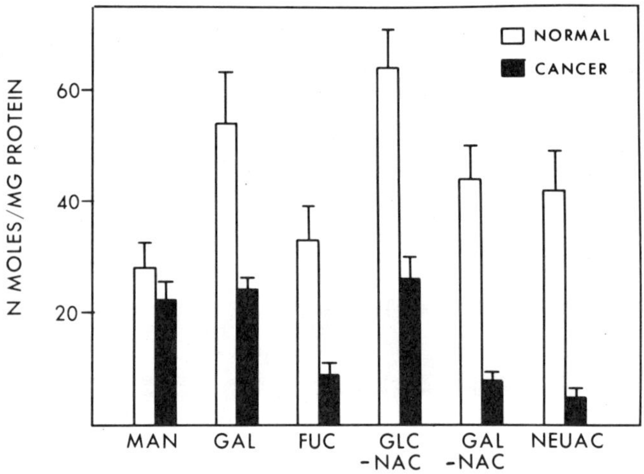

Figure 1. Carbohydrate composition of the membrane fraction of normal and cancerous colonic mucosa

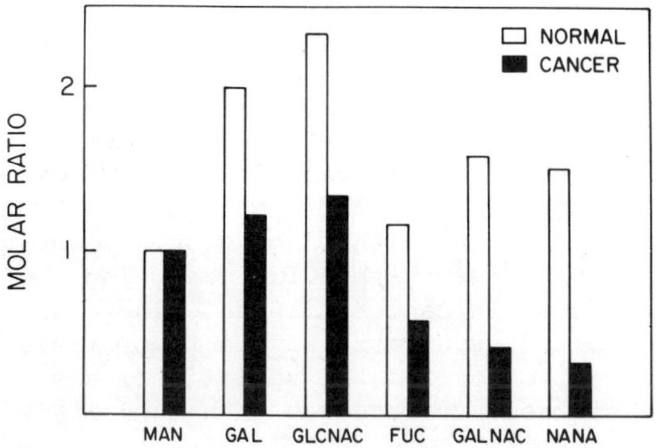

Figure 2. Molar ratio of carbohydrate relative to mannose in membrane fraction of normal and cancerous colonic mucosa

Table II

Glycosyltransferase Activities in Human Colonic Mucosa

Glycosyltransferase	Blood Type A (6)	AB (4)	B (4)	O (6)
	cpm/mg protein/hr ($X\ 10^{-3}$)			
N-acetylgalacto-saminyl-transferase (A)	210 ± 97	176 ± 30	0	0
Galactosyl-transferase (B)	0	59 ± 12	98 ± 27	0
Galactosyl-transferase I	13 ± 3	12 ± 4	15 ± 4	9 ± 4
Galactosyl-transferase II	77 ± 17	90 ± 24	110 ± 32	79 ± 29
Sialyl-transferase	54 ± 8	45 ± 6	32 ± 6	44 ± 8
Fucosyl-transferase	117 ± 32	110 ± 30	90 ± 23	119 ± 27

Mean ± S.D.

normal groups.

In view of the absence of detectable A antigen in cancerous tissues, it might be expected that the level of the transferase responsible for its synthesis might be low. Indeed the tumors had only 20% of the activity found in the normal tissues. Although not all of the enzymes are shown, two fucosyltransferases and two galactosyltransferases were found to be lower in the cancer tissues, but the difference was statistically significant only for one galactosyltransferase. The level of a sialyltransferase was not significantly different in normal and cancerous tissues. In two patients with blood group B, the galactosyltransferase activity responsible for producing the B determinant was markedly reduced in cancer tissues. These data suggest that the absence of blood group activity in tumor tissues is a consequence of a reduction or loss of specific glycosyltransferases. The sialyltransferase activity was unchanged in the colonic cancer tissues despite the significantly lower level of N-acetylneuraminic acid in the colonic tumor tissues.

Several sialyltransferases in mammalian tissues have been described; the one shown in Figure 3 utilizes desialyzed α_1-acid glycoprotein as an acceptor. This enzyme catalyzes the addition of sialic acid (N-acetylneuraminic acid) to terminal galactose residues of oligosaccharides. With desialyzed fetuin as an acceptor, it is likely that the same enzyme is being measured. However, with desialyzed ovine submaxillary mucin as an acceptor, an eznyme which catalyzes the transfer of sialic acid to N-acetylgalactosamine is measured. Warren and his coworkers observed that virally transformed cells have a higher level of a sialyltransferase than the normal cells when the membrane glycopeptides, obtained from transformed fibroblasts and from Novikoff ascites tumor cells, were used as acceptors. Table III shows the activities of four sialyltransferases measured in tumors, adjacent normal tissue and colonic mucosa from patients with inflammatory bowel disease (8). Sialyltransferase activity was essentially the same using three of the acceptors including glycopeptides from Novikoff ascites tumor cells. However, when desialyzed OSM was the acceptor, the sialyltransferase activity was substantially lower in the cancer tissues. From these data, only the activity of a sialyltransferase involved in the addition of N-acetylneuramine acid to a mucin acceptor was reduced in tumors.

Experimentally determined activities of the glycosyltransferases in mixing experiments agreed with the theoretical predictions, demonstrating that no inhibitors or activators of these enzymes are present in tumor or normal tissues. Thus, the observed reduction in glycosyltransferase activity appears to be due to an actual reduction of the enzymes.

Alterations in antigenic determinants and carbohydrate compositions of cancerous tissue could also be due to changes in glycosidase activity. Five glycosidase activities were measured

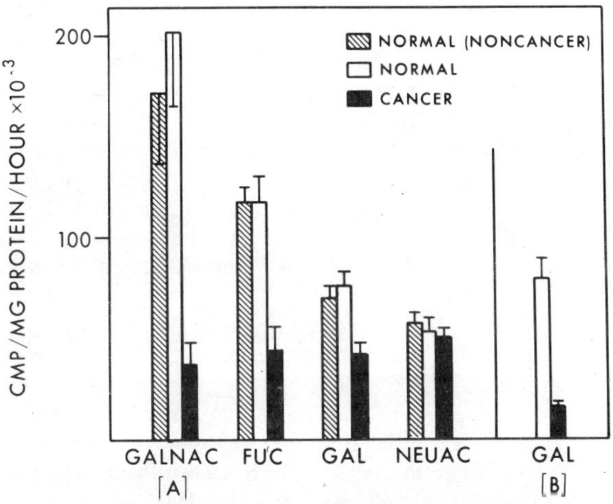

Figure 3. Glycosyltransferase activities in normal and cancerous colonic mucosa

Table III

Sialyltransferases of Normal and Cancerous Colonic Mucosa

Acceptors	Normal (6)	Cancer (6)	Inflammatory Bowel Disease (3)
	cpm/mg protein/hr (X 10^{-3})		
α_1-acid glycoprotein	34.2 ± 16.1	33.4 ± 11.2	36.4 ± 12.4
Fetuin	41.4 ± 16.9	44.1 ± 18.8	40.7 ± 14.6
Tumor glycopeptides	4.9 ± 1.8	5.0 ± 1.6	5.1 ± 1.5
Ovine submaxillary mucin	14.2 ± 9.6	3.7 ± 1.7**	17.6 ± 4.9

*All the acceptors are desialyzed. **$P < 0.025$.

Cancer Research

using 4-methylumbelliferone glycosides as substrates. To deter- the level of blood group A degrading glycosidases, a radioactively labeled blood group A oligosacchride was prepared by incubating fucosyllactose with ^{14}C-labeled UDP-N-acetylgalactosaminyltransferase. Of 6 glycosides shown in Figure 4, no appreciable difference was seen between normal and cancerous tissue. In fact, the mean level of the enzyme capable of degrading the A antigen was lower in the cancer tissues. From this information, the lower level of A antigen in cancerous tissue probably is due to increased glycosidase activity. Similar results were also obtained with tissues from patients with O and B blood types. One should be cautious in correlating the carbohydrate content of the tissues or cells with the levels of activities of these enzymes since the glycoprotein or oligosaccharide acceptors or synthetic substrates used may not reflect accurately the structure of natural acceptors or substrates in the tissues or cells.

Figure 5 depicts schematically what may occur in normal-to-tumor cell transformation. The terminal trisaccharide sequence at the non-reducing end of the carbohydrate side chain is the blood group A determinant. When galactose is substituted for N-acetylgalactosamine, the B determinant is formed.

Our data show that A or B blood group activities which were present in the normal adjacent mucosa disappeared or were markedly reduced in colonic cancer cells and that there was an increased exposure of terminal galactose residues. This suggests that the carbohydrate side chains of blood group active membrane glycoproteins may be incomplete in the colonic cancer tissue.

A similar deletion or reduction in the antigenicity of ABH blood group active in gastrointestinal tumors has been reported previously by Dr. Hakomori's group (9) and recently by Dr. Siddiqui in our laboratory (10). Table IV gives the glycolipid-associated blood group ABH activities obtained from normal and cancerous colonic tissues (10). In the eight specimens examined, there was a decrease or absence of glycolipid associated blood group activity in all cancer tissues compared to adjacent normal tissues.

An incompletion of the carbohydrate side chains of blood group active glycolipids and glycoproteins resulting in the appearance of new, precursor, blood group antigens such as T and i has been reported to occur in gastrointestinal and breast tumors by several investigators including Drs. Springer (11) and Feizi (12). These data and earlier work by Drs. Hakomori (13), Fishman (5) and Robbins (14) using transformed fibroblasts have been the basis for the hypothesis that a deletion or reduction in carbohydrate antigens is accompanied by an increase in its less complex precursors.

Our recent studies on glycolipid compositions of normal and cancerous colonic tissues indicate that a simplification of

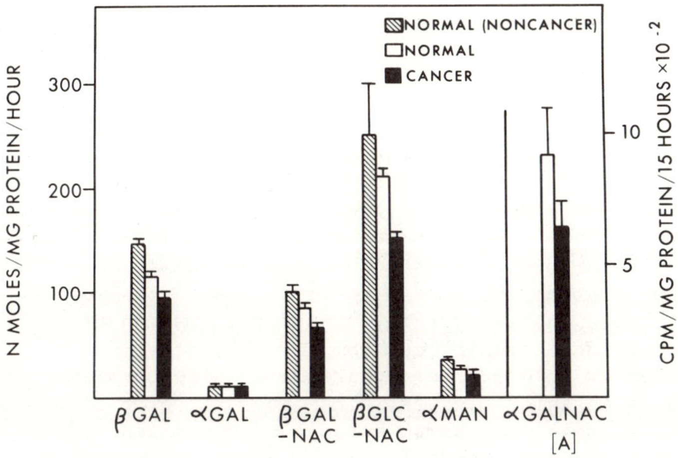

Figure 4. Glycosidase activities in normal and cancerous colonic mucosa

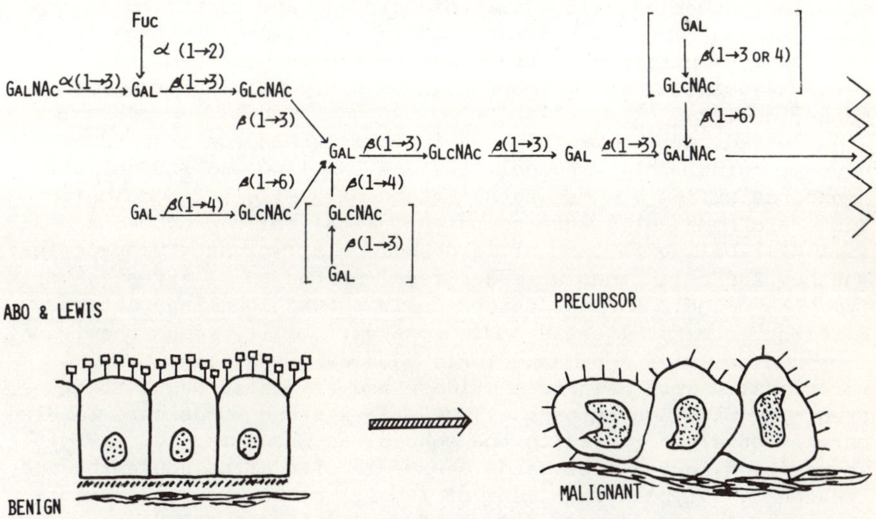

Figure 5. Glycoprotein-associated ABO(H) glycoproteins in colonic cancer

carbohydrate side chains does not occur in all classes of glycolipids (10)

Gangliosides, both simple and complex sialic acid-containing glycolipids such as GM_3, GD_3, GM_1 and bD_{1a}, were increased over two-fold in 3 out of 8 cancer tissues (10). In 5 cases no difference was observed. Of the latter group, none of the tumors showed lymph node involvement while two of the former had substantial lymph node metastasis. Pathological examination showed the third tumor of this group to be an extensive mucoid carcinoma but examination of lymph node involvement was incomplete. Thus, more complex glycolipids seem to be abundant in some human colonic cancer tissues, unlike virally-transformed malignant fibroblasts which become enriched in less complex precursors.

Another example of the appearance of more complex carbohydrate side chains in tumor tissues is the appearance of Forssman antigenic activity in some human colonic cancer tissues from individuals who are Forssman negative (15). These data will be described by Dr. Hakomori. To understand tumor behavior, it is important to compare structures and processes between the tumor and normal tissues. However, some problems or studies are best approached by examining tumor cells grown in culture.

To study some aspects of tumor cell biochemistry, we have begun to examine human colonic adenocarcinoma cell lines maintained in monolayer culture.

Cell Surface Glycoproteins of Human Colonic Adenocarcinoma Cells

We have used a method for specifically labeling galactose and galactosamine residues which are present at non-reducing termini of oligosaccharides with galactose oxidase and tritiated sodium borohydride. This method was used to investigate the organization and chemical nature of surface membrane carbohydrates, such as those associated with glycoproteins and glycolipids of colonic cancer cells.

The radioactive profiles of tritium-labeled cell surface glycoproteins, following cell surface labeling and SDS-gel electrophoresis from 5 human fetal intestinal cells and 6 colonic adenocarcinoma cells were compared. Fetal cells produced a similar labeling pattern of large molecular weight glycoproteins, whether the cell lines were derived from the small intestine or from the colon. Colonic cancer cells showed labeling patterns distinct from fetal cells with some variability among them.

When cells of a human colonic adenocarcinoma line, SKCO-1 were labeled with galactose oxidase and tritiated $NaBH_4$ after treatment with neuraminidase, the pre-existing bands were labeled more intensely compared to the control as shown in Figure 6 (16). In addition, a few more bands became labeled after neuraminidase treatment. Trypsin treatment of labeled cells produced a diminution of only few bands. Some of the tritiated galactoproteins were also found to be radioiodinated by lactoperoxidase.

Table IV

Glycolipid Blood Group Activities in Colonic Mucosa (16)

Patient	Blood Type	ABH blood group activity Normal	ABH blood group activity Cancer
1	A	++*	-
2	AB	++++	+
3	O	++++	-
4	O	++++	++
5	O	++	+
6	O	++++	-
7	O	++++	-
8	O	++++	-

*++++, highest dilution of glycolipid which inhibits hemagglutination; -, no inhibition.

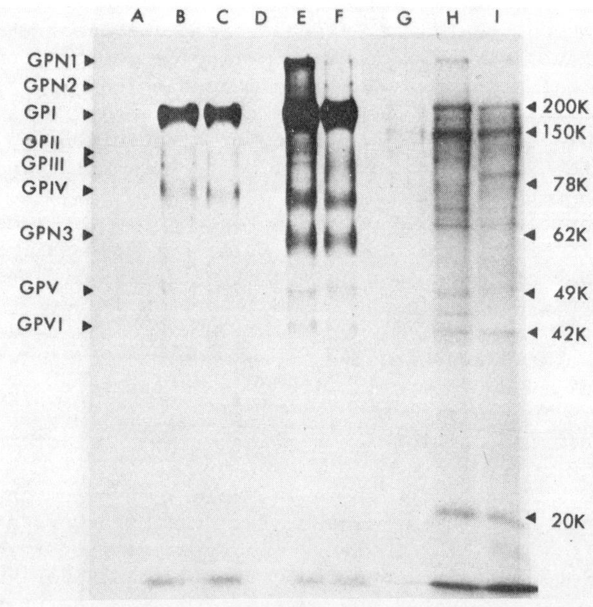

Figure 6. Autoradiographic pattern of ^{3}H (A to F) and ^{125}I-labeled (G to I) SKCO-1 cell membrane proteins separated in SDS-polyacrylamide slab gel. A, no treatment; B, galactose oxidase treatment; C, galactose oxidase followed by trypsin treatment; D, treatment with neuraminidase; E, treatment with neuraminidase followed by galactose oxidase; F, treatment with neuraminidase followed by galactose oxidase with subsequent trypsin treatment; G, lactoperoxidase alone; H, lactoperoxidase and H_2O_2; I, lactoperoxidase and H_2O_2 with subsequent trypsin treatment (16).

Additional ^{125}I labeled membrane proteins not labeled by tritium could also be detected.

Following external cell surface labeling with galactose oxidase and triated borohydride, the membrane fraction of SKCO-1 cells was solubilized with detergent and fractionated by a series of chromatographic methods consisting of affinity chromatography on *Ricinus communis* and Concanavalin A-Sepharose columns, and gel filtration using Sephadex G-200.

Using this procedure, we succeeded in isolating a labeled glycoprotein which we designated as Galactoprotein I. This protein corresponded to the slow migrating major band which was present only in SKCO-1 cells.

The isolated Galactoprotein I was found to be homogenous on SDS-gel electrophoresis when stained for protein or carbohydrate. When compared to carcinoembryonic antigen on SDS-gel electrophoresis, this galactoprotein was found to have an electrophoretic mobility identical to that of ^{125}I labeled CEA and had an apparent molecular weight of 200,000.

When immunodiffusion of Galactoprotein I against CEA antiserum was carried out, a single precipitin line of fusion formed indicating that Galactoprotein I shares immunological identity with CEA. No precipitin line was formed between anti-CEA antiserum and other galactoproteins. The results of quantitative precipitation indicate that Galactoprotein I could be completely precipitated by CEA antiserum..

Galactoprotein I could not be released from the plasma membranes of SKCO-1 cells with either EDTA, buffered saline or sonication but could be solubilized effectively with detergents. These properties indicate that Galactoprotein I is an "integral" protein of the membrane.

The amino acid and carbohydrate composition of Galactoprotein I was determined and compared, with those obtained by various investigators, for carcinoembryonic antigen. Although not shown, the amino acid composition was similar between galactoprotein I and CEA.

Carbohydrate analysis, shown in Table V revealed that Galactoprotein I contained the same neutral and amino sugars as CEA reported by others (1). However, the relative proportion of each sugar in Galactoprotein I and CEA differed considerably. For instance, much less fucose and more sialic acid was present in Galactoprotein I than the published values for CEA. This difference may, in part, be due to the difference in tissues from which these materials are prepared. CEA materials are likely to have been prepared from colonic cancer metastasized to the liver, while SKCO-1 cells were derived from a primary colonic tumor. It is possible that the decreased sialic acid and increased fucose content of the surface membrane glycoprotein of these cancer cells may be related to the capacity of colonic tumor cells to invade and metastasize to other organs. Alternatively, the difference in the proportion of various carbohydrates may be due to the

Table V

Carbohydrate Content of Galactoprotein I (16)

	GPI	CEA			
		(Weight of monosaccharide/weight total carbohydrate) X 100			
		A	B	C	D
Fucose	6.1	11.3	12.0	15.2	21.0
Mannose	20.5	12.8	21.7	14.3	11.3
Galactose	24.8	19.9	15.6	17.9	26.6
Sialic acid	18.8	8.1	7.9	10.7	3.9
N acetylgalactosamine	6.3	2.6	N.S.*	1.8	1.1
N-acetylglucosamine	23.6	43.3	43.3	40.2	35.7

*N.S. = not significant

‡Carbohydrate analyses of CEA were reported by: A, Banjo *et al.*, B, Kupchik *et al.*, C, Coligan *et al.*, and D, Westwood *et al.*

difference between tissues and cultured cells.

We also examined the disappearance from cell membrane of radioactivity associated with Galactoprotein I as a function of time after galactose oxidase and tritiated sodium borohydride labeling. The turnover rate of labeled Galactoprotein I indicated that it was released from the membrane slowly into the medium with a half-life of about 5 days. When labeled Galactoprotein I isolated from cells immediately after and 5 days after galactose oxidase labeling was compared, no difference in the mobility on SDS gel electrophoresis was observed (16).

When the subcellular distribution of the carcinoembryonic antigen (CEA) was examined in 5 human colonic cancer cell lines, 85-90% of the activity was found to be associated with the membrane fraction of each cell. Considerable variation in the amount of CEA content was observed among the 5 cell lines. Only one cell line produced an appreciable amount of CEA, namely SKCO-1. This line contained 14 ug of CEA per mg protein in the cell homogenate. It is also of interest that one cell line secreted into the media only 6% of the cellular CEA while another line secreted over 300% of the cellular CEA in 3 days. Although the numbers of cell lines examined are small, these results suggest that membrane-bound CEA from each colonic cancer cell line has a different turnover or shedding rate. It may be significant that the SKCO-1 line was the least tumorigenic when injected to nude mice and produced the largest amount of CEA.

Inflammatory bowel disease such as ulcerative and granulomatous colitis are precancerous conditions, and patients with such diseases are at higher risk for developing colon cancer than are healthy individuals. The more extensive the involvement of the bowel and the longer the duration of the inflammatory diseases, the higher the incidence of colonic cancer.

Although considerable histochemical changes have been observed in the mucosal cells indicating alteration in glycoconjugates, only limited data are available on many of the biochemical parameters in these disease states. In the few patients with inflammatory bowel diseases that we have studied, we were unable to observe significant alterations in the activities of glycosyltransferases or glycosidases seen in patients with colonic tumors (8). Clearly, there is a need to carefully examine glycoconjugates and their associated enzymes in the intestinal mucosa of these patients. By understanding how and which membrane glycoproteins and glycolipids are synthesized and degraded in normal, premalignant and cancerous mucosa, we should be in a much better position to understand carcinogenesis.

Summary

As information is accumulated, it is becoming increasingly apparent that alterations in carbohydrate content and metabolism are associated with the malignant process. Our approach has been to examine glycoproteins and glycolipids, especially those on the cell surface, in tissues and cultured cells and the enzymes involved in the synthesis and degradation of their oligosaccharide side chains. With regard to the studies comparing tumors with adjacent normal tissues, we found a deletion or marked reduction in A and B blood group activities in tumor tissues with a concomitant increase in *Ricinus communis* binding components. The carbohydrate contents of the glycoprotein fraction from membranes and cytoplasm were generally reduced in cancerous tissues. Sugars associated with mucin type glycoproteins were those most affected. The activities of some glycosyltransferases responsible for the formation of the A and B blood group determinants were markedly lower in the tumors while glycosidase activities were relatively unchanged.

Regarding studies of tumor cells grown in culture, we have isolated a membrane glycoprotein antigen from the surface of a cultured human adenocarcinoma cell line which is immunologically identical to CEA. This glycoprotein is a 200,000 molecular weight integral membrane protein with a carbohydrate composition qualitatively similar, but quantitatively distinct from CEA. Studies of the subcellular distribution of CEA have indicated that CEA in several cultured human colonic tumor cells is a membrane protein and that the rate of appearance of CEA in the culture medium was different among these cell lines.

Acknowledgement

This work was supported in part by United States Public Health Service Grant CA-14905 from the National Cancer Institute through the National Large Bowel Cancer Project and by the Veterans Administration Medical Research Service. Reprint requests should be addressed to Young S. Kim, M.D.

Literature Cited

1. Kim, Y. S., Tsao, D., and Whitehead J. Cancer (1977) 40:2473-2478.
2. Meezan, E., Wu, H. C., Black, P. H. and Robbins, P. W. Biochemistry (1969) 8:2518-2524.
3. Warren, L., Fuhrer, J. P., and Buck, C. A. Fed. Proc. (1973) 32:80-85.
4. Gahmberg, C. G., and Hakomori, S. Proc. Natl. Acad. Sci. U.S. (1973) 70:3329-3333.
5. Fishman, P. H., Brady, R. O., Bradley, R. M., Aronson, S. A., and Todaro, G. J. Proc. Natl. Acad. Sci. U.S. (1974) 71: 298-301.
6. Davidsohn, I., Kovarik, S., and Lee, C. L. Arch. Pathol. (1966) 81:381-390.
7. Kim, Y. S., Issacs, R., and Perdomo, J. M. Proc. Natl. Acad. Sci. U.S. (1974) 71:4869-4873.
8. Kim, Y. S., and Issacs, R. Cancer Res. (1975) 35:2092-2097.
9. Stellner, K. and Hakomori, S. Biochem. Biophys. Res. Commun. 55:439-445.
10. Siddiqui, B., Whitehead, J., and Kim, Y. S. J. Biol. Chem. (1978) 253:2168-2175.
11. Springer, G. F., and Desai, P. R., Yang, H. J., and Murthy, M. S. Clin. Immun. and Immunopath. (1977) 7:426-441.
12. Feizi, T., Turberville, C., and Westwood, J. H. Lancet (1975) 2:391-393.
13. Nakomori, S., and Murakami, W. T. Proc. Natl. Acad. Sci. U.S. (1968) 59:254-261.
14. Robbins, P. W., and MacPherson, I. Nature (1971) 229:569-570.
15. Hakomori, S., Wang, S. W., and Young, Jr., W. W. Proc. Natl. Acad. Sci. U.S. (1977) 74:3023-3027.
16. Tsao, D. and Kim, Y. S. J. Biol. Chem. (1978) 253:2271-2278.

RECEIVED March 14, 1978.

17

Human Carcinoma-Associated Precursors of the Blood Group MN Antigens

G. F. SPRINGER, P. R. DESAI, and M. S. MURTHY

Department of Immunochemistry Research, Evanston Hospital, 2650 Ridge Ave., Evanston, IL 60201

E. F. SCANLON

Department of Surgery and Microbiology—Immunology, Evanston Hospital, 2650 Ridge Ave., Evanston, IL 60201

M and N are the major antigens of the second human blood group system (1). Glycoproteins are the predominant carriers of the human red blood cell MN specificities which are determined by carbohydrates (cf. 2,3). A look at the structural basis of the interrelation of blood group MN specificities, which also occur on human epithelial tissues (4,5,6), and their carbohydrate precursor specificities T and Tn is important for an understanding of the human carcinoma-associated antigenic specificities T and Tn, as is also the appreciation that all humans possess anti-T and anti-Tn but not anti-M and anti-N antibodies. The biosynthesis of T-, N- and M- specific structures and the surprising significance of the carcinoma-associated precursor antigens T and Tn for the human immune response, humoral as well as cell-mediated, against adenocarcinoma are the main topics of this paper.

Materials and Methods

Bacteria and Vaccines. Bacteria were obtained from the American Type Culture Collection (Washington, DC); bacterial antigens were prepared in this laboratory or given by Drs. D.A.L. Davies, O. Lüderitz and E. Rietschel. Gram-negative bacteria were grown on fully defined media using procedures described previously (7). Vaccines were from the individuals and Drug Houses listed in Table IV.

Carcinomatous and Control Tissue. Mammary gland tissues originated from ductal adenocarcinomata Stage I (International Nomenclature) which had not invaded regional lymph nodes and from Stage II and III ductal carcinomata as well as from nonmalignant breast lesions (fibroadenomatosis and fibrocystic disease) of U.S. women 18 to 86 years old. Glandular tissue of the primary tumor as well as distant metastases were obtained under aseptic conditions during surgery. Specimens left in the Pathology Department for more than a few minutes and autopsy tissues were not used because of autolytic changes. No specimen showed obvious

0-8412-0452-7/78/47-080-311$05.00/0

bacterial contamination and none tested was pyrogenic. The glandular structures were sliced and freed of other tissues. Histologic sections were made to assure that only predominantly malignant regions were used in case of carcinoma. Cell membranes were prepared by hypotonic lysis as described previously following procedures used by others (8,9).

The two human colon carcinoma-derived tissue culture cell lines were given by Dr. B. Tom (10); cells derived from normal colon epithelium were not available.

Preparation of T antigen from Human Blood Group O,MN Red Cells. The mode of preparation has been described earlier (11,12, 13). From red blood cell stroma the MN antigens were extracted with aqueous 45% phenol plus electrolyte at 23-25°C. Active material was isolated by fractional ultracentrifugation and ethanol fractionation. The glycoprotein which precipitated between 45 and 70% ethanol concentration was exhaustively dialyzed, freeze-dried and used for T antigen preparation. T antigen was uncovered by the removal of NAN from the isolated O,MN antigens with *Vibrio cholerae* neuraminidase (RDE) at pH 6.8; RDE was then inactivated by heating at 100°C for 5 min. T antigen was prepared aseptically.

Absorption of Antisera. Because of their strict intraspecies specificity human antisera only were used against blood group antigens M, N, T, and Tn, unless stated otherwise; all sera used have been described before (14,37). The antisera were absorbed once with the membrane-cytoplasm preparations or homogenized glandular tissue or colon carcinoma-derived tissue culture cells or bacteria under standard conditions (7,14,15). Control absorptions of human anti-blood group B sera with blood group H(O)-specific human breast glandular tissue, benign or malignant, resulted in no titer decrease, nor did absorption of human anti-Rh_o(D) sera with benign or malignant tissues from blood groups A_1 and H(O) persons (see also Table I).

Blood Collection and Serologic Procedures. Bloods for determination of anti-T titers were collected 24 to 48 hr before or immediately after surgery, unless stated otherwise. The clotted bloods were coded and recorded by individuals who were not involved in the assays. Sera were prepared and used at once or stored in aliquots at -20°C.

For serological assays O,MM and O,NN erythrocytes were obtained, stored, washed, and employed as such or after T-activation with RDE (2). Tn red blood cells were donated by Drs. G. Leonard (Cla.Ric.) and W.D. Bowman (Car.Lip.) (*cf.* 16,17). Hemagglutination and hemagglutination inhibition tests including controls and standards were performed and interpreted by routine procedures (7,13). Tests were scored (*cf.* 18) independently by three individuals to whom nature of tissues used for absorption

Table I. Percent Absorption of Anti-Blood Group M, N, and Precursor Agglutinins with Human Breast Gland Membrane-Cytoplasm Preparations, Colon Carcinomata and Controls

Human sera	Breast glandular tissue			Colon carcinoma (Tissue culture)	OMM erythrocytes	"Buffy coat"
	Benign	Stage I carcinoma	Stages II & III carcinoma			
	6[a]	3	15	2[b]	3	2
Anti-M	52[c] (30 - 75)	72 (60 - 80)	66 (0[d] - > 90)	49 - 59	100	
Anti-N	58 (50 - 66)	53 (20 - 70)[e]	77 (50 - > 80)	53 - 100	10 - 25	
Anti-T	< 1 (0 - 6)	53 (33 - 75)	70 (30 - > 90)	50 - 83	0	0
Anti-Tn	< 10 (0 - 25)	42 (25 - 50)	64 (50 - > 80)	58 - 98	0	0

[a] Number of cases. [b] No normal epithelial cells available. [c] Arithmetic averages if > 2 persons tested, figures in parentheses = range. [d] M activities were found, with one exception, if M antigen was expressed on the subject's red cells (see text). [e] Membranes from MM individual.

and source of the sera were unknown; three serum standards were included as controls in all determinations. Specimens were decoded only after completion of the tests.

Biosynthetic Procedures. For enzymatic transformation of Tn to T group B red cells of Cla.Ric., which are completely Tn-transformed, were used (19,20). Sera of blood group B and A_1B persons freed of anti-T and -Tn agglutinins served as β-galactosyltransferase source, UDP-Gal as Gal donor, and ATP, $MnCl_2$ and $MgCl_2$ as activators. Analytical grade chemicals were used; UDP-Gal and ATP were from Sigma Chemicals. Tn red cells incubated in absence of one or more of the other reactants served as controls; they did not acquire T specificity, nor did normal MN red blood cells when incubated under experimental conditions with transferase-containing serum.

For biosynthesis of N- and M- specific determinants, T antigens prepared by desialation of three different preparations, each from isolated human red blood cell NN and MM glycoproteins, were used as substrates, and for M-activation of N, three different human red cell NN glycoprotein preparations isolated under gentle conditions (13) were employed. Sera from three to six different donors of NN as well as MM blood types served as sialyltransferase source and CMP-NAN as NAN donor. CMP-NAN was prepared in Dr. H. Schachter's laboratory as described by others (21, 22) and ^{14}C-CMP-NAN was from New England Nuclear Corp. In "cold" experiments, the reaction mixtures consisted of 5 mg (0.5% final conc.) of antigen with 2.0 μmoles CMP-NAN and 500 μl serum as transferase source. In "hot" experiments, all reactants were reduced to 12.5% of those in the "cold" experiment. Incubation and work up were as described previously (19,20). Control samples were run throughout exactly as the experiment proper; they consisted of reaction mixtures lacking in either substrate or CMP-NAN. Substrates incubated with transferase in absence of CMP-NAN showed no change in specificity. In all instances recovery of substrate was $< 50\%$ at the end of the experiment. Controls without added substrate occasionally acquired traces of activity.

In vivo Measurement of Delayed-Type "Recall" Hypersensitivity. T antigen was dissolved at 1% final concentration in buffered saline containing 0.25% phenol; controls were solvent alone and solvent containing 1% MN antigen. All reagents were sterile and free of pyrogens and Au-antigen. These solutions (0.1 ml each) were injected i.d. into the upper arm. Patients with mastectomy were injected in the contralateral arm. Induration and erythema were measured ca. 24 and 48 hr later and scored as by Hollinshead *et al.* (23). Reactions, if any, given by the controls were subtracted from those due to T antigen.

In vitro Measurement of Cell-Mediated Immune Response. Cellular immunity towards T antigen was assessed by the leukocyte migration inhibition (LMI) assay in agarose plates essentially as by Clausen (15,24). As control, leukocytes of an apparently healthy person were tested in parallel on the same plates with those of the patient for MI by T antigen (5 or 10 μg T antigen per 10 μl) in each agarose plate. Each test was run in triplicate, and arithmetic averages of the results were used. Control tests with MN instead of T antigen were performed under like conditions. For healthy individuals, the average migration index at 5 and 10 μg of T antigen was 1.00; a migration index of $\leq$ 0.90 is considered a positive reaction (25).

Results

T-specific haptenic structures on erythrocytes were readily biosynthesized by the procedure outlined in Materials and Methods. Specific reactivity of T-activated Tn cells in agglutination and absorption tests amounted to 20-35% of that obtained with optimally T-activated normal human red cells. Hemagglutination of T-activated Tn cells by four doses of anti-T was completely inhibited by 0.005-0.02 mg/ml T antigen, depending on the serum used, but not by 100-times as much of MN antigens.

We have also transformed T antigens prepared from isolated NN or MM antigens into those carrying N- and M-specificities. The specificity of this transformation depended solely on the MN-type of the transferase donor. Sialyltransferases from MN and MM donors produced N- and M- activation, those from NN donors only N specificity. Isolated NN antigen was transformed to that possessing M specificity with sera from MN and MM but not NN donors. In hemagglutination inhibition assays, activities of the partially purified products, when compared to the corresponding isolated erythrocyte antigen, ranged from two to > 75% for N-activation of T, from four to > 34% for M-activation of T and from two to > 75% for M-activation of N antigen depending on the human or animal antiserum inhibited. Incorporation of ^{14}C-NAN into T- and N-antigens during incubation with transferase-containing sera was in keeping with the serological activities of the products obtained. Serological results were similar for M-activation of the α-1 glycopeptide isolated from NN erythrocytes. The biosynthetic experiments are summarized in Figure 1.

We were struck by the structural similarity of the N- and M-specific groups of the human red cell glycoproteins and glycoprotein I (epiglycanin) (26) of the TA3 mouse mammary adenocarcinoma; we found that this glycoprotein indeed possessed blood group N-like specificity (27). TA3 glycoprotein I of strain A mice is carcinoma-associated and does not occur in normal strain A mouse tissue (28). Thus, a link was established for the first time between an animal carcinoma-associated antigen and a human blood group substance (N and precursor T). The human anti-T agglutinin of Thomsen and Friedenreich readily killed TA3-St carcinoma cells

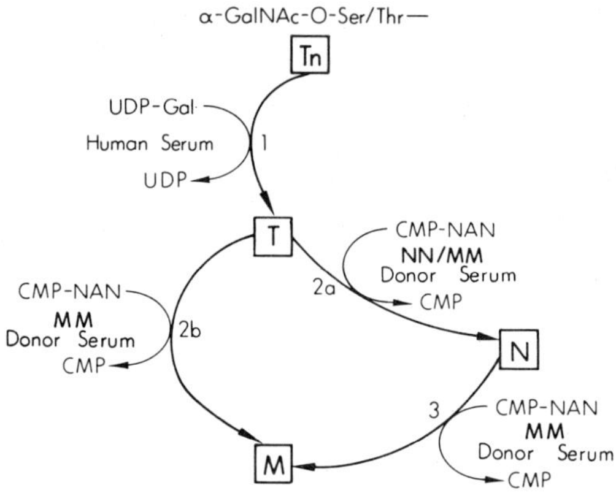

Figure 1. Biosynthetic pathway of the M- and N-specific immunodeterminant structures of human blood group M and N glycoproteins

in the presence of guinea pig complement (28).

We therefore turned our attention from murine mammary glands and the TA3 carcinoma to human breast tissue and its malignancies. Human breast gland cell membrane preparations were used to absorb human antisera. The results of some of these studies are listed in Table I. M- and N- specific structures are present in both benign and malignant human breast glands, T- and Tn- specificities occurred in all of the human breast (primary) and colon carcinomata tested but not in corresponding preparations from benign or normal breast tissues. T activity was low in one invasive Stage III breast carcinoma. All but one gland membrane preparation specifically absorbed anti-M and anti-N antibodies in agreement with the representation of these antigens on the patient's red cells. There was one exception where N but not M was expressed on the patient's carcinoma cell membranes, while both antigens were present on her red cells.

We have tested metastases removed up to 60 months after surgery for primary ductal carcinoma of the breast from four patients. All contained T- and the two tested for Tn contained also Tn-specific structures. This promising observation indicates that T- and Tn- specificities are not eliminated by either somatic mutation or modulation and thus remain accessible to potential immunodiagnosis and immunotherapy.

The T-specificity on adenocarcinomata prompted our determination of anti-T titer scores of human sera, which ordinarily range from 22 to 24. They were severely depressed in a highly significant number of a large population of breast carcinoma patients as compared to 270 patients with benign breast disorders and 470 controls all together (Table II). Similar results were obtained in patients with lung and gastrointestinal (G.I.) carcinomata (15; to be published).

Of the just described patients with breast disease, 32 with mastectomy for carcinoma and 32 with breast biopsy who suffered from benign disease were rebled once within 14 months of surgery. Of the 32 breast carcinoma patients rebled, 21 (65.6%) showed an increase of > 25 to 90% in anti-T titer, while only one of 32 (3.1%) benign breast disease patients and none of 11 patients with major operations for non-malignant disorders showed a significant increase in anti-T. This difference between the carcinoma patients and the controls was significant ($p < 0.001$). In some patients, bled a third time, clinical recurrence of the carcinoma was associated with renewed lowering of serum anti-T levels (to be published).

Table III summarizes the findings on delayed-type skin hypersensitivity (DTH) observed on i.d. injection of T antigen, and those of the leukocyte migration inhibition assay (LMI). It can be clearly seen that the immune response towards T antigen of breast carcinoma patients extends from humoral to cell-mediated reactivity and is demonstrable *in vivo* as well as *in vitro*. Forty-four breast carcinoma patients were skin tested (Table III), 36 of

Table II. Anti-T Agglutinin Titer Score[a] in Patients with Breast Disease and Control Hospital Population (*cf.* 12)

Test group	Individuals with score: 10 or < (%)	10 or < (No. of persons)	12 or < (%)	12 or < (No. of persons)
Breast carcinoma	21.16[b]	40/189	32.28[b]	61/189
Benign breast disease (A)	5.19	14[c]/270	8.15	22[c]/270
All noncarcinomatous persons studied (B)	3.62	17[c]/470	8.30	39[c]/470

[a] Blind studies; standards: three pools of 36-1000 different sera each; arithmetic average titer scores: 22-24 (U.S. population). [b] *p*: that difference between anti-T scores of breast carcinoma patients and control group (A) as well as (B) is due to chance is < 0.001 throughout (χ^2 test).
[c] Two of the 14 patients have since developed carcinoma.

Table III. Delayed-Type Recall Skin Reactivity upon i.d. Injection of T and *in vitro* Cell-Mediated Immune Response to T Antigen in Patients with Breast Carcinoma, Benign Breast Disease and in Apparently Healthy Controls[a]

Disease stage	DTH Persons positive[b]/Persons tested	LMI Persons positive/Persons tested[c]
IV	6/6	7/14
III	8[d]/9	10/21
II	13/13	17/31
I	14[e]/16	8/27
Benign	2[f]/38	11/84
"Healthy"	0/23	1[g]/95

[a] Only patients > 3 weeks after surgery or chemotherapy and > 12 months after radiation therapy were studied, because of the known immunosuppression due to these interventions (46). Individuals of Chinese and Japanese descent not included. [b] DTH positive = > 4.5 mm induration after 24-48 hr or erythema at 48 hr > 25 mm (average of 2 diameters measured perpendicularly) (23). [c] Leukocyte migration inhibition on agarose plate; 10 μg T antigen/1.5 x 10^6 leukocytes; not toxic up to > 45 μg. [d] Negative was the only infiltrating lobular carcinoma. [e] The two negative persons had *in situ* lobular carcinoma. [f] 2/38 positive. Intraductal papilloma was the most ominous structure in both. [g] Vietnam war veteran with typhoid, cholera and tetanus toxoid vaccinations.

these suffered from carcinoma of ductal origin, 2 were non-invasive. The remaining 8 patients had lobular carcinoma. All patients with ductal carcinoma gave a positive DTH reaction regardless of Stage and invasiveness (15,29). Five with lobular carcinomata also gave a positive reaction but one of the patients with invasive and two with non-invasive lobular tumors did not react (29). In contrast, none of the 23 healthy individuals tested had any delayed hypersensitivity reaction toward T antigen, and 36 of the 38 patients diagnosed histologically to have fibro-adenoma-fibrocystic disease had a negative reaction.

The right hand side of Table III shows that in LMI 34 (51.5%) of the 66 patients with breast carcinoma Stages II, III and IV had a positive reaction, while of 27 patients with Stage I disease, eight (29.6%) were positive. Seven of the eight patients with lobular carcinoma referred to above were tested in LMI. The leukocytes of only two of these patients, both with infiltrating tumor showed migration inhibition.

Of 84 patients with benign breast disease, 11 (13.1%) were positive in LMI with T antigen. No effect of T antigen was found on the peripheral leukocytes of a large number of presumably healthy people. LMI was positive in only one of 95 healthy individuals, a Vietnam war veteran vaccinated repeatedly with S.typhi, V.cholerae and Tetanus toxoid all of which possess T-specificity (see Table IV). T antigen up to 50 μg did not affect viability (trypan blue) or migration of the white cells of healthy persons, MN antigen at the same high concentration had no effect on migration of leukocytes of any of the patients and controls tested.

Discussion

In this communication we report that we have reversed the degradative steps leading from blood group M-specific structures to its precursors and synthesized T-specific groups (19,20) on Tn erythrocytes as well as N- and M- specific groups on isolated T antigens. The peptide core of T antigen in M- differs in 2 amino acids from that in N- individuals and M antigenic determinants possess more carbohydrate (30,31,32,33,34). Since glycosyltransferases generally recognize only limited areas (one to two monosaccharides) underlying the acceptor molecule, sialyltransferases from MM and MN donors produced N and M activation, whereas those from NN donors produced only N regardless of the MN type origin of T antigen. Isolated N antigen could be transformed to M only with sera from MM and MN donors. The genetical sequence of these immunological specificities is therefore: Tn→ T→ N→ M (M could conceivably arise directly from T). These structures have also been found on lipidic carriers, including those from a carcinoma (35).

The discovery that human adenocarcinoma but not healthy human tissues and benign structures possess T- and Tn- specificities in reactive, unmasked form establishes that these precursors are

Table IV. T-Specific Substances in Microbes, their Lipopolysaccharides (LPS) and Microbial Vaccines[a]

T Specificity					
Present			Absent		
BACTERIA:	9/9	*Escherichia coli*	BACTERIA:		*Aerobacter aerogenes*
	3/3	*E. freundii*			*Klebsiella pneumoniae*
		Salmonella poona		3/3	*Proteus* strains
	1/3	*Pseudomonas* strains			
LPS:	4/8	*E. coli*	LPS:		*Pasteurella pseudotuberculosis*[b]
	6/9	*Salmonella* strains			*Pneumococcus* XIV polysaccharide[c]
		S. typhimurium			
		S. abortus equi	VACCINES:		BCG, Tice strain (Univ. of Illinois)[d,e]
		S. typhi			Tuberculin, PPD (Tubersol)[e,f]
		S. minnesota			Influenza virus:
	2/3	*Shigella* strains			A/Victoria + B/Hong Kong (Parke-Davis)[e]
		Serratia marcescens			*Pseudomonas*[e,g]
		Bacterium tularensis			
	2/3	*Chromobacterium violaceum*			
VACCINES:		*Salmonella typhi* (Wyeth, E. Lilly)			
		Corynebacterium parvum (Coparvax)[e,h]			
		Clostridium tetani toxoid (Wyeth)[e,b]			
		Vibrio cholerae (Wyeth)[e,b]			
		Poliomyelitis virus[e,i]			

[a] Non-dialyzable part. [b] Active with some but not all human anti-T. [c] Donated by Dr. M. Heidelberger. [d] Donated by Dr. R. G. Crispen. [e] A possible contribution to activities by growth medium has not been excluded. [f] From Connaught Laboratories. [g] Donated by Dr. J. U. Gutterman. [h] Donated by Wellcome Laboratories. [i] From Lederle, Orimune trivalent.

characteristic of adenocarcinomata of human breast, G.I. tract and probably lung. The occurrence of fully reactive T- and Tn-antigenic specificities in these cancers is probably due to incomplete biosynthesis of normal cell-surface components, as indicated by the presence of all these specificities except M, in the breast carcinoma of one patient whose red cells carried M and N antigens as well as by the extensive absorption (70%) of anti-N by breast carcinoma membranes from one patient of erythrocyte group MM. T- and Tn- specific antigenic determinants have recently been demonstrated directly with immunofluorescent lectins in all carcinomatous breast glands tested but not in non-carcinomatous glandular areas of the same sections (36; Dr. Cr. McNeil, personal communication).

Ordinarily humans have no anti-N or anti-M antibodies since all possess blood group N, M, or NM, and individuals homozygous for M uniformly have some N as well (1,37). Healthy mammals do not form antibodies against their own tissues which are in contact with their immune machinery. Furthermore, N and M specificities are not ubiquitous as are A and B specificities (3,7,38). On the other hand, T- and Tn- specific structures are always masked in healthy humans, but are widespread among intestinal and airborne bacteria (see below); consequently antibodies against these structures are found in all humans (16,39). Levels of these antibodies are quite constant under ordinary circumstances (29,39). Anti-T was severely depressed in patients with carcinoma of the breast, lung and G.I. tract much more frequently when compared to populations of comparable age and socio-economic status who had either benign disease or were apparently healthy. Of all individuals with severely depressed anti-T, > 80% suffered from carcinoma. This depression was not due to IgG, IgM, or IgA decrease. Similar observations on anti-T of patients with breast cancer have recently been reported by others (40). Anti-T antibody may be depressed because of its interaction with T-specific substances either attached to cancer cell-surface and/or released from the cell-surface into the circulation. This hypothesis is supported by the observation of release of T-specific antigens from human mammary carcinoma-derived tissue culture cells into the culture fluid (41). We did not find depressed anti-T in patients with sarcoma, melanoma or with brain tumors tested so far, nor T-specific structures on these tumors.

As mentioned above, there is evidence that humoral anti-T as well as anti-Tn result largely from immunization by the host's own intestinal flora (7,14,38,42,43,44,45). We studied numerous, predominantly gram-negative bacteria and/or their lipopolysaccharides, and some vaccines. Table IV shows that many microbes indeed possess T-specific structures as determined by hemagglutination inhibition and by absorption assays. Three of the 12 active *Escherichia coli* and one of the two active *Chromobacterium violaceum* strains, *Salmonella milwaukee* and *S. abortus equi* inhibited human anti-T only when tested after boiling. It is noteworthy that *Corynebacterium parvum* vaccine, in wide use to stimulate "unspecific" resistance against cancer, possessed significant T specificity, while the one commercial BCG tested and the *Pseudomonas* vaccine (also used as "unspecific" stimulant against cancer) and PPD tuberculin had no T activity.

Tn activity of bacterial preparations was also found in 25 of 30 *Enterobacteriaceae* (including *S. typhi*), *Bacterium tularensis*, and in *C. parvum*, typhoid, tetanus and poliomyelitis vaccines. Tn activity was not demonstrable in BCG, PPD tuberculin, three *Pseudomonas* strains and some *Enterobacteriaceae*.

The discovery of T- and Tn- specific immunologically reactive determinants in the cell membranes of breast and colon adenocarcinomata but not in healthy tissues establishes for the first time a physically and chemically well-defined immunospecific structure characteristically associated with human carcinoma. These precursor specificities T and Tn in reactive form appear to be due to incomplete biosynthesis (or accelerated degradation) of normal cell-surface components. It remains to be seen if T antigen also occurs in other malignancies and occasionally in some permanently benign human tumors.

The T antigen quite likely can be considered a paradigm for other precursor antigens, e.g. Tn, and also for the Forssman antigens and related blood group antigens, with similar functions in malignant processes. T antigen is available in unlimited quantity, is not contaminated with HL-A- and Au- antigens, and is obtainable pyrogen-free. It can be readily prepared by removal of NAN from isolated human red cell blood group NM antigens. T antigen and the anti-T antibodies which are present in all humans (conceivably coupled with radioactive isotopes) may be useful in diagnosis, prognosis, and possibly therapy and immunoprophylaxis of some adenocarcinomata.

Acknowledgments

G. F. S. is Julia S. Michels Investigator in Surgical Oncology. This work was supported by the National Institutes of Health grants CA 19083 and CA 22540. We thank surgeons and pathologists of Evanston and St. Francis Hospitals, Evanston, IL and the Berkshire Medical Center, Pittsfield, MA for fresh tissues. We are grateful to Mrs. H. Tegtmeyer, Mrs. T. Mo, Mr. M. Coleman, Mrs. V. Scanlon, Mrs. E. Mitchell and Mrs. J. Kellogg for excellent assistance.

Literature Cited

1. Landsteiner, K.; Levine, P. J. Exp. Med. (1928), 47, 757-775.
2. Springer, G. F.; Ansell, N. J. Proc. Natl Acad. Sci. USA (1958), 44, 182-189.
3. Springer, G. F. Naturwissenschaften (1970), 57, 162-171.
4. Kosjakov, P. N.; Tribulev, G. P. J. Immunol. (1939), 37, 283-295.
5. Boorman, K. E.; Dodd, B. E. J. Pathol. Bacteriol. (1943), 55, 329-339.
6. Stalder, K.; Springer, G. F. Fed. Proc. (1960), 19, 70.
7. Springer, G. F.; Horton, R. E. J. Clin. Invest. (1969), 48, 1280-1291.
8. Davies, D. A. L. Immunology (1966), 11, 115-125.
9. Mann, D. L.; Rogentine, G. N., Jr.; Fahey, J. L.; Nathenson, S. G. Nature (1968), 217, 1180-1181.
10. Tom, B. H.; Rutzky, L. P.; Jakstys, M. M.; Oyasu, R.; Kaye, C. I.; Kahan, B. D. In Vitro (1976), 12, 180-191.
11. Springer, G. F.; Desai, P. R. Carbohydr. Res. (1975), 40, 183-192.
12. Springer, G. F.; Desai, P. R.; Scanlon, E. F. Cancer (1976), 37, 169-176.
13. Springer, G. F.; Nagai, Y.; Tegtmeyer, H. Biochemistry (1966), 5, 3254-3272.
14. Springer, G. F.; Desai, P. R.; Banatwala, I. J. Natl. Cancer Inst. (1975), 54, 335-339.
15. Springer, G. F.; Desai, P. R.; Murthy, M. S.; Scanlon, E. F. J. Surg. Oncol. (1978), in press.
16. Moreau, R.; Dausset, J.; Bernard, J.; Moullec, J. Bull. Soc. Méd. Hôp. (1957), 73, 569-587.
17. Sturgeon, P.; McQuiston, D. T.; Taswell, H. F.; Allan, C. J. Vox Sang. (1973), 25, 481-497.
18. Stratton, F.; Renton, P. H. "Practical Blood Grouping", p. 34, Blackwell, Oxford, 1958.
19. Springer, G. F.; Desai, P. R.; Schachter, H.; Narasimhan, S. Naturwissenschaften (1976), 63, 488-489.
20. Desai, P. R.; Springer, G. F. Proc. 4th Int. Symp. Glycoconjugates (1977), in press.
21. Comb, D. G.; Roseman, S. J. Biol. Chem. (1960), 235, 2529-2537.
22. Kean, E. L. J. Biol. Chem. (1970), 245, 2301-2308.
23. Hollinshead, A. C.; Stewart, T. H. M.; Herberman, R. B. J. Natl. Cancer Inst. (1974), 52, 327-338.
24. Clausen, J. E. Acta Allergol. (Kbh.) (1971), 26, 56-80.
25. Springer, G. F.; Desai, P. R. Transplant. Proc. (1977), 9, 1105-1111.
26. Codington, J. F.; Sanford, B. H.; Jeanloz, R. W. Biochemistry (1972), 11, 2559-2564.
27. Springer, G. F.; Codington, J. F.; Jeanloz, R. W. J. Natl. Cancer Inst. (1972), 49, 1469-1470.

28. Springer, G. F.; Desai, P. R. Ann. Clin. Lab. Sci. (1974), 4, 294-298.
29. Springer, G. F.; Desai, P. R.; Murthy, M. S.; Scanlon, E. F. Naturwissenschaften (1978), in press.
30. Springer, G. F.; Yang, H. J.; Huang, I.-Y. Naturwissenschaften (1977), 64, 393.
31. Springer, G. F.; Yang, H. J. Immunochemistry (1977), 14, 497-502.
32. Yang, H. J.; Springer, G. F. Proc. 4th Int. Symp. Glycoconjugates (1977), in press.
33. Waśniowska, K.; Drzeniek, Z.; Lisowska, E. Biochem. Biophys. Res. Commun. (1977), 76, 385-390.
34. Friedensen, B.; Yang, H. J.; Springer, G. F., in preparation.
35. Cantrell, J. L.; Springer, G. F.; Desai, P. R., in preparation.
36. Fischer, K.; Poschmann, A.; Stegner, H. E. Dtsch. Med. Wochenschr. (1977), 102, 1226-1227.
37. Springer, G. F.; Tegtmeyer, H.; Huprikar, S. V. Vox Sang. (1972), 22, 325-343.
38. Hoskins, L. C.; Boulding, E. T. J. Clin. Invest. (1976), 57, 74-82.
39. Friedenreich, V. "The Thomsen Hemagglutination Phenomenon" Levin and Munksgaard, Copenhagen, 1930.
40. Luner, S. J.; Wile, A. G.; Sparks, F. C. Proc. 68th Annu. Meet. Am. Assoc. Cancer Res. (1977), Abstract 375.
41. Anglin, J. H.; Lerner, M. P.; Nordquist, R. E. Nature (1977), 269, 254-255.
42. Springer, G. F.; Desai, P. R. Annu. Meet. Am. Soc. Microbiol. (1977), Abstracts, E42, p. 88.
43. Springer, G. F.; Desai, P. R.; Murthy, S. M.; Scanlon, E. F. Proc. 4th Int. Symp. Glycoconjugates (1977), in press.
44. Boccardi, V.; Attinà, D.; Girelli, G. Vox Sang. (1974), 27, 268-272.
45. Springer, G. F.; Horton, R. E.; Forbes, M. J. Exp. Med. (1959), 110, 221-244.
46. Daban, A.; Schneider, M.; Calle, R. Bull. du Cancer (1975), 62, 21-28.

RECEIVED March 30, 1978.

18

Alpha-fetoprotein as an Indicator of Early Events in Chemical Carcinogenesis

STEWART SELL

Department of Pathology, M-012, University of California, San Diego, La Jolla, CA 92093

Alphafetoprotein (AFP), a serum glycoprotein present in high concentration in fetal sera, but in only trace amounts in adult sera, has attracted increasing attention among oncologists and developmental biologists. AFP was discovered by the great Soviet scientist, Garri Israelevich Abelev, in 1963, when he noted the presence of a specific serum protein in neonatal mice and in mice bearing transplantable hepatomas (1). Several other laboratories soon were able to confirm this observation in other species, including man (2, 3, 4, 5, 6). However, further advances in understanding the biologic and diagnostic significance of AFP were hindered by lack of a sensitive accurate technique for measuring AFP. Accurate radioimmunoassays for AFP were established in the early 1970's (7, 8, 9, 10). This was followed by development of specific immunoassays for AFP synthesis by labeled amino acid incorporation and immunofluorescent labeling techniques. Using these assays, AFP production and cellular localization both in vivo and in vitro under a variety of experimental systems has been examined (11, 12). These experimental studies led to understandings which have been expanded clinically in regard to the use of AFP in the diagnosis of congenital anomalies, the differentiation of benign from malignant liver disease, and the determination of the effects of therapy on AFP-producing tumors, as well as to insights into the significance of AFP in normal development and of the sequence of events culminating in development of hepatocellular carcinomas after exposure of rats to chemical hepatocarcinogens.

Properties of AFP

AFP is a single protein chain with physical chemical properties similar to albumin (Table I). In native configuration AFP is immunochemically distinct from albumin, but cross inhibition of antibody binding between AFP and albumin is observed after unfolding of the molecules by disulfide reduction and carboxyaminylmethylation (13). In addition, preliminary amino acid

0-8412-0452-7/78/47-080-326$05.00/0

sequencing indicates a 45% homology (14). AFP of murine species has a high estrogen binding affinity (10^{-9}M), whereas the AFP of other species does not (15, 16).

TABLE I
PROPERTIES OF RAT AFP AND ALBUMIN

	AFP		ALBUMIN
Molecular Weight	70,000		60,000
Isoelectric Point	4.9		5.6
CHO Content	4%		None
Serum Concentration (ug/ml)			
Adult	.03		60,000
Fetal	8,000		100
Estrogen Binding			
Ka	10^{-9}		10^{-5}
pm/mg Protein	50		0.2
Sequence Homology		45%	
Cross Reactivity			
Native		0	
Unfolded*		+	

*Disulfide reduction plus carboxamidomethylation.

A series of systematic experimental studies on the nature and significance of alphafetoprotein (AFP) in normal development, during restitutive hepatocellular proliferation, during tumor growth and therapy, and after exposure to hepatocarcinogens has been carried out in our laboratory. A summary of the serum concentrations of AFP under various conditions is given in Chart 1.

Normal Development

Alphafetoprotein is synthesized by the fetal liver and yolk sac (18, 19), crosses the placenta into the maternal circulation and is rapidly catabolized by the mother (20). There is a reciprocal relationship between the serum concentration of AFP and albumin. Early in development AFP concentrations are high and albumin concentrations low (4, 5, 21). When the serum concentrations of AFP fall during neonatal development, the concentration of albumin rises. The delineation of the feto-maternal synthesis and distribution in the rat has established a base for understanding elevations of AFP concentrations in aminotic fluid or maternal serum associated with congenital anomalies in the human (Table II) (22, 25).

TABLE II

INCREASED AMNIOTIC FLUID AFP IN ABNORMAL PREGNANCY

ABNORMALITY	MECHANISM
Fetal Death	Release of Fetal AFP into AMF
Neural Tube Defects	Transfer of AFP from Fetal CSF
Esophageal Atresia	Reduced Turnover, Lack of Swallowing
Congenital Nephrosis	Loss of Fetal AFP from Renal Defect
Multiple Pregnancy	Increased Production
Rh Incompatibility	?Increased Production

Alphafetoprotein is synthesized by the liver of the newborn rat until four weeks of age where there is an abrupt cessation of synthesis (11). This striking cessation of AFP synthesis is associated with the termination of neonatal liver cell proliferation and change in the chromosome number from diploid to tetraploid.

There is a close relationship between the fetal tissues that synthesize AFP and tumors in adults that produce AFP. Tissue sites of synthesis of AFP, measured by incorporation of radiolabeled aminoacids, were determined using different fetal tissues (19). The fetal liver and the yolk sac elements are the fetal tissues which produce the highest amounts of AFP; in adults liver and teratocarcinoma (yolk sac) fetal tumors produce AFP with the highest frequency (26, 27, 28). The fetal GI tract (stomach, small intestine and pancreas) also produces AFP although at much lower amounts than liver or yolk sac; human tumors of the GI tract may produce AFP but the frequence of AFP production by tumors of these tissues is much less than the production by hepatomas or teratocarcinomas (29). Small amounts of AFP are produced by fetal lung and a small number of bronchiogenic carcinomas produce (21, 29). Finally, although not produced detectably by fetal mesenchymal tissues, it is possible that mesenchymal tumors may also produce AFP (30). These results support the concept of different levels of expression of developmental products by tumors (31, 32). Such exposures also applies to ectopic hormone production and carcinoembryonic antigen by tumors (Table III).

TABLE III
LEVELS OF EXPRESSION OF SOME ONCODEVELOPMENTAL GENE PRODUCTS BY TUMORS

	AFP	CEA	HORMONE (e.g. ACTH)
FIRST LEVEL			
Fetal or Adult Tissue Which Normally Produce It	Hepatoma, Teratocarcinoma With Yolk Sac Elements	Colonic Carcinoma	Pituitary Adenoma
SECOND LEVEL			
Closely Related Embryologically (Same Cell Line)	Pancreas, GI CA	Pancreas, Gastric Liver CA	Medullary CA of Thyroid
THIRD LEVEL			
More Distantly Related (Same Cell Line)	Pulmonary CA	Pulmonary, Breast CA	Oat Cell CA of Lung
FOURTH LEVEL			
Different Cell Line	?Sarcoma Lymphoma	Sarcoma, Lymphoma	Pulmonary CA

The control of AFP synthesis by fetal hepatocytes has been determined by quantitative assay of AFP and albumin synthesis in a fetal hepatocytes culture system developed by Hyam Leffert of the Salk Institute (18). The data indicate that albumin is synthesized throughout the cell cycle, whereas AFP is largely synthesized prior to S and released prior to M (19). Factors such as hormones, which may affect AFP synthesis, appear to do so through alterations in the growth state of fetal hepatocytes. When fetal hepatocytes are not in an active growth state, production of AFP is not detectable.

Hepatotoxic Liver Injury

The liver of adult rats subjected to 70% partial hepatectomy (16) or following exposure to chemical hepatotoxic agents such as CCl_4 or galactosamine (35, 36) will synthesize AFP during the restitutive phase of liver regeneration. The serum concentrations of AFP become elevated shortly after proliferation is observed and decline to normal shortly after proliferation stops (16). This effect is reproducible in vitro as adult hepatocytes cultures will re-express fetal phenotype during the active growth phase but re-establish adult phenotype when becoming confluent (37). The absence of an elevation in serum AFP following liver injury is a poor prognostic sign, as all rats not demonstrating an AFP elevation died. In humans, AFP concentrations may be used in a similar manner to determine the extent of liver damage following liver necrosis or hepatitis and to determine the occurrence of restitutive proliferation (38,39). By immunofluorescence, AFP may be identified in hepatocytes undergoing proliferation (40).

A specific function for AFP has not been convincingly demonstrated. Some of the postulated functions of AFP are listed in Table IV. Several of these deserve further comment. Those functions related to the high binding capacity of AFP for estrogen seem not to be generally applicable since only the AFP of rat and mouse have this property. The possibility of AFP protecting the fetus from maternal immune rejection is an attractive one, but conflicting experimental data have been obtained (41, 42, 43, 44). It is clear that AFP does carry out some of the osmotic and carrier functions of albumin in the fetus since AFP is the major serum protein during a considerable time in the fetus (Table I).

TABLE IV

SOME PROPOSED FUNCTIONS OF AFP DURING FETAL DEVELOPMENT

OBSERVATION	POSSIBLE FUNCTION
Structure Homology with Albumin	Fetal Albumin - Carrier and Osmotic
Inverse Serum Concentration to ALB	
Immunosuppressive In Vitro	Blocks Maternal Rejection of Fetus
Binds Estrogen	Protects Fetus From Maternal Estrogen
Produced During Hepatocyte Proliferation	Growth Control
Produced During Liver Lobule Formation	Developmental Tissue Organizing Signal

The correlation of AFP production with normal neonatal liver development and with restitutive liver cell proliferation has suggested that the function of AFP may be to serve as a tissue organization signal for proper alignment of different cell types in the liver (31). In both neonatal development and resotration of the adult liver after partial hepatectomy, not only must the number of hepatocytes be rapidly increased, but also these cells must assume the proper alignment with sinusoidal cells, bile canniculi and vessels. The role of AFP as an organization signal is supported by analogy to other "oncodevelopmental" markers such as the T locus markers of the mouse and the T cell antigens of lymphocytes, one being responsible for cell-cell recognition during normal development of the early embryo, the other believed to be required for proper cell-cell interactions in immune responses. It is attractive to assume other "oncodevelopmental" products such as AFP might function in a similar manner (31).

Tumor Growth

The elevations of serum concentrations of AFP following transplantation of a variety of hepatomas in syngeneic rats has been analysed (45, 46, 47). Not all transplantable hepatomas produce elevated serum concentrations of AFP. The level of production of AFP is related to multiple factors including degree of aneuploidy, growth rate and degree of histologic differentiation. Upon transplantation of tumors that produce AFP an exponential continuous elevation of serum concentrations of AFP occurs (48) (see Chart 1). This is in contrast to the transient, usually much lower, elevations found following liver cell necrosis and proliferation. Thus, this model suggests that serial determinations of AFP in patients with undiagnosed hepatomas may be useful in differentiating cirrhosis or hepatitis (with liver cell proliferation) from hepatoma (49, 31).

The serum concentration of AFP in animals bearing hepatomas may be used to follow the effects of therapy (48, 50, 51). In a model system, surgical removal of an AFP producing tumor results in an immediate decline of serum AFP concentrations. A prolonged return to normal is found in long-term survivors; re-elevation associated wtih development of pulmonary or lymph node metastases. Radiation to the lung shortly following tumor removal results in a significant increase in the number of long-term survivors (51). Again, this animal model points out the potential usefulness of AFP in clinical diagnosis and prognosis (31).

Tumor Development

For over twenty years the induction of rat hepatocellular carcinomas has been used as a model for chemical carcinogenesis (52, 53, 54). In this system a sequence of morphologic alterations is observed during exposure of rats culminating in the appearance of hepatocellular carcinomas after 16-20 weeks. The sequence of morphologic changes proceeds from the appearance of small foci of a few hepatocytes with altered staining characteristics through larger microscopic foci of cells which impinge upon the surrounding parenchyma and grossly visible neoplastic nodules to frank hepatocellular carcinomas (55, 56, 57, 58). The hypothesis that this sequence is related and that the nodules represent pre-malignant but reversible lesions is supported by the presence of altered hepatic enzyme expression in the neoplastic nodules that is also found in hepatocellular carcinomas (59, 60).

Data obtained from the application of AFP measurements and cellular localization as an early indicator of events in chemical carcinogenesis may result in significant changes in the current concept of the sequence of events leading to development of hepatocellular carcinomas. This use of AFP has been made in collaboration with F.F. Becker, E. Smuckler, B. Lombardi and H. Shinozuka. Exposure of rats to hepatocarcinogens such as ethionine (61), acetylaminofluorene (62) and 3-methyl-4-dimethylaminoazobenzene (63) results in elevation of the serum concentrations of AFP within a few hours or days even though (?premalignant) morphologic changes may not be observed until after 6-8 weeks of exposure and frank hepatomas do not develop until 16-20 weeks after the initial exposure (64). Discontinuation of exposure to the carcinogen after AFP elevations have occurred does not cause a cessation of AFP production. For instance, serum concentrations of AFP remain elevated for 12-18 weeks after one 3-week pulse of a non-carcinogenic dose of 2-AAF, and four such pulses are required to produce tumors (62). Therefore, the kinetics of AFP following administration of non-hepatotoxic doses of hepatocarcinogens is quite different than the kinetics after administration of agents which induce liver cell necrosis. On the other hand, continued feeding of 2-AAF does not maintain the serum AFP elevation as the serum AFP concentration of rats fed a continuous diet of 2-AAF will fall back to just about normal with essentially the same pattern as is observed when exposure is discontinued. A striking observation is that the serum concentration of AFP is actually falling at the time of maximal development of neoplastic nodules (65) (Chart 2). The serum AFP rises sharply with the appearance of an AFP producing hepatocellular carcinoma at 16-20 weeks. About two-thirds of the hepatocellular carcinomas that are produced by 2-AAF exposure actually produce AFP (65).

The identification of the cells producing AFP after exposure to chemical hepatocarcinogens has been the subject of much unrewarding study, but immunofluorescent studies recently carried out have clarified the situation. Several groups reported that some of the "oval" or transitional cells which appear several weeks after exposure to diaminoazobenzene (DAB) contained AFP (66-71). However, DAB is a potent necrotizing agent for liver cells so that the relationship between oval cells and carcinogenesis is difficult to interpret. On the other hand, Okita et al. reported that not oval cells but some of the noeplastic nodules that appear after exposure to ethionine contain AFP (72). They use this evidence to support Farber's concept of a sequence of metabolic and morphologic changes linking the neoplastic nodule to the hepatoma as a premalignant lesion. Other data does not support this concept. As mentioned above, there is a reverse correlation between serum AFP concentrations and the appearance of neoplastic nodules in rats fed FAA (65). In addition, the livers of rats treated with ethionine or DEN and fed a choline deficient diet display a marked oval cell proliferation that occurs within 2-3 weeks in the absence of detectable liver cell necrosis. This "oval cell" proliferation is associated with serum AFP concentrations two logs higher than that found with ethionine or DEN alone (60 ug/ml vs. 0.6-1.0 ug/ml). Immunofluorescent labeling reveals that many of these oval cells contain albumin and approximately 3% also contain AFP (27, and unpublished observations). Those elements recognizable as bile ducts in the proliferative lesions are negative for AFP and albumin. Double immunofluorescent

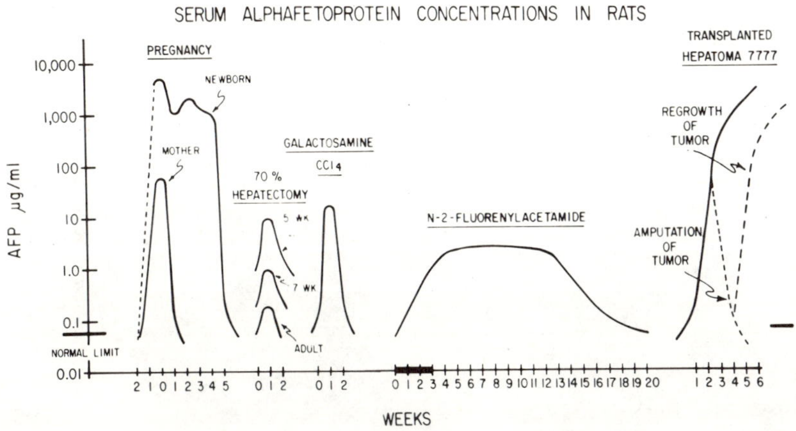

Chart 1. Serum alphafetoprotein concentrations in rats. The serum AFP concentrations of rats during normal gestation, after partial hepatectomy and chemically induced liver injury, after exposure to chemical carcinogens, and during growth of an AFP producing transplantable hepatoma are presented (from Ref. 17).

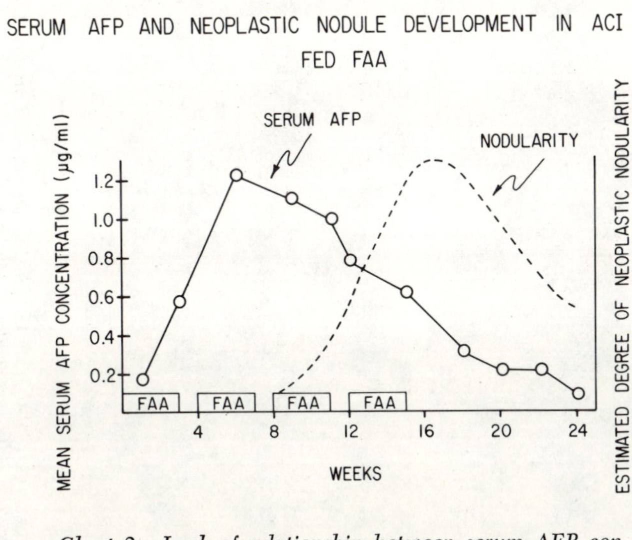

Chart 2. Lack of relationship between serum AFP concentrations and development of neoplastic nodule after FAA feeding of ACI rats. Following feeding of 0.06% FAA the serum concentrations of AFP become elevated with the first week of feeding, attain a maximum after six weeks, and then fall during the third and fourth feeding cycles when neoplastic nodule formation is most evident (see *Ref.* 65).

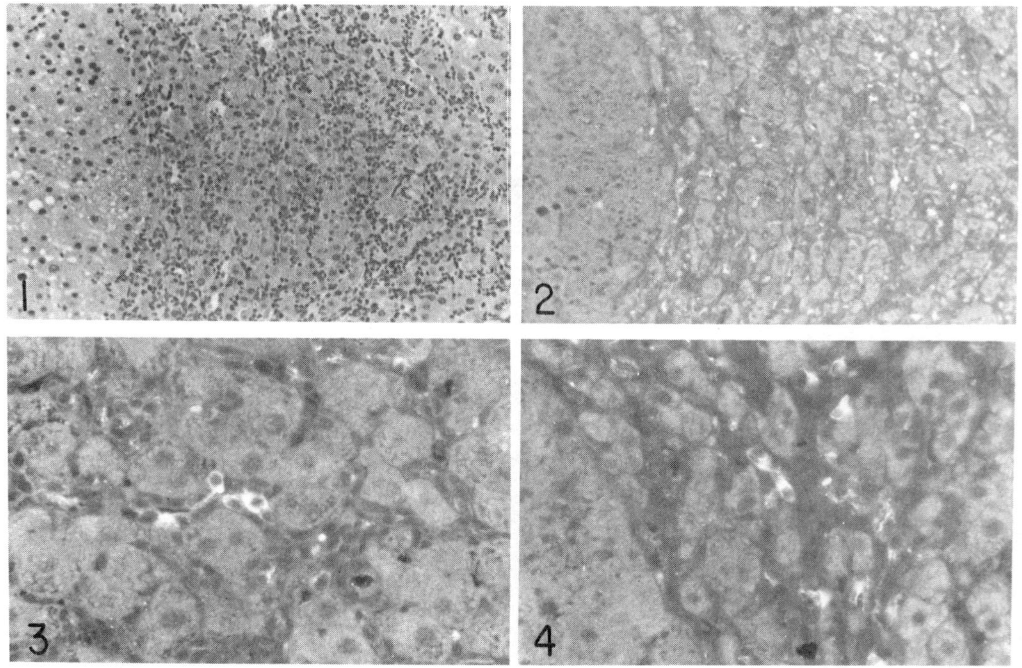

Figure 1. Edge of neoplastic nodule (left margin) and adjacent liver with marked increase in "oval cells." The oval cells are the smaller cells which alternate with larger hepatocytes. H & E ×245.
Figure 2. Same section as Figure 1 labeled for AFP by immunofluorescence. Note AFP containing oval cells. ×245. Figures 3 and 4. Oval cells containing AFP. ×245.

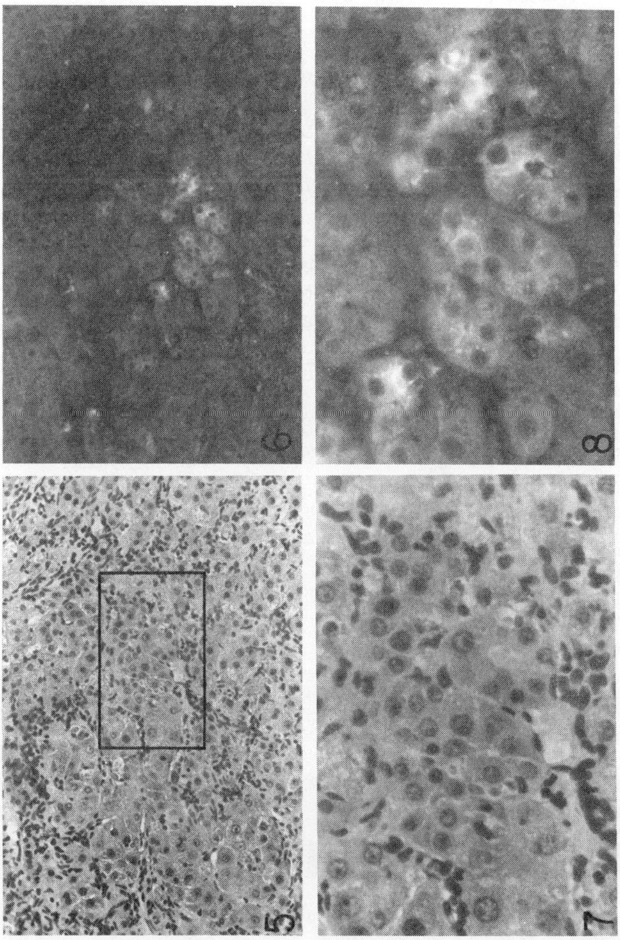

Figure 5. Area of "atypical" hyperplasia of liver. Note gland-like collection of atypical hepatocytes. H & E ×230. The rectangle outlines the area shown in Figures 7 and 8. Figure 6. AFP immunofluorescence staining of area of "atypical" hyperplasia. ×90. Figure 7 (H & E) and 8 (AFP immunofluorescence). Area shown within Figure 5. AFP containing cells are in gland-like area of "atypical" hyperplasia. ×230.

staining for albumin and AFP reveals that the AFP cells are also positive for albumin. Early foci of cellular alteration and small neoplastic nodules are AFP negative although many of the cells in these hepatocellular lesions contain albumin. Extensive examination of the livers of choline supplemented rats receiving 1 injection of DEN (75 mg/kg) reveals that strongly AFP positive "oval" celsl (approximately 1 per 5,000 liver cells) may be identified as single cells in the portal zones only one week after injection, at a time when little or no morphologic change is recognizable.

This finding has stimulated a more extensive analysis of the relationship of AFP production to the development of hepatocellular carcinomas after exposure of rats to chemical carcinogens (74). Preliminary observations show that the neoplastic nodules that develop after feeding 2-FAA to Fisher male rats for four, 2-week on 1-week off, cycles do not contain cells with AFP although many of the cells of these nodules are positive for albumin. At the time of this writing, over 1,000 neoplastic nodules varying in size from eight cells to over 12 mm in diameter have been examined, and none of these contained cells with AFP. Neoplastic nodules are recognizable as discrete well-demarcated collections of liver cells similar to normal hepatocytes without normal sinusoid (74). In zones of these livers where there are not identifiable neoplastic nodules the sinusoids contain a diffuse increase in "oval" cells. Approximately 3-5% of these oval cells are positive for AFP. In addition, some zones of "atypical hyperplasia" of hepatocytes also contain cells with AFP. Representative immunofluorescent staining patterns are shown in Figures 1-8. Since none of the "neoplastic nodules" contain AFP but many of the tumors that arise from carcinogen exposure produce AFP (65), the relationship between neoplastic nodules and hepatocellular carcinoma must be re-evaluated (75). The dogma among workers in chemical carcinogenesis has been that hepatocellular carcinomas arise from neoplastic nodules. On the basis of the above observations as well as the preliminary observations made in collaboration with Dr. Leffert, that cells from neoplastic nodules do not proliferate as well in vitro as normal hepatocytes, it is proposed that hepatocellular carcinomas do not arise from neoplastic nodules, but from oval cells which are not part of the nodules. The production of AFP by these cells after exposure to chemical carcinogens most likely represents inappropriate synthesis. Chemical carcinogens may select a very limited cell population (?stem cells) to produce AFP and this signal may be misread by the normally controlling environment or by the developing stem cells so that the stem cells proliferate and differentiate rather than form normal liver lobules, and eventually progress to hepatocellular carcinomas (Figure 9).

In conclusion, by correlation and analogy, it is hypothesized that the function of AFP is to serve as a signal for normal tissue (liver) organization (31). The correlation is that AFP is produced by fetal liver and by the adult liver following partial

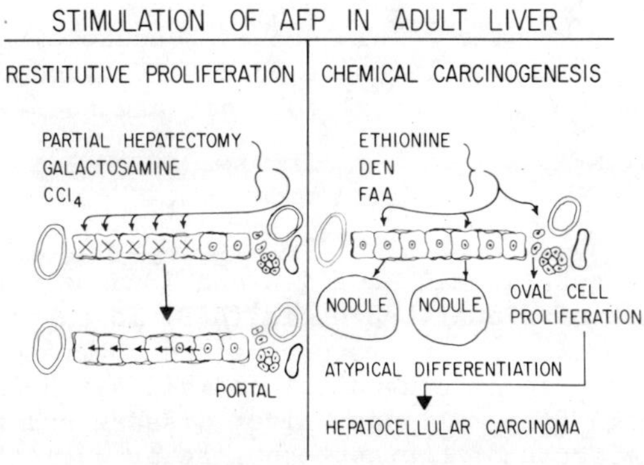

Figure 9. Representation of hypothesized targets for chemical toxins and chemical carcinogens in relationship to target cell and resulting AFP synthesis (for explanation, see text)

hepatectomy or chemically induced necrosis. In these situations, the developing or growing liver must align a number of cell types including hepatocytes, sinusoids, bile canniculi as well as blood and lymphatic vessels. The analogy is with other markers shared by fetal tissues and tumors in adults (oncodevelopmental gene products) that have been demonstrated to play a role in cell-cell interactions required for normal development and tissue organization. It is postulated that chemical carcinogens act on a small number of undifferentiated hepatocellular stem cells (?oval cells) causing the inappropriate synthesis of AFP. Under the continuing influence of the carcinogen, these cells develop into hepatocellular carcinomas. This occurs because either the progeny of the stem cells are unable to respond to environmental signals or the appropriate environmental controls are not available.

Abstract

Systematic quantitative sutdies and immunofluorescent cellular localization of alphafetoprotein (AFP) in experimental systems has provided a basis for understanding and applying AFP quantitation to the diagnosis of congenital anomalies, to the differentiation of benign from malignant liver disease and to the determination of the effects of therapy upon AFP producing tumors. In addition, these studies have contributed to new insights into the significance of AFP in normal development and of the sequence of events in induction of hepatocellular carcinoma by chemical carcinogens. An hypothesis has been formulated that AFP functions during fetal development and in the adult during liver undergoing regeneration as a marker for organization of growing liver tissue into normal lobules. Production of AFP during exposrue to chemical hepatocarcinogens represents an inappropriate signal. Carcinogens induce a small number of immature hepatocytes (?stem cells) to proliferate. Under inappropriate signals for organization, some of these stem cells, perhaps only one or two per liver, eventually produce hepatocellular carcinomas. The previously accepted sequential evolution of hepatocellular carcinomas from foci of altered hepatocytes through neoplastic nodules to malignant tumors is not supported by the results of ongoing experiments. Neoplastic nodules may represent a non-cancerous response of differentiated hepatocytes to carcinogen exposure.

Literature Cited

1. Abelev, G.I., Perova, S.D., Khramkova, N.I. et al., Transplantation (1956) 1:174.
2. Tatarinov, Y. Vopr. Med. Khim. (1964) 10:90.
3. Stanislawski-Birencwajg, M., Uriel, J., Grabar, P. Canc. Res. (1967) 27:1990.
4. Abelev, G.I. Cancer Res. (1968) 28:1334.
5. Abelev, G.I. Adv. Cancer Res. (1971) 14:295.
6. Sell, S., Wepsic, H.T. "The Liver. The Molecular Biology of Its Diseases" (Becker, F.F. Ed.). pp. 773, Marcel Dekker, New York, 1975.
7. Ruoslahti, E., Seppala, M. Int. J. Cancer (1971) 8:374.
8. Oakes, D.D., Shuster, J., Gold, P. Cancer Res. (1972) 32: 2753.
9. Sell, S., Gord, D. Immunochem. (1973) 10:439.
10. Sizaret, P., Breslaw, N., Anderson, S.G. J. Biol. Stand. (1975) 3:201.
11. Sell, S., Nichols, M., Becker, F.F. et al. Cancer Res. (1974) 34:865.
12. Sell, S., Becker, F.F., Leffert, H.L. et al. Cancer Res. (1976) 36:4239.
13. Ruoslahti, E., Engvall, E. Proc. Natl. Acad. Sci. USA (1976) 73:4641.
14. Ruoslahti, E., Terry, W.D. Nature (1976) 260:804.
15. Liang-Tang, L.K., Soloff, M.S. Biochim. Biophys. Acta (1972) 263:753.
16. Nunez, E., Vallette, G., Benassayag, C., Jayle, M-F. Biochem. Biophys. Res. Commun. (1974) 57:126.
17. Sell, S., Becker, F.F. J. Natl. Cancer Inst. (1978) 60:19.
18. Gitlin, D., Boesman, M. J. Clin. Invest. (1967) 46:1010.
19. Sell, S., Skelly, H. J. Natl. Cancer Inst. (1976) 56:645.
20. Sell, S., Alexander, D. J. Natl. Cancer Inst. (1974) 52:1489.
21. Gitlin, D. Ann. N.Y. Acad. Sci. (1975) 259:7.
22. Seppala, M. Ann. N.Y. Acad. Sci. (1975) 259:59.
23. Adinolfi, A., Adinolfi, M., Lessof, M.H. J. Med. Genet. (1975) 12:138.
24. Norgaard-Pedersen, B. Scand. J. Immunol. (1976) 5:1.
25. Brock, D.J., Scrimgeour, J.B., Nelson, M.M. Clin. Genet. (1975) 7:163.
26. Masseyeff, R. Pathol. Biol. (Paris) (1972) 20:703.
27. Waldmann, T.A., McIntire, K.R. Cancer Res. (1974) 34:1510.
28. JcIntire, K.R., Bogel, C.L., Princler, G.L., et al. Cancer Res. (1972) 32:1941.
29. McIntire, K.R., Waldmann, T.A., Moertel, C.G., et al. Cancer Res. (1975) 35:991.
30. Mihalev, A., Tzingilev, D., Sirakov, L.M. Neoplasma (1976) 23:103.
31. Sell, S. "Handbook of Cancer and Immunology" (Waters, H. Ed.). Garland Publishing Co., in press.

32. Sell, S. "Basic Immunologic Mechanisms in Cancer" (Hanna, M. G., Jr., Ed.). Academic Press, New York, in press.
33. Leffert, H.L., Sell, S. J. Cell Biol. (1974) 61:823.
34. Sell, S., Leffert, H.L., Skelly, H., et al. Ann. N.Y. Acad. Sci. (1975) 259:45.
35. Sell, S., Stillman, D., Gochman, N. Am. J. Clin. Pathol. (1976) 66:847.
36. Watanabe, A., Miyazaki, M., Taket, K. Gann (1975)67:279.
37. Leffert, H.L. et al. Proc. Natl. Acad. Sci., in press.
38. Endo, Y., Kanai, K., Oda, T., et al. Ann. N.Y. Acad. Sci. ((1975) 259:234.
39. Murray-Lyon, A.M., Orr, A.H., Gazzard, B., Kohn, J., Williams, R. Gut (1976) 17:576.
40. Engelhardt, N.V., Lazareva, M.N., Abelev, G.I., et al. Nature (1976) 263:146.
41. Murgita, R.A., Tomasi, T.B., Jr. J. Exp. Med. (1975) 141: 269.
42. Murgita, R.A., Tomasi, T.B., Jr. J. Exp. Med. (1975) 141: 440.
43. Sheppard, H.W., Jr., Sell, S., Trefts, P., et al. J. Immunol. (1977) 119:91.
44. Sell, S., Sheppard, H.W., Jr., Poler, M. J. Immunol. (1977) 119:98.
45. Sell, S., Morris, H.P. Cancer Res. (1974) 34:1413.
46. Becker, F.F., Klein, K.M., Wolman, S.R., et al. Cancer Res. (1973) 33:3330.
47. Becker, F.F., Wolman, S.R., Asofsky, R. et al. Cancer Res. (1975) 35:3021.
48. Sell, S., Wepsic, H.T., Nickel, R. et al. J. Natl. Cancer Inst. (1974) 52:133.
49. Sell, S. "Proceedings of the Seventh Annual Conference on Environmental Toxicology" (Fries, S.L., Ed.). pp. 278, National Technical Information Service, Springfield, 1977.
50. Sell, S., Sheppard, H.W., Jr., Nickel, R. et al. Cancer Res. (1976) 36:476.
51. Sell, S., Stillman, D., Michaelsen, M., et al. Radiat. Res. (1977) 69:54.
52. Farber, E. Cancer Res. (1956) 16:142.
53. Orr, T.W. Path. Bact. (1940) 50:393.
54. Price, J.M., Harmon, J.W., Miller, E.C., Miller, J.A. Cancer Res. (1952) 12:192.
55. Epstein, S., Ito, N., Merkow, L., Farber, E. Cancer Res. (1967) 27:1702.
56. Farber, E. Adv. Cancer Res. (1963) 7:383.
57. Farber. E. Cancer Res. (1973) 33:2537.
58. Farber, E., Parker, S., Gruenstein, M. Cancer Res.(1976) 36:3879.
59. Knox, W.E. "Enzyme Patterns in Fetal, Adult and Neoplastic Rat Tissues." S. Karger, Basel, 1976.

60. Pitot, H.C., Barsness, L., Goldsworthy, T., Kitagawa, T. Nature (1978) 271:456.
61. Smuckler, E.A., Koplitz, M., Sell, S. Cancer Res. (1976) 36:4558.
62. Becker, F.F., Sell, S. Cancer Res. (1974) 34:2489.
63. Watabe, H. Cancer Res. (1971) 31:1192.
64. Teebor, G.W., Becker, G.W. Cancer Res. (1971) 31:1.
65. Becker, F.F., Sell, S. Submitted.
66. Bannikov, G.A., Gel'Shtein, V.I., Chipysheva, T.A. Vestn. Akad. Med. Nank. SSSR (1977) 3:39.
67. Dempo, K.N., Chisaka, N., Yoshida, Y., Kaneko, A., Onoe, T. Cancer Res. (1975) 35:1282.
68. Fujita, S., Ishizuka, H., Kamimura, N., Kaneda, H., Ariga, K. Ann. N.Y. Acad. Sci. (1975) 259:217.
69. Onda, H. Gann (1976) 67:253.
70. Onoe, T., Kaneko, A., Dempo, K., Ogawa, K., Minase, T. Ann. N.Y. Acad. Sci. (1975) 259:168.
71. Tchipysheva, T.A., Guelstein, V.I., Bannikov, G.A. Intl. J. Cancer (1977) 20:388.
72. Okita, K., Gruenstein, M., Klaiber, M., Farber, E. Cancer Res. (1974) 34:2758.
73. Shinozuka, H., Lombardi, B., Sell, S., Iammarino, R.M. Cancer Res. in press.
74. Squire, R.A., Levitt, M.H. Cancer Res. (1975) 35:3214.
75. Sell, S. submitted.

RECEIVED December 29, 1977.

19

Carcinoembryonic Antigen—A Marker of Human Colonic Cancer

CHARLES W. TODD and JOHN E. SHIVELY

Division of Immunology, City of Hope National Medical Center, Duarte, CA 91010

The current literature has numerous examples of qualitative and quantitative changes in antigens associated with human malignancy. Despite the potential that these antigens offer for the control of cancer through detection, therapy, or prevention by immunization, in most cases the antigens are poorly defined and practical applications are lacking. One antigen which has achieved considerable application in the diagnosis and monitoring of cancer therapy is the carcinoembryonic antigen (CEA). Our study of the chemical nature of this complex molecule, or perhaps more accurately stated molecular complex, was undertaken with the goal of improving its clinical utility in the management of cancer.

The initial reports by Gold and Freedman (1,2) suggested that CEA was synthesized by the rudimentary digestive system in the human fetus, that synthesis ceased prior to birth, but that it resumed with the onset of colonic adenocarcinoma in the adult. This sequence lead them to name the substance the carcinoembryonic antigen. Their development of a radioimmunoassay for CEA (3) and the potential importance of such a tumor marker stimulated studies of the diagnostic value of CEA assays in a variety of pathological conditions, both benign and malignant. The literature on these clinical studies has been recently reviewed (4). These studies demonstrated that CEA-reactivity was not universally elevated in adenocarcinoma of the digestive system, that it was often elevated in other malignancies notably of breast, bladder, and lung, and that moderate elevations are often present in nonmalignant inflammatory diseases. Moreover, normal sera showed variable low levels of CEA-reactivity. Clearly the situation was more complex than originally visualized. Since radioimmunoassays are not completely specific (5), more precise chemical knowledge of the substances being detected was necessary.

Most of the studies done on CEA have employed material isolated from hepatic metastases of colonic adenocarcinoma using essentially the method originally described by Krupey *et al.* (6), although the final block electrophoresis step is usually omitted. The CEA employed in most of our studies was obtained by homogeni-

0-8412-0452-7/78/47-080-342$05.00/0

zation of liver metastases of colon adenocarcinoma in water, addition of an equal volume of 2M perchloric acid, centrifugation, dialysis of the supernatant, concentration, chromatography on Sepharose 4B, and rechromatography on Sephadex G200 (7). The CEA activity during these isolations was followed by a triple isotope double antibody radioimmunoassay (8), which has been recently improved by the substitution of ^{57}Co for ^{22}Na as a volume marker for the supernatant (9).

CEA prepared in this way when examined by electron microscopy consisted of twisted rod or cruller shaped particles with dimensions of 9 x 40 nm (10). Further purification of CEA by affinity chromatography on concanavalin A Sepharose removed a small amount (15%) of contaminants, which were not retained by the affinity column. This material was found to contain mucopolysaccharides as identified by cellulose acetate strip electrophoresis, depolymerization with bovine testicular hyaluronidase, and identification of glucuronic acid as the trimethylsilyl derivative of its methyl glycoside by gas chromatography-mass spectrometry (12).

Early studies by Krupey *et al.* (13) revealed CEA to be a glycoprotein with a molecular weight about 180,000, comprising about 60% carbohydrate and 40% protein. This high percentage of carbohydrate may explain the fact that the molecular dimensions as visualized by electron microscopy indicate a molecular volume about 16 times greater than that calculated for a closely packed molecule of 180,000 molecular weight. Presumably the frequency and size of the carbohydrate groupings interfere with compact folding of the chain. There appears to be only one polypeptide chain present in CEA. Consistent with this view is the finding that reduction and alkylation of the disulfide bonds prior to electron microscopic examination of CEA leads to the appearance of thread-like structures. The maximum extended length observed, 220 nm, approaches that expected for a protein of the requisite size allowing 0.364 nm extension per amino acid residue (14).

Characterization of CEA by isoelectric focusing (15), ion exchange chromatography (15,16) and affinity chromatography on concanavalin A Sepharose (11,17,18) demonstrated considerable heterogeneity, as might be expected for a highly glycosylated protein. Indeed, variation is seen in the relative amounts of the carbohydrate components of CEA isolates from different tumors (19). Typical results from a CEA preparation purified by affinity chromatography on concanavalin A Sepharose (11) are presented in Table I. The carbohydrate compositions were determined by the method of Clamp (20) for neutral sugars, by the method of Warren (21) for sialic acid, and by the method of Liu and Chang (22) with the amino acid analyzer for the amino sugars. The sialic acid has been identified as N-acetylneuraminic acid (23).

These analyses required samples of approximately 250 μg. More recently we have developed a method approximately 2 orders of magnitude more sensitive employing methanolysis and trifluoroacetylation to generate methyl N,O-trifluoroacetyl glycosides, which are separated by gas chromatography on 7m x 0.5 mm packed capillary columns and quantitated by electron capture detection (24).

In contrast to the variability seen for the carbohydrate composition, the amino acid composition of various CEA isolates remains quite constant (25). A typical analysis is presented in Table 1.

Using the results in Table 1 some deductions regarding the structure of CEA can be made. The total number of amino acid residues is 829. A single polypeptide chain with 829 amino acid residues with a length of 0.364 nm per residue (14) would have a length of 302 nm, if fully extended. This compares favorably with the 220 maximum observed by electron microscopy considering the approximations involved. The low content of N-acetylgalactosamine (1.5% of the sugar residues) suggests that the oligosaccharide units are not bound to the protein backbone through N-acetylgalactosamine linked to serine or threonine. The failure to accomplish β-elimination supports this view (19). Alternatively, if the attachment is through N-acetylglucosamine linked to asparagine, the total aspartic acid/asparagine content limits the number of oligosaccharide units to 128 with an average size of 4 residues. Subsequently we shall see evidence that there are about half as many units with an average of twice this size. The amount of N-acetyglucosamine indicates some of these residues must be distal from the attachment site since the ratio of N-acetylglucosamine to aspartic acid/asparagine is 1.5. The sum of serine (91 residues) and threonine (74 residues) is 165. Thus there are ample residues to establish the asp,x,ser/thr sequences in the protein chain, believed to code for N-acetylglucosamine-asparagine attachment (26).

More precise information on the structure of the oligosaccharide units was obtained by methylation analysis (27) employing gas chromatography and mass spectrometry as depicted in *Figure 1*. Samples were methylated by the method of Hakomori (28) in which the methylsulfinyl anion (29) was used to generate the polysaccharide alkoxide before the addition of methyl iodide. Acetolysis, hydrolysis, reduction, and acetylation of the permethylated polysaccharides were performed using the procedures described by Stellner *et al.* (30). Identification of the partially methylated alditol acetates was performed by the method of Björndal *et al.* (31) for the neutral sugar derivatives and by the method of Stellner *et al.* (30) for the amino sugar derivatives. The products are identified by their retention times in the gas chromatograph and their fragments as produced in the mass spectrometer. There are certain preferred cleavage points that occur on electron impact in the mass spectrometer. The order of the preference of these cleavages is indicated by the numbers 1, 2, and 3 (*Figure 1*). A detailed procedure has been published (32).

The results of these studies (27) on 4 CEA samples are presented in Table 2 together with the results on a sample analyzed by Hammarström *et al.* (33). It is apparent that considerable

TABLE 1

COMPOSITION OF CEA

Components	*Moles per 180,000 MW*
Sugars	
N-Acetylgalactosamine	7.5
N-Acetylglucosamine	188.2
N-Acetylneuraminic acid	12.1
Fucose	102.4
Galactose	110.9
Mannose	82.4
Amino Acids	
Alanine	49.3
Arginine	20.1
Aspartic acid/Asparagine	127.8
Cysteine	15.8
Glutamic acid/glutamine	84.6
Glycine	47.7
Histidine	20.1
Isoleucine	31.3
Leucine	67.8
Lysine	20.3
Methionine	Trace
Phenylalanine	18.9
Proline	58.6
Serine	91.2
Threonine	74.0
Tryptophan	11.9
Tyrosine	36.1
Valine	53.5
Carbohydrate	49.9%
Protein	50.1%

These valves were calculated from the data for Fraction V, Table 2, Reference 11.

variation exists among these samples, but that a basic pattern is being followed. As expected from patterns previously observed in mammalian glycoproteins (34), all the fucose is terminal. Similarly all the N-acetylneuraminic acid is terminal. This follows not from the methylation analysis, but from the observation that removal of the N-acetylneuraminic acid with neuraminidase exposed to attack by periodate additional moles of galactose equivalent to the moles of N-acetylneuraminic acid removed (23). This observation together with the negligible amounts of branching galactose observed by methylation analysis (Table 2) also fixes the point of attachment of the N-acetylneuraminic acid at the 3-position of galactose (23). Three-fourths of the mannose residues in all the CEA samples were linked to 3 other sugar residues (branching mannose). Except for the small amounts of 3,4- and 2,3-linked galactose, the only other branching sugar is N-acetylglucosamine of which 32-41% is present at branching points.

The data in Table 2 can also be used to calculate the number of oligosaccharide units linked to the protein chain and their average size (35). The number of oligosaccharide units equals the sum of the nonreducing terminal residues minus the sum of the branching residues. From this the average size can be obtained by dividing the total number of residues by the number of oligosaccharide units. Thus the numbers of oligosaccharide units for the CEA samples in Table 2 are 51, 77, 124, 64, and 69 respectively. On this basis the comparable oligosaccharide units contain 9, 7, 4, 8, and 6 monosaccharide residues each. Considering the cumulative errors and the expected inherent variation, the data are in reasonably good accord. The picture that emerges is many small oligosaccharide units attached to about 1 out of 10 of the amino acid residues utilizing about 2/3 of the aspartic acid/asparagine residues.

The high carbohydrate content of CEA suggested the possibility that the antigenic site as detected in the radioimmunoassay might reside there. Although the precise nature of this site is still undefined and may vary depending on the antisera used to define it, the cumulative weight of evidence indicates that the protein portion of the molecule plays the major, if not exclusive, role in determining the site. This does not exclude the possibility that many CEA antisera contain antibodies to carbohydrate. Under the conditions of the usual radioimmunoassay only a rather narrow population of high affinity antibodies plays a significant role (4). A consequence of this selection is that the antiserum used may contain large populations of antibodies playing an insignificant role in the immunoassay titration, but quite capable of reacting with the antigen under the more concentrated conditions prevailing in immunodiffusion tests. For this reason it is dangerous to assume that phenomena observed by various immunodiffusion techniques are necessarily involved in immunoassay titrations(4).

TABLE 2

STRUCTURAL UNITS OF THE POLYSACCHARIDE PORTION OF CEA

		Moles/1.8·10^{5}g of CEA				
Carbohydrate residues	*Glycosidic linkage*	CEA_1[a]	CEA_2	CEA_3	CEA_4	CEA-14[b]
N-Acetylneuraminic acid	Terminal	23	11	34	13	18
Fucose	Terminal	103	103	104	95	110
Galactose	Terminal	43	72	58	63	36
	3	61	22	45	18	11
	2	5	18	18	9	36
	6	11	2	11	4	13
	3,4 and 2,3	5	2	2	4	
Total		125	116	134	98	96
Mannose	Terminal	9	14	9	13	4
	2	9	7	9	7	14
	6	Trace	Trace	Trace	Trace	Trace
	2,4	20	18	22	13	18
	2,6	16	18	20	18	23
	3,6	23	25	11	22	18
Total		77	83	71	73	77
N-Acetylglucosamine	Terminal	Trace	16	32	20	5
	4	92	94	88	110	83
	3,4	63	76	58	83	41
Total		155	186	178	213	129

[a]The subscripts 1 to 4 indicate CEA purified from different tumors. All tumors were liver metastases that had originated in the colon.

[b]Data from Hammarström *et al.* (33).

The evidence supporting the role of the protein portion in determining the antigenic site may be summarized as follows. Serial periodate oxidation, Smith degradation (36), of CEA demonstrated persistence of antigenic activity even after removal of 90% of the carbohydrate (23,33,37). Antisera to CEA can detect as little as 10^{-11} M of CEA. Such high binding affinities are characteristically achieved against protein antigens but not against carbohydrates (38). These high binding affinities can not be explained by bivalent antibody binding to adjacent sites on the CEA molecule, since equivalent concentrations of binding sites of monovalent Fab' fragments of CEA antibodies were equally effective as the intact antibody in the CEA radioimmunoassay (39). Dilute alkali treatment of CEA destroys its antigenic activity, but does not affect its lectin binding ability (33). It is well known that protein structure is rapidly degraded by alkali, whereas the carbohydrate linkages of asparagine linked glycoproteins are relatively insensitive to base. Westwood *et al.* (40) have shown that controlled acid hydrolysis of CEA destroys its antigenic activity, producing N-terminal proline peptides with intact carbohydrate chains except for the loss of sialic acid and fucose. It had previously been shown that destruction of sialic acid and fucose by periodate oxidation does not alter the antigenic activity (23). Reduction and alkylation of the cystine disulfide bonds in CEA destroys its antigenic activity (22,33,41,42). Clearly the secondary and tertiary structures of CEA are critical in maintaining its antigenic nature.

The sequence of the first 24 amino acids at the N-terminus of CEA was readily determined by standard automated Edman techniques. This sequence is presented in *Figure 2* (43). It has been confirmed several times (44) and extended an additional 6 residues (45), although the choice between glutamic acid and glutamine at certain residues has not been uniform (46).

Additional sequencing encountered technical problems arising from the high degree of glycosylation. These difficulties began with the resistance to specific cleavage, presumably due to steric hinderance by the carbohydrate groups. Since the carbohydrate substituents are highly heterogeneous in degree of substitution among a population of CEA molecules, specific cleavage methods did not give uniform or stoichiometric yields of expected peptides. Once peptides were obtained, their purification was rendered difficult by these heterogeneous carbohydrate substituents. Glycopeptides have broad, ill defined peaks in separations based on conventional sizing techniques. Edman degradations of glycopeptides frequently come to an abrupt halt when an amino acid to which carbohydrate is attached is encountered.

The first of these difficulties was solved by the use of the detergent Triton X100 (0.25%), which rendered CEA susceptible to trypsin cleavage (47). Although CEA contains sufficient arginine and lysine residues to produce about 40 tryptic peptides,

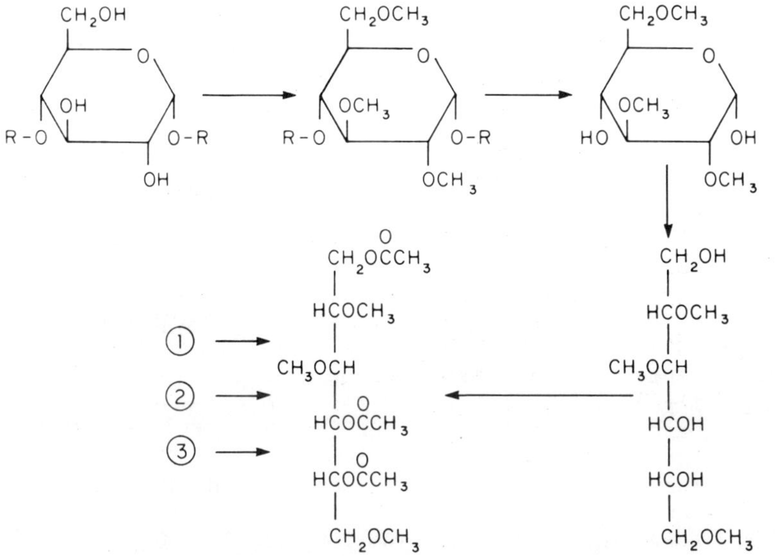

Cancer Research

Figure 1. Methylation analysis. Numbers indicate preferred order of bond rupture under electron bombardment in the mass spectrometer.

CARCINOEMBRYONIC ANTIGEN
Amino - Terminal Residues

Lys Leu Thr Ile Glu[5] Ser Thr Pro Phe Asn[10]
Val Ala Glu Gly Lys[15] Glu Val Leu Leu Leu
Val His Asn Leu

Figure 2. Sequence of amino acids at the N-*terminus of CEA*

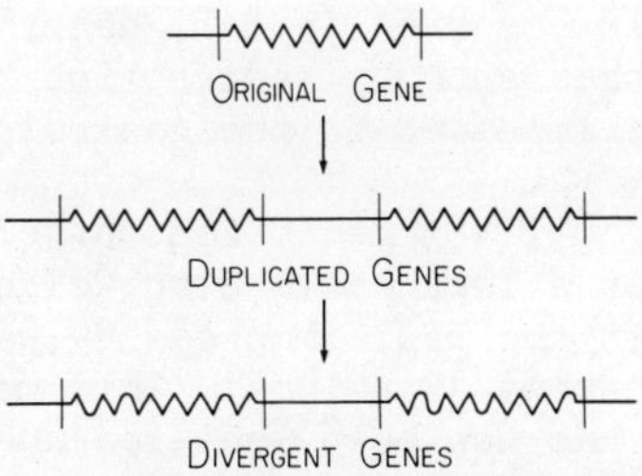

Figure 3. Proposed genetic evolution for the protein chains of CEA and related antigens

only 10 were isolated. These were fractionated by a combination of anion and cation exchange resins using high pressure liquid chromatography. In some cases these peptides could be sequenced through 20 Edman degradation cycles. Details are presented elsewhere (47). Currently methods of deglycosylating CEA are being evaluated with the objective of simplifying the sequencing problem.

It has long been recognized that radioimmunoassay of serum from normal individuals for CEA always showed some variable low level of CEA reactivity. The demonstration in normal tissues of material crossreactive with but distinguishable from CEA appeared to explain the observed normal levels. These molecules included NCA (48), NGP (49), CCEA-2 (50), CEX (51), and β_E protein (52). These molecules all have lower molecular weights than does CEA. Subsequently other groups (53-55) reported the presence in normal serum and normal tissue extracts of material with CEA-like activity and similar to CEA in elution from gel filtration columns. Unfortunately the concentration of this substance in normal tissue was so low that sufficient material could not be obtained for chemical characterization. In 1974 Go *et al.* (56) measured the secretion in various portions of the gastrointestinal tract of material that inhibited in the assay for CEA. The colon secreted the largest quantity of the material. It eluted from a Sephadex G-200 column in the same position as CEA.

The presence of these quantities of material with CEA-like activity in the colon lavages of healthy individuals made it possible to isolate sufficient material for chemical characterization and direct comparison with CEA (57). The material with CEA-like activity from colonic lavages was purified by gel filtration on Sepharose 6B and Sephadex G200 followed by affinity chromatography on concanavalin A linked to Sepharose. The purified material migrated in polyacrylamide-sodium dodecyl sulfate electrophoresis as a single diffuse band with mobility identical to that of tumor CEA. It possessed the same specific activity as CEA in radioimmunoassay. The carbohydrate and amino acid compositions were similar to those of CEA. Moreover, methylation analysis demonstrated that the monosaccharide linkages were similar to those in CEA (57), and its N-terminal amino acid sequence was homologous to CEA (58). Thus by all criteria applied it was indistinguishable from CEA. It thus seems likely that CEA is a normal tissue product. This raises the question of whether the elevated serum levels observed in malignancy result from increased cellular synthesis or by redirection to the blood stream of material normally eliminated in the gastrointestinal tract.

During the isolation of CEA from liver metastases of colonic adenocarcinoma, it was observed that a lower molecular weight component possessing CEA activity in the radioimmunoassay was present in the eluate from the Sepharose 4B column. This material, which we have designated TEX, has now been isolated and purified (59). Like CEA, TEX is a glycoprotein. It binds to concanavalin A Sepharose from which it can be eluted by displacement with methyl α-D-mannoside. By Sephadex G200 gel filtration and by sodium dodecyl sulfate polyacrylamide gel electrophoresis its molecular weight was shown to be about 110,000 daltons.

TEX contained 35% carbohydrate, considerably less than that present in CEA. Linkage studies of the carbohydrate by methylation analysis revealed that TEX contained substantially less terminal galactose, as well as less 4-linked intrachain and 3,4-branched N-acetylglucosamine than CEA. A single treatment with periodate destroyed all of the sialic acid and fucose and 25% of the N-acetylglucosamine as with CEA, but in addition twice as much mannose and galactose are destroyed as compared with CEA (59).

When the size of the protein component is calculated from the molecular weight and percent protein, it is found that this size is nearly identical to that of CEA. Similarly the amino acid composition of TEX is nearly identical to that of CEA, except for the presence of small but reproducible amounts of methionine in TEX, but not in CEA. The sequence of the first 24 N-terminal amino acids of TEX differs from CEA only in the presence of alanine, at position 21 in lieu of valine in CEA. The combined evidence, both immunological and chemical, indicates that TEX is closely related to CEA, but differs in its mode and degree of glycosylation and has at least one and probably more amino acid alterations in its polypeptide chain (59).

Since TEX, which was isolated from a tumor, may be similar to NCA and related materials isolated from normal tissue, a preparation of NCA was obtained from normal spleen tissue by an immunochemical method utilizing insolubilized antibodies to CEA and NCA (60). The molecular weight of the NCA obtained was found to be about 100,000 daltons. The carbohydrate composition was 30%. Unfortunately the material available was insufficient for methylation analysis and periodate oxidation studies. The sample possessed nearly the same amino acid analysis as CEA, but like TEX contained a small but demonstrable amount of methionine. The sequence of the first 26 N-terminal amino acids was identical to that of CEA, except at position 21, where as in TEX alanine is present instead of the valine of CEA (60).

The immunological data show that NCA and TEX are identical if tested with either goat anti-NCA or rabbit anti-TEX, but TEX crossreacts with CEA when tested with monkey anti-CEA (59), an antiserum which was previously shown not to crossreact with spleen NCA (61). An alternative interpretation is that TEX contains trace amounts of CEA_{low}, recently described as identical to CEA in all respects except, perhaps, in degree of glycosylation (62).

Whatever may be the interrelationships of the various antigens crossreactive with CEA (48-52,62), it appears that from the standpoint of primary structure, *i.e.* sequence, of the protein component, CEA, NCA, and TEX are closely related. NCA and TEX may or may not differ from one another in this respect, but they both differ from CEA. We believe that they arise from distinct but related genes which have a common evolutionary origin in a single gene. This primoridal gene (*Figure 3*) has duplicated, or replicated to more than 2 genes, which have subsequently undergone divergent evolution. A more profound understanding of the interaction of these genes and their gene products in benign and malignant disease will hopefully lead to improved clinical utility of this system.

ACKNOWLEDGMENTS

This review was supported in part by National Cancer Institute grants CA 16434 and CA 19163 from the National Large Bowel Cancer Program.

REFERENCES

1. Gold, P. and Freedman, S.O.: Specific carcinoembryonic antigens of the human digestive system. *J. Exp. Med.*, 122: 467-481, 1965.
2. Gold, P. and Freedman, S.O.: Demonstration of tumor specific antigens in human colonic carcinomata by immunological tolerance and absorption techniques. *J. Exp. Med.*, 121: 439-462, 1965.
3. Thompson, D.M.P., Krupey, J., Freedman, S.O. and Gold, P.: The radioimmunoassay of circulating carcinoembryonic antigen of the human digestive system. *Proc. Nat. Acad. Sci. USA*, 64: 161-167, 1969.
4. Egan, M.L., Engvall, E., Ruoslahti, E.I. and Todd, C.W.: Detection of circulating tumor antigens. *Cancer*, 40: 458-466, 1977.
5. Egan, M.L., Coligan, J.E. and Todd, C.W.: Radioimmunoassay for the diagnosis of human cancer. *Cancer*, 34: 1504-1509, 1974.
6. Krupey, J., Gold, P. and Freedman, S.O.: Purification and characterization of carcinoembryonic antigens of the human digestive system. *Nature*, 215: 67-68, 1967.
7. Coligan, J.E., Lautenschleger, J.T., Egan, M.L. and Todd, C.W.: Isolation and characterization of carcinoembryonic antigen. *Immunochemistry*, 9: 377-386, 1972.
8. Egan, M.L., Lautenschleger, J.T., Coligan, J.E. and Todd, C.W.: Radioimmune assay of carcinoembryonic antigen. *Immunochemistry*, 9: 289-299, 1972.
9. Egan, M.L., Todd, C.W. and Knight, W.S.: ^{57}Co: A volume marker for the triple isotope double antibody radioimmune assay. *Immunochemistry*, 14: 611-613, 1977.

10. Slayter, H.S. and Coligan, J.E.: Electron microscopy and physical characterization of the carcinoembryonic antigen. *Biochemistry*, 14: 2323-2330, 1975.
11. Slayter, H.S. and Coligan, J.E.: Characterization of carcinoembryonic antigen fractionated by concanavalin A chromatography. *Cancer Res.*, 36: 1696-1704, 1976.
12. Pritchard, D.G. and Todd, C.W.: Purification of carcinoembryonic antigen by removal of contaminating polysaccharides. *Cancer Res.*, 36: 4699-4701, 1976.
13. Krupey, J., Gold, P. and Freedman, S.O.: Physicochemical studies of the carcinoembryonic antigens of the human digestive system. *J. Exp. Med.*, 128: 387-398, 1968.
14. Pauling, L., Corey, R.B. and Branson, H.R.: The structure of proteins: Two hydrogen-bonded helical configurations of the polypeptide chain. *Proc. Nat. Acad. Sci. USA*, 37: 205-240, 1951.
15. Coligan, J.E., Henkart, P.A., Todd, C.W. and Terry, W.D.: Heterogeneity of the carcinoembryonic antigen. *Immunochemistry*, 10: 591-599, 1973.
16. Eveleigh, J.W.: Heterogeneity of carcinoembryonic antigen. *Cancer Res.*, 34: 2122-2124, 1974.
17. Chism, S.E., Bell, P.M. and Warner, N.L.: Heterogeneity of CEA and 'CEA-like' preparations determined by Farr assays for lectin binding. *J. Immunol. Meth.*, 13: 83-89, 1976.
18. Harvey, S.R. and Chu, T.M.: Demonstration of two molecular variants of carcinoembryonic antigen by concanavalin A Sepharose affinity chromatography. *Cancer Res.*, 35: 3001-3008, 1975.
19. Egan, M.L., Coligan, J.E., Morris, J.E., Schnute, W.C., Jr. and Todd, C.W.: Physical characterization and structural studies of the carcinoembryonic antigen. *Cancer Res.*, 36: 3482-3485, 1976.
20. Clamp, J.R., Bhatti, T. and Chambers, R.E.: The examination of carbohydrate in glycoproteins by gas-liquid chromatography. In: Glycoproteins. A. Gottschalk, (ed.), *Elsevier Publishing Co.*, 5 (part A): 300-321, 1972.
21. Warren, L.: The thiobarbituric acid assay of sialic acids. *J. Biol. Chem.*, 234: 1971-1975, 1959.
22. Liu, T.-Y. and Chang, Y.H.: Hydrolysis of proteins with p-toluenesulfonic acid. *J. Biol. Chem.*, 246: 2842-2848, 1971.
23. Coligan, J.E. and Todd, C.W.: Structural studies on carcinoembryonic antigen: Periodate oxidation. *Biochemistry*, 14: 805-810, 1975.
24. Wrann, M.M. and Todd, C.W.: Sensitive determination of sugars utilizing packed capillary columns and electron capture detection. *J. Chromatog.* 147: 309-316, 1978.

25. Coligan, J.E., Egan, M.L., Guyer, R.L., Schnute, W.C., Jr. and Todd, C.W.: Structural studies on the carcinoembryonic antigen. *Ann. N.Y. Acad. Sci.*, 259: 355-365, 1975.
26. Spiro, R.G.: Glycoproteins. *Ann. Rev. Biochem.*, 39: 599-638, 1970.
27. Coligan, J.E., Pritchard, D.G., Schnute, W.C., Jr. and Todd, C.W.: Methylation analysis of the carbohydrate portion of carcinoembryonic antigen. *Cancer Res.*, 36: 1915-1917, 1976.
28. Hakomori, S.: A rapid permethylation of glycolipid and polysaccharide catalyzed by methylsulfinyl carbanion in dimethylsulfoxide. *J. Biochem. Tokyo*, 55: 205-208, 1964.
29. Sandford, P.A. and Conrad, H.E.: The structure of the *Aerobacter aerogenes* A3(S1) polysaccharide. I. A reexamination using improved procedures for methylation analysis. *Biochemistry*, 5: 1508-1517, 1971.
30. Stellner, K., Saito, H. and Hakomori, S.: Determination of aminosugar linkages in glycolipids by methylation. *Arch. Biochem. Biophys.*, 155: 464-472, 1973.
31. Björndal, H., Hellerqvist, C.G., Lindberg, B. and Svensson, S.: Gas-liquid chromatography and mass spectrometry in methylation analysis of polysaccharides. *Angew. Chem. Intern. Ed. Engl.*, 9: 610-619, 1970.
32. Pritchard, D.G., Todd, C.W. and Egan, M.L.: Chemistry of carcinoembryonic antigen. *Methods in Cancer Research*, 14: 55-85, 1977.
33. Hammarström, S., Engvall, E., Johansson, B.G., Svensson, S., Sundblad, G. and Goldstein, I.J.: Nature of the tumor-associated determinant(s) of carcinoembryonic antigen. *Proc. Nat. Acad. Sci. USA*, 72: 1528-1532, 1975.
34. Heath, E.C.: Complex polysaccharides. *Ann. Rev. Biochem.*, 40: 29-56, 1971.
35. Shively, J.E. and Todd, C.W.: Carcinoembryonic antigen. *Scand. J. Immunol.*, in press.
36. Smith, F. and Unrau, A.M.: On the presence of 1→6 linkages in laminarin. *Chem. Ind. London*, 881, 1959.
37. Bessel, E.M., Thomas, P. and Westwood, J.H.: Multiple Smith degradations of carcinoembryonic antigen (CEA) and of asialo CEA. *Carbohyd. Res.*, 45: 257-268, 1975.
38. Eisen, H.N., pp 376-377 in Davis, B.D., Dulbecco, R., Eisen, H.N., Ginsberg, H.S. and Wood, W.B., *Microbiology*, Harper and Row, New York, 1968.
39. Morris, J.E., Egan, M.L. and Todd, C.W.: The binding of carcinoembryonic antigen by antibody and its fragments. *Cancer Res.*, 35: 1804-1808, 1975.
40. Westwood, J.H., Bessel, E.M., Bukhari, M.A., Thomas, P. and Walker, J.M.: Studies on the structure of the carcinoembryonic antigen. I. Some deductions on the basis of chemical degradations. *Immunochemistry*, 11: 811-818, 1974.

41. Westwood, J.H. and Thomas, P.: Studies on the structure and immunologic activity of carcinoembryonic antigen -- the role of disulfide bonds. *Br. J. Cancer,* 32: 708-719, 1975.
42. Westwood, J.H., Thomas, P. and Foster, A.B.: The treatment of carcinoembryonic antigen with sodium metaperiodate. *Biochem. Soc. Trans.,* 2: 1250-1251, 1974.
43. Terry, W.D., Henkart, P.A., Coligan, J.E. and Todd, C.W.: Structural studies of the major glycoprotein in a preparation with carcinoembryonic antigen activity. *J. Exp. Med.,* 136: 200-204, 1972.
44. Terry, W.D., Henkart, P.A., Das, S., Coligan, J.E. and Todd, C.W.: Characterization of human carcinoembryonic antigens. *Proceed. of Miles Symposium,* pp 241-247, 1973.
45. Chu, T.M., Bhargava, A.K. and Harvey, S.R.; Structure studies of the glycoproteins associated with carcinoembryonic antigen (CEA). *Fed. Proc.,* 33: 1561, 1974.
46. Coligan, J.E., Egan, M.L., Guyer, R.L., Schnute, W.C., Jr. and Todd, C.W.: Structural studies on the carcinoembryonic antigen. *Ann. N.Y. Acad. Sci.,* 259: 355-365, 1975.
47. Shively, J.E., Kessler, M. and Todd, C.W.: The antigenic determinants of carcinoembryonic antigen: The formation, isolation, composition, and N-terminal sequences of its tryptic peptides. *Cancer Res.,* submitted for publication.
48. von Kleist, S., Chavanel, G. and Burtin, P.: Identification of an antigen from normal human tissue that crossreacts with the carcinoembryonic antigen. *Proc. Nat. Acad. Sci. USA,* 69: 2492-2494, 1972.
49. Mach, J.-P. and Pusztaszeri, G.: Carcinoembryonic antigen (CEA): Demonstration of a partial identity between CEA and a normal glycoprotein. *Immunochemistry,* 9: 1031-1034, 1972.
50. Turberville, C., Darcy, D.A., Laurence, D.J.R., Jones, E.W. and Neville, A.M.: Studies on carcinoembryonic antigen (CEA) and a related glycoprotein, CCEA-2. Preparation and chemical characterisation. *Immunochemistry* 10: 841-843, 1973.
51. Darcy, D.A., Turberville, C. and James, R.: Immunological study of carcinoembryonic antigen (CEA) and a related glycoprotein. *Br. J. Cancer,* 28: 147-160, 1973.
52. Ørjasaeter, H.: Study of substances related to carcinoembryonic antigens, CEA-NCA and association with α_1-antichymotrypsin. *Acta Path. Microbiol. Scand.,* 84: 235-244, 1976.
53. Chu, T.M., Reynoso, G. and Hansen, H.J.: Demonstration of carcinoembryonic antigen in normal human plasma. *Nature,* 238: 152-153, 1972.
54. Kupchik, H.Z. and Zamcheck, N.: Carcinoembryonic antigen(s) in liver disease. *Gastroenterology,* 63: 95-101, 1972.

55. Pusztaszeri, G. and Mach, J.: Carcinoembryonic antigen (CEA) in nondigestive cancerous and normal tissues. *Immunochemistry*, 10: 197-204, 1973.
56. Go, V.L.W., Ammon, H.V., Holtermuller, K.H., Krag, E. and Phillips, S.F.: Quantification of carcinoembryonic antigen-like activities in normal human gastrointestinal secretions. *Cancer*, 36: 2346-2350, 1975.
57. Egan, M.L., Pritchard, D.G., Todd, C.W. and Go, V.L.W.: Isolation and immunochemical and chemical characterization of carcinoembryonic antigen-like substances in colon lavages of healthy individuals. *Cancer Res.*, 37: 2638-2643, 1977.
58. Shively, J.E., Todd, C.W., Go, V.L.W. and Egan, M.L.: Amino terminal sequence of a carcinoembryonic antigen-like glycoprotein isolated from the colonic lavages of healthy individuals. *Cancer Res.*, in press.
59. Kessler, M.J., Shively, J.E., Pritchard, D.G. and Todd, C.W.: Isolation, immunologic characterization, and structural studies of a tumor antigen (TEX) related to carcinoembryonic antigen. *Cancer Res.*, in press.
60. Engvall, E., Shively, J.E. and Wrann, W.: Isolation and characterization of the normal crossreacting antigen (NCA). Homology of its N-terminal amino acid sequence with that of carcinoembryonic antigen (CEA). *Proc. Nat. Acad. Sci. USA*, in press.
61. Ruoslahti, E., Engvall, E., Vuento, M. and Wigzell, H.: Monkey antisera with increased specificity to carcinoembryonic antigen (CEA). *Int. J. Cancer*, 17: 358-361, 1976.
62. Hammarström, S., Svenberg, T., Hedin, A. and Sundblad, G.: Antigens related to CEA. *Scand. J. Immunology*, in press.

RECEIVED December 29, 1977.

20

Status of Blood Group Carbohydrate Chains in Human Tumors

WILLIAM W. YOUNG, JR. and SEN-ITIROH HAKOMORI

Biochemical Oncology, Fred Hutchinson Cancer Research Center, and Departments of Pathobiology and Microbiology/Immunology, University of Washington, 1124 Columbia, Seattle, WA 98104

1. Introduction

Clear changes of glycolipid composition and glycoprotein profiles in cell surface membranes associated with oncogenic transformation have been described during the past ten years. Possibly the first clear documentation for this topic was the decrease of hematoside and increase of lactosylceramide in polyoma transformed BHK cells as described in 1968 (1). A number of studies carried out by different investigators agreed that two kinds of changes in carbohydrate chains may occur during oncogenic transformation (see for review ref. 2). One is the block of synthesis certain glycolipids resulting in the associated accumulation of precursor structures and the second is induction of synthesis of a unique glycolipid for the transformed cells which most probably depends on activation of a new glycosyltransferase.

Since *in vitro* cellular systems are only models for *in vivo* tumors, it is important to observe the changes of membrane glycolipids in tumors, particularly in human cancer. Among the great number of carbohydrate chains in human cells and tissues, the most well-defined carbohydrate chains, in terms of their chemistry, enzymology and genetics, are the blood group determinants. For this reason, our major effort has been directed to characterizing the changes of blood group determinants in human cancer as compared to comparable normal tissue; however, it should be noted that such comparisons may not be ideal as the precursor cell of human cancer may represent only a small population of cells present in normal tissue. However, for practical purposes such as diagnosis and therapy of human tumor using a specific cell surface marker, a gross difference of carbohydrate determinants observable between tumor and normal tissue may offer useful information.

The first clear change of blood group determinants in human cancer was described in the laboratory of the late Dr. Masamune, who was a good friend of the late Dr. Ward Pigman to whom this Symposium is dedicated. Oh-uti in 1949, while working in Masamune's laboratory, described the deletion of blood group activity in "blood group polysaccharide" prepared from human gastric cancer (3). A number of subsequent studies were made in

0-8412-0452-7/78/47-080-357$05.00/0

Masamune's laboratory comparing the immunological and chemical differences of blood group glycoproteins from tumors and normal mucosa (4,5). Independently, Davidsohn and his co-workers (6,7), Kay and Wallace (8), Prendergast (9), and Dabelsteen and Fulling (10), have studied the status of blood group A and B in human tumors as compared to normal tissue. Although a few discrepencies were described, they essentially agreed that blood group A or B determinants are deleted in various human tumors and that the degree of deletion could be correlated with the degree of malignancy (6,7). Most of these studies were made before the chemical structures of A and B determinants were established. Since the structural relationship between A, B, H, Le^a, Le^b, I and i determinants was established late in the 1960's, studies have focused on the status of the precursors of ABH determinants.

As seen in Table I, accumulation of precursor carbohydrate chains is quite obvious for some human cancers, if not all of the cases. A similar precursor accumulation has been observed in *in vivo* transformed cells and in experimental animal tumors and such precursors could be useful surface markers of tumor cells (13).

Another category of changes in blood group determinants that has been observed is the appearance of new blood group determinants in some human tumor tissue. Examples include the possible synthesis of P and P_1 antigens in the gastric tumor of a patient belonging to the extremely rare blood group p (14,15) and induction of A-like antigen in tumors of host blood group O (16,17). The A-like antigen appearing in blood group O or B could have been the result of the new synthesis of Forssman antigen in tumors (18, 19). These items will be discussed extensively in subsequent sections.

2. Accumulation of Precursor Carbohydrates of Blood Group Determinants

Since I-, and i-determinants have been assigned as precursors of blood group ABH determinants (26), we have studied the i/I reactivity of water-soluble blood group glycoproteins extracted from gastrointestinal tumors and normal mucosal tissue. The glycoprotein fraction was extracted according to the procedure described by Masamune *et al* (5) and by the perchloric acid extraction method described by Krupey *et al* (27). As shown in Table II, the perchloric acid extractable fraction of tumor tissue showed significant i-activity in contrast to the normal mucosa in 4 cases, whereas, 3 cases showed higher i-activity in normal mucosa, and one case did not show any significant difference (Watanabe, Hakomori and Warner, unpublished data). Recently, the hexasaccharide structure, Galβ1→4GlcNAcβ1→3Galβ1→4GlcNAcβ1→3 Galβ1→4Glc was found to be a determinant for i-specificities (28). Such a structure may appear or increase in some human cancer.

In order to study the status and distribution of blood group carrier carbohydrate chains in normal and cancer tissues, the reactivity of glycolipids directed against three structures, as shown in Figure 1, were compared using their respective antibodies (24). The complement fixing reactivities of glycolipids

Table I. Status of Blood Group Determinants in Human Cancer.

Status	References
1. Deletion of AB-determinants	
Gastrointestinal tumors	(Oh-uti 1949; Masamune *et al* 1952; 1958; Davidsohn *et al* 1966; Iseki *et al* 1962; Stellner *et al* 1973)
Cervical cancer	(Davidsohn *et al* 1969)
Bladder cancer	(Kay and Wallace 1961)
Oral carcinoma	(Prendergast 1968; Dabelsteen and Fulling 1971)
2. Precursor accumulation	
Association of CEA with I (Ma) antigen	(Simmons and Perlman 1973)
Presence of I (Step) antigen	(Feizi *et al* 1975)
Presence of i (Dench) antigen	(Watanabe and Hakomori unpublished)
Accumulation of lacto-N-triosylceramide (GlcNAcCTH)	(Watanabe and Hakomori 1977; Karlsson 1976)
3. Induction of "incompatible" determinants	
P_1, P, P^k antigen in tumor of p-individual	(Levine *et al* 1951)
A-like antigen in tumors of O-, B-individuals	(Hakomori *et al* 1967; Häkkinen 1971)
4. Heterophile F-antigen as a possible new isoantigen in human and the presence of F-antigen in tumors of F^--tissues	(Hakomori, Wang and Young 1977; Makita 1977)

Table II. The i-activity of the perchloric-acid-extractable fraction of normal mucosa and tumor tissue*

Case	Blood group	Dench i activity in	
		Normal mucosa	Tumor tissue
MIM	O	>50	3
Tn	A. MN	>50	6
V. His	ND	25	>50
G. Gutt.	O, Le^a, MN	>50	12
A. Dalz.	O, N, Le^a	>50	6
C. Walt.	O, Le^a, MN	6	25
M. Hutch.	ND	3	12
Anonym.	ND	>50	>50

*Activity is expressed by doses of glycoprotein which showed obvious hemagglutination inhibition in µg. The test system contained anti-i (Dench) serum (3 hemagglutination doses) and 1% cord erythrocytes K. Watanabe, S. Hakomori and G.A. Warner, unpublished data).

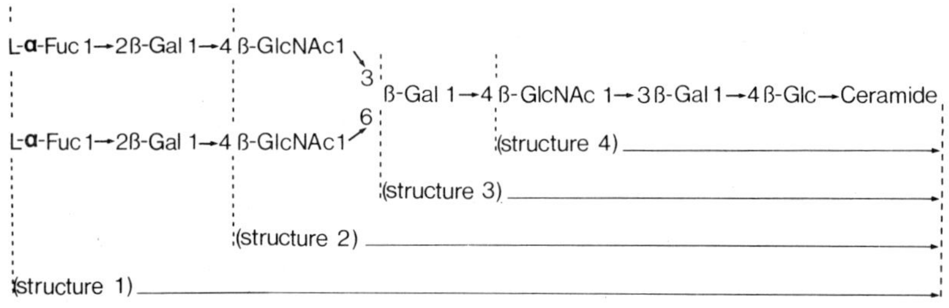

Journal of Experimental Medicine

Figure 1. Structure of H_3 glycolipid and its degradation products to which antibodies were directed

extracted from normal and colon mucosa with antibodies directed against structures 1, 2 and 4 are shown in Table III. The average reactivity with anti-structure 4 antisera was 1 to 160 for normal mucosal tissue and 1 to 750 for cancer tissue. While the antibody directed against structure 4 showed this remarkable differential reactivity between normal and tumor glycolipids, the antibodies directed against structure 1 and 2 showed similar reactivity towards normal and tumor glycolipids. The glycolipids of several cases of colon carcinoma were compared with those of normal mucosal tissue by thin-layer chromatography (TLC). An intensified spot corresponding to structure 4 (GlcNAcβ1→3Galβ1→4 Glcβ1→1ceramide) was observed in the glycolipid fraction of colon carcinoma as compared to normal mucosa. Such a remarkable accumulation of this structure in some human cancers can be regarded as a precursor accumulation. Recently, Karlsson briefly mentioned the accumulation of the same glycolipid in human melanoma tissue (25). Since the glycolipid is the common structure for all blood group ABH, Le^a, Le^b, and Ii antigens, the accumulation of this glycolipid can be correlated to a precursor accumulation due to a blocked synthesis of these antigens.

3. Induction of "incompatible" determinants and Forssman glycolipid.

The appearance of incompatible blood group determinants of both the P and ABH systems have been reported (*see* Table I). Because of the well-known cross-reactions of the A and Forssman antigens, there was the possibility that the apparent "neo-A" reactivity that was detected in tumors of type O patients (16,17) could have been caused by the presence of Forssman antigen.

The heterogenetic Forssman antigen (29) is a glycosphingolipid whose structure has been identified as GalNAcα1→3GalNAcβ1→3 Galα1→4Galβ1→4Glc→ceramide (30,31). In the early literature, reviewed by Buchbinder in 1935 (32), the various animal species were categorized as being either Forssman-positive or - negative. It is now clear that all Forssman-positive animal species possess a glycolipid with a carbohydrate structure identical to that described above (33-36). In addition the immunodeterminant of the Forssman antigen, namely a terminal disaccharide GalNAcα1→3Gal NAcβ1→R, has been found in other "carrier" structures as well, all of which are Forssman active: (1) a ceramide tetrasaccharide of hamster fibroblasts, GalNAcα1→3GalNAcβ1→3Galα1→4Galβ1→ceramide (37); a polysaccharide of *Streptococcus* type c (38); and a ceramide-heptasaccharide and -octasaccharide of dog gastric mucosa (39,40). It is also possible that certain glycoproteins may be Forssman-active, since glycoproteins have been recently detected that contain globoside and ganglioside reactivities (41).

In the early literature most primates were found to be Forssman-negative, although humans of blood groups A and AB were found to be immunologically Forssman-reactive. These results could have been due to cross-reactions with blood group A structures. However, Forssman has been detected in certain human

Table III. Complement fixing reactivities of glycolipids extracted from normal and colon mucosa.

	Case 1		Case 2		Case 3		Case 4	
	Normal	Cancer	Normal	Cancer	Normal	Cancer	Normal	Cancer
Reactivity with anti-structure 4	320	1280	160	640	320	1280	320	1280
Reactivity with anti-structure 2	80	320	160	160	160	160	160	160
Reactivity with anti-structure 1 (anti-H_3)	160	80	160	80	80	80	80	80

Complement fixing reactivities of glycolipids extracted from normal colon mucosa and colon tumors with antibodies that are directed against structure 1, 2, and 4 of Fig. 1. Numbers are reciprocals of the dilution of antisera that could fix complement by 1.2 μg/50 μl of glycolipid antigen complexed with two times amount of lecithin and cholesterol.

Journal of Experimental Medicine

tissues in the following recent studies: (1) by Kawanami in a human metastatic tumor of biliary adenocarcinoma in liver (42); (2) by Makita in human lung tumor tissue but not in normal lung (18); and (3) by Hakomori *et al* (19) in gastrointestinal tissue. In the latter report Forssman was detected as a normal component of the gastrointestinal mucosa (F^+ population) in only 5 of 21 cases studied, while in the remaining cases Forssman was not detectable in the normal mucosa (F^- population).

In tumors derived from these cases the following striking pattern was found: (1) all tumors derived from F^- mucosa possessed Forssman glycolipid, while (2) none of the tumors originating in F^+ mucosa contained Forssman glycolipid. Globoside (GalNAcβ1→3Galα1→4Galβ1→4Glc→Ceramide), the immediate precursor of Forssman, was found to be present in large amounts in both the normal mucosa and tumors of all cases studied. Thus, the distribution of Forssman in the normal mucosa appeared to be like that of an isoantigen. In addition the appearance of Forssman in tumors of F^- individuals indicated that it might represent a tumor-specific antigen. A summary of the cases studied to date, all of which come from Taiwan, appears in Table IV.

To further explore this potentially exciting area, we have begun a thorough investigation of the Forssman status of humans. Relevant aspects of this problem include the following: (1) The surgical cases described above all come from Taiwan. Does such a distribution occur in other human populations? (2) In F^- individuals is Forssman absent in all normal tissues besides the gastrointestinal tract? This question is crucial to any possible role of Forssman as a tumor-specific antigen. Related to this is the question of whether Forssman is expressed as an embryonic antigen.

To conduct such investigations requires testing tiny amounts of tissues in some cases. This necessitates the use of immunological rather than chemical methods. Lipid antigens possess certain immunochemical properties that greatly effect their reactivity (see for reviews 49 and 50). Individual glycolipid molecules are univalent and amphipathic. They do not form "true" solutions in water but rather form micellar solutions on dispersion in an aqueous solution by sonication and heating. Micellar aggregates of glycolipids represent multivalent structures which display minimal reactivity with antibodies. This reactivity can be greatly stimulated by the addition of "auxiliary lipids" (most often a phospholipid and cholesterol) to the glycolipid solution prior to drying and subsequent addition of buffer. The lipid vesicles or liposomes that are produced create an obvious multivalent structure. However, a disadvantage of such liposomes is that because of their multi-lamellar nature only those antigen molecules on the external surface of the outermost bilayer (perhaps only 5% of the total antigen) can react with antibody. To avoid this problem the liposomes can be sonicated until they reach a nearly unilamellar state. Alternatively, the dried lipids can be dissolved first in hot MeOH or EtOH followed by buffer, which appears to result in maximum antigen exposure as well.

An additional problem of glycolipid antigens is their apparent "masking" when in the presence of a large amount of other glycolipids. Forssman reactivity could not be detected in the total neutral glycolipid fractions of the surgical cases mentioned above but became quite strong after purification of the ceramide pentasaccharide fraction by Iatrobeads chromatography (19). A similar situation occurred with the H-activity of H-glycolipid (43) and Ii activity of Ii-active glycolipids (unpublished observation).

The immunochemical methods most frequently used for glycolipid analysis include the following. Double diffusion in agar gel is the simplest but least sensitive method and occasionally produces spurious precipitin bands. Inhibition of hemagglutination is simple, and of moderate sensitivity. Complement fixation can detect about 1 nanogram of glycolipid but requires considerable effort. Release of trapped markers from glycolipid-liposomes by antibody-complement (44) has a sensitivity limit of about 50 nanograms.

Recently we have developed a radioimmunoassay (RIA) for detecting glycolipid antigens. Initial attempts to use radiolabeled glycolipids were unsuccessful because of transfer of label between "hot" and "cold" liposomes. The present assay utilizes a glycolipid-polymer synthesized as shown in Fig. 2. First, the olefinic double bond in the ceramide moiety of the glycolipid is converted to a carboxylic acid by a recent improvement (45) of our earlier procedure (46). Then the glycolipid is coupled to a polyacrylic hydrazide (PAH) polymer (47). Radiolabels can be attached to both the glycolipid and PAH moieties as shown in Fig. 3. It should be noted in passing that such a "polyglycolipid" should prove useful not only for detection of glycolipid antigens as described below but also for quantitation of anti-glycolipid sera and as a polymeric antigen in immunization protocols.

The RIA procedure is summarized in Fig. 4. "Cold" glycolipid antigen in the form of lipid vesicles is preincubated with diluted rabbit anti-Forssman immunoglobulin. Then the labeled Forssman-PAH complex is added and allowed to bind to the remaining available antibodies. Addition of intact, fixed *Staphylococcus aureus*, which has "Protein A" binding sites for immunoglobulin (48), and centrifugation results in precipitation of any antigen-antibody complexes. The supernatant can then be counted to determine the amount of inhibition of precipitation of labeled antigen by the unlabeled competitor.

Figure 5 demonstrates that this assay can readily detect 1 nanogram of Forssman, while inhibition by related structures (globoside and blood group A glycolipids) is negligible. Figure 6 compares the reactivity of the purified ceramide

Table IV. Summary of Forssman status of Taiwan tumor cases

Blood type	F^- cases	F^+ cases
A	4	1
B	10	1
O	6	4
Total	20	6

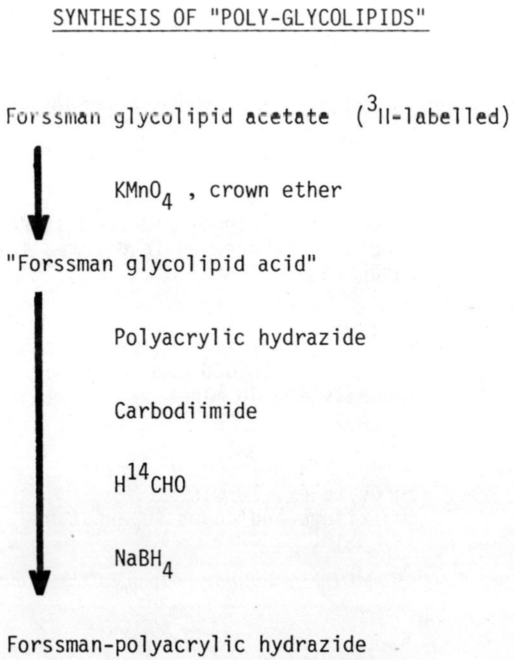

Figure 2. Synthesis of a "poly-glycolipid," Forssman-polyacrylic hydrazide

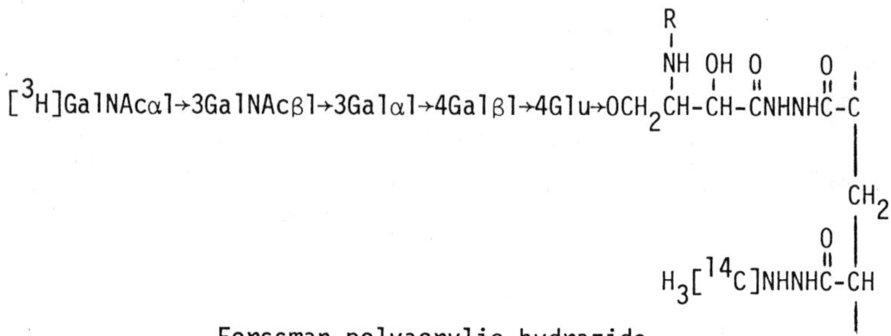

Forssman polyacrylic hydrazide

Figure 3. Structure of Forssman-polyacrylic hydrazide

Step 1. Glycolipid/phospholipid/cholesterol vesicles
Rabbit anti-Forssman Ig diluted 1:1000
Incubate 4°, 2 hr

Step 2. [$^{14}C/^{3}H$] Forssman-polyacrylic hydrazide
(10,000 cpm per assay tube)
Incubate 4°, 30 min

Step 3. Intact Staphylococcus aureus
Incubate 4°, 10 min
Centrifuge and count supernatant

Figure 4. Protocol for radioimmunoassay of Forssman glycolipid

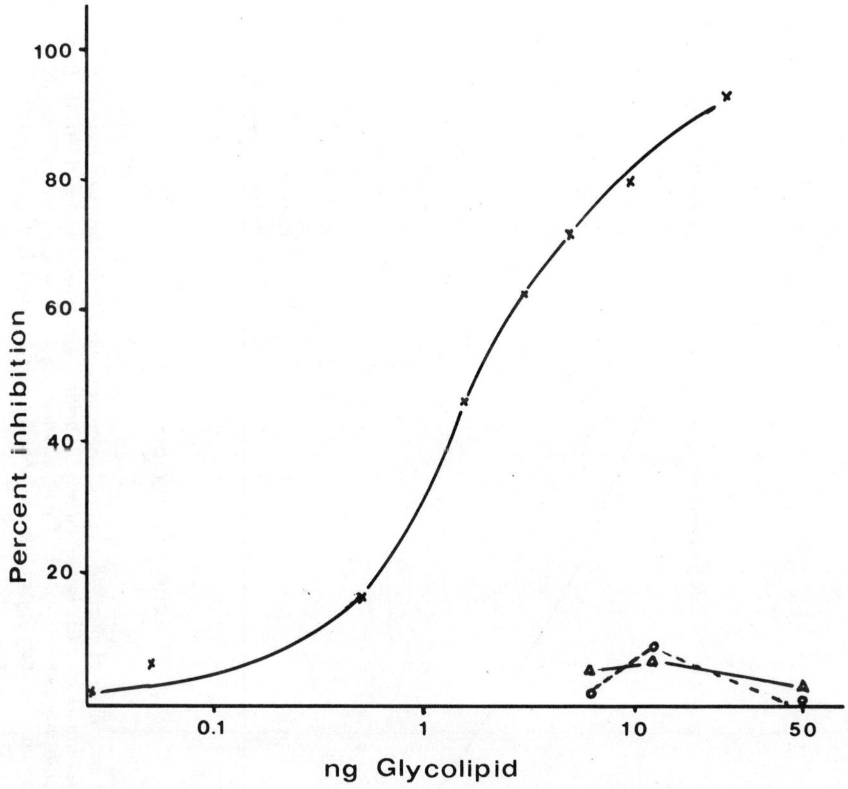

Figure 5. Inhibition of precipitation of [^{14}C/^{3}H]-labeled Forssman-PAH by unlabeled glycolipid antigens. Lipid vesicles of the indicated composition were incubated with anti-Forssman (10 μL, 1/1000 dilution) for 2 hr at 4°C in a total volume of 100 μL. [^{14}C/^{3}H]-labeled Forssman-PAH (10,000 cpm, 100 μL) was added and incubation continued for 30 min, 4°C. S. aureus *(7 mg, 700 μL) was added and after 10 min at 4°C removed by centrifugation at 100 g for 10 min. 450 μL of supernatant was counted, and the results are expressed as follows:*

$$\% \text{ Inhibition} = \left(\frac{C - B}{A - B}\right) \times 100$$

where A = supernatant cpm of control tube without anti-Forssman; B = supernatant cpm of control tube with anti-Forssman but no inhibitor; C = supernatant cpm of experimental tube. (×—×) Forssman/SM/CHOL; (○—○) Globoside/SM/CHOL; and (△—△) A^b glycolipid/SM/CHOL. Mole ratio glycolipid/SM/CHOL = 1/250/200.

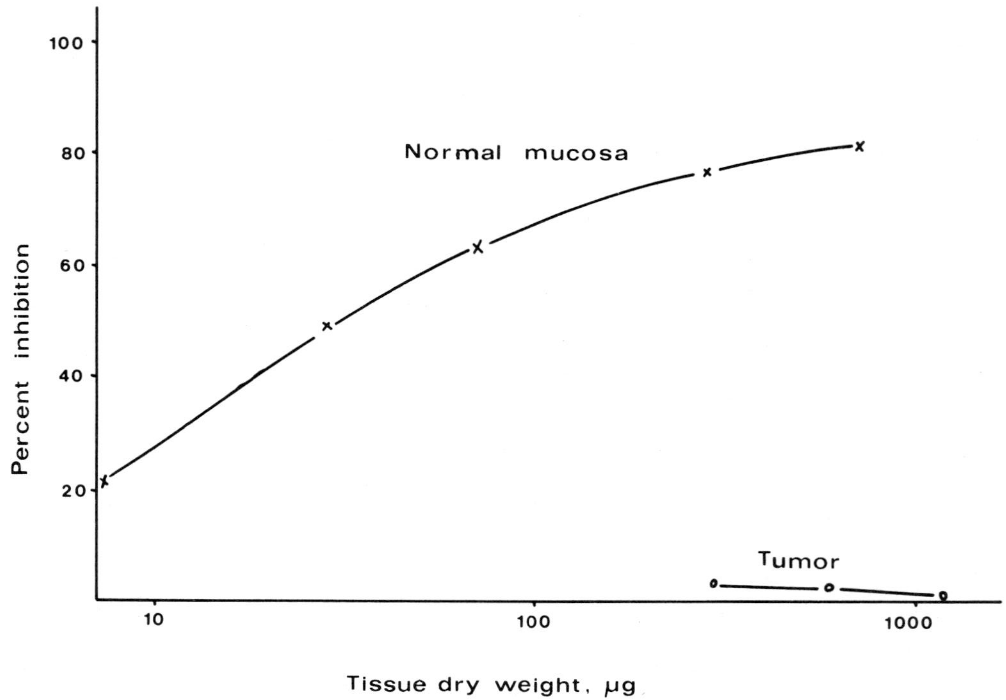

Figure 6. Forssman reactivity of purified ceramide pentasaccharide fractions of human gastrointestinal tissues as determined by radioimmunoassay. Purified ceramide pentasaccharide fractions from the normal colonic mucosa (×—×) and colon tumor tissue (○—○) from a Taiwan patient were tested for Forssman reactivity by radioimmunoassay as described in the legend to Figure 5.

pentasaccharide fraction from the normal mucosa and tumor tissue of one Taiwan case which by chemical analysis had been found to be F^+, as defined above.

4. Concluding remarks

In conclusion, examples have been presented in which striking alterations of blood group determinants occur in human tumor tissue apparently due to one of two mechanisms: (a) accumulation of precursor structures due to a block in synthesis of more complex determinants such as A or B; and (b) the appearance of apparently new determinants due to the possible activation of allelomorphic genes. Many examples of the first possibility have been revealed, not only in tumors but also in *in vitro* model systems as well, and is a rather well-accepted mechanism. In contrast, the second process has only limited experimental evidence. In fact until recently the only example was a single case of apparent expression of P blood group antigens in the tumor of a p individual (14,15). Now, however, the appearance of the Forssman antigen in F^- individuals (19) may result from this process as well.

Both processes may represent the expression of onco-fetal antigens as both the "structure 4" ceramide trisaccharide (24) and Forssman glycolipid (K. Watanabe, unpublished observation) have been detected in human fetal tissue. In addition both processes generate tumor cell surface determinants which could serve as specific targets for immunotherapy or for immunochemotherapy.

This study has been supported by National Cancer Institute Grant CA 19224, and partially supported by a grant from the Cancer Research Institute, Inc. W.W.Y. is recipient of a fellowship from the Cancer Research Institute, Inc.

REFERENCES

1. Hakomori, S., and Murakami, W.T. (1968) Proc. Nat. Acad. Sci. U.S.A. 59, 254.
2. Hakomori, S. (1975) Bioch. Biophys. Acta 417, 55.
3. Oh-uti, K. (1949) Tohoku J. Exp. Med. 51, 297.
4. Masamune, H., Yosizawa, Z., Oh-uti, K., Matuda, Y., and Masukawa, A. (1952) Tohoku J. Exp. Med. 56, 37.
5. Masamune, H., Kawasaki, H., Abe, S., Oyama, K., and Yamaguchi, Y. (1958) Tohoku J. Exp. Med. 68, 81.
6. Davidsohn, I., Kovarik, S., and Lee, C.L. (1966) Arch. Path. 81, 381.
7. Davidsohn, I., Kovarik, S., and Ni, Y. (1969) Arch. Path. 87, 306.
8. Kay, H.E.H., and Wallace, B.H. (1961) J. Natl. Cancer Inst. 26, 1349.
9. Prendergast, R.C., Tot, P.D., and Garguilo, A.W. (1968) J. Dent. Res. 47, 306.
10. Dabelsteen, E., and Fulling, J.H. (1971) Scand. J. Dent. Res. 79, 387.
11. Iseki, S., Furukawa, K., and Ishihara, K. (1962) Proc. Jap. Acad. Sci. 38, 556.
12. Simmons, D.A.R., and Perlman, P. (1973) Cancer Res. 33, 313.
13. Rosenfelder, G., Young, W.W., Jr., and Hakomori, S. (1977) Cancer Res. 37, 1333.
14. Levine, P., Bobbitt, O.B., Waller, R.K., and Kuhmichel, A. (1951) Proc. Soc. Exp. Biol. Med. 77, 403.
15. Levine, P. (1976) Ann. N.Y. Acad. Sci. 277, 428.
16. Hakomori, S., Koscielak, J., Bloch, K.J., and Jeanloz, R.W. (1967) J. Immunol. 98, 31.
17. Häkkinen, I. (1970) J. Natl. Cancer Inst. 44, 1183.
18. Makita, A. (1977) United States-Japan Joint Seminar "Structural and Functional Significance of Membrane Glycolipids".
19. Hakomori, S., Wang, S.-M., and Young, W.W., Jr. (1977) Proc. Natl. Acad. Sci. U.S.A. 74, 3023.
20. Stellner, K., Hakomori, S., and Warner, G.A. (1973) Biochem. Biophys. Res. Commun. 55, 439.
21. Iseki, S., Furukawa, K., and Ishihara, K. (1962) Proc. Jap. Acad. Sci. 38, 556.
22. Cooper, A.G. (1974) J. Immunol. 113, 1246.
23. Feizi, T., Turberville, C., and Westwood, J.H. (1975) Lancet 1, 391.
24. Watanabe, K., and Hakomori, S. (1976) J. Exp. Med. 144, 644.
25. Karlsson, K.A. In Structure of Biological Membranes (1976), S. Abrahamsson and I. Pascher (eds.) 34th Nobel Symposium, New York, Plenum Publishing.
26. Feizi, T., Kabat, E.A., Vicar, G., Anderson, B., and Marsh, W.L. (1971) J. Immunol. 106, 1578.
27. Krupey, J., Wilson, T., Freedman, S.O., and Gold, P. (1972) Immunochemistry 9, 617.

28. Niemann, H., Watanabe, K., Hakomori, S., Childs, R.A., and Feizi, T. submitted for publication.
29. Forssman, J. (1911) Biochem. Z. 37, 78.
30. Siddiqui, B., and Hakomori, S. (1971) J. Biol. Chem. 246, 5766.
31. Stellner, K., Saito, H., and Hakomori, S. (1973) Arch. Biochem. Biophys. 155, 464.
32. Buchbinder, L. (1935) Arch. Pathol. 19, 841.
33. Sung, S.J., Esselman, W.J., and Sweeley, C.C. (1973) J. Biol. Chem. 248, 6528.
34. Taketomi, T., Hara, A., Kawamura, N., and Hayashi, M. (1974) J. Biochem. (Tokyo) 75, 197.
35. Ziolkowski, C.H.J., Fraser, B.A., and Mallette, M.F. (1975) Immunochemistry 12, 297.
36. Karlsson, K.-A., Pascher, I., Pimlott, W., and Samuelsson, B.E. (1974) Biomed. Mass Spectrometry 1, 49.
37. Gahmberg, C.G., and Hakomori, S. (1975) J. Biol. Chem. 250, 2438.
38. Coligan, J.E., Fraser, B.A., and Kindt, T.J. (1977) J. Immunol. 118, 6.
39. Slomiany, A., Slomiany, B.L., and Annese, C. (1977) FEBS Letters 81, 157.
40. Slomiany, A., and Slomiany, B.L. (1977) Eur. J. Biochem. 76, 491.
41. Tonegawa, Y., and Hakomori, S. (1977) Biochem. Biophys. Res. Commun. 76, 9.
42. Kawanami, J. (1972) J. Biochem. (Tokyo) 72, 783.
43. Stellner, K., Watanabe, K., and Hakomori, S. (1973) Biochemistry 12, 656.
44. Six, H.R., Young, W.W., Jr., Uemura, K., and Kinsky, S.C. (1974) Biochemistry 13, 4050.
45. Young, W.W., Jr., Laine, R.A., and Hakomori, S. Meth. Enzymol. in press.
46. Laine, R.A., Yogeeswaren, G., and Hakomori, S. (1974) J. Biol. Chem. 249, 4460.
47. Wilchek, M., and Miron, T. (1974) Meth. Enzymol. 34, 72.
48. Kessler, S.W. (1976) J. Immunol. 117, 1482.
49. Rapport, M.M., and Graf, L. (1969) Prog. Allergy 13, 273.
50. Marcus, D.M., and Schwarting, G.A. (1976) Advances in Immunology 23, 203.

RECEIVED March 15, 1978.

Alterations of Membrane-Associated Glycolipids and Glycoproteins during Malignant Transformations

Introduction

JOHN F. CODINGTON

Laboratory for Carbohydrate Research, Departments of Biological Chemistry and Medicine, Harvard Medical School and Massachusetts General Hospital, Boston, MA 02114

Current interest in comparing the physicochemical properties and concentrations of particular glycoproteins and glycolipids at the surfaces of malignant and normal cells derived from the same tissues stems mainly from a few major discoveries of the last three decades. The rationale for such studies is the same as that which motivated biochemists in their search for differences in enzyme activities or nucleic acid compositions in tumor and normal cells. Only the stage (the cell surface or the isolated plasma membrane) and the players (the glycoproteins and the glycolipids) on that stage are different. "The play is the thing", however, and "there are more things" in the varied and intricate plots unfolding at the cell surface "than are dreamt of in your philosophy" (apologies to Shakespeare).

Paul Ehrlich in 1909 suggested that effective host resistance to malignancy was due to an immune response to tumor cells (1). It was not until much later, however, that the presence of certain antigens on tumor cell surfaces, but not on the surfaces of normal cells derived from the same tissue, were effectively demonstrated by the use of cells from inbred mouse strains (2). Since murine, and later human, histocompatibility antigens, H-2 and HL-A, respectively, were shown to be glycoprotein in character (3), it was believed by many investigators that tumor associated antigens were also glycoproteins. Such predictions have been proven to be correct, and during recent years a wide variety of cell-surface tumor antigens, virion antigens, virus-induced antigens, fetal antigens, and antigens induced by chemicals, have been shown to be glycoproteins.

The suggestion made by Hirst in 1945 (4) that a carbohydrate component, later termed sialic acid, served as a substrate for influenza virus sialidase (later called neuraminidase) represented a milestone in cell-surface biochemistry. The presence of sialic acid at the cell surface was later shown by Cook and coworkers (5) to be mainly responsible for the electrophoretic mobility of erythrocytes. This finding explained the prior results of Ambrose and his colleagues (6), who demonstrated that certain malignant cells exhibited greater electrophoretic mobility (presumably due to a greater concentration of cell-surface sialic acid) than their normal counterparts.

In 1962 Abercrombie and Ambrose (7) published their results on contact inhibition. These indicated that normal cells in culture exhibit a greater degree of growth control than do malignant cells. Since contact between cells is first made *via* macromolecular structures at the cell periphery, it was suggested that the degree of contact inhibition manifested by a cell was due to the type or concentration of cell-surface glycoproteins or glycolipids.

A report by Aub and coworkers on an agglutinin present in wheat germ (8) initiated an era of the use of lectins as probes for cell-surface macromolecular structures. The upsurge of activity in this field was due mainly to the work of Burger and his coworkers (9), who reported that wheat germ agglutinin had the capacity to clump malignant or transformed cells significantly more than normal cells. These results suggested that carbohydrate-containing cell-surface macromolecules were qualitatively or quantitatively different in normal and malignant cells.

Because of the complexity of the surface macromolecular structures and their interactions with both internal (cytoplasmic) and external (cellular and molecular) environments, it is not surprising that the conclusions resulting from many of the early findings had to be modified. Current studies of the mammalian cell-surface glycoproteins have achieved a degree of sophistication not believed possible only a few years ago. This development has resulted from an application of the latest fractionation, immunological, immunochemical, electron microscopic, and organic chemical techniques. It is anticipated that during the next few years studies of the chemical, physical, and biological properties of cell-surface macromolecules will lead to a far greater understanding of the malignant processes than we have today and will open new doors to the possible control of human cancer.

Literature Cited

1. Klein, G., and Klein, E., *Proc. Natl. Acad. Sci.* U.S.A. (1977) *74*, 2121.
2. Foley, E. J., *Cancer Res.* (1953) *13*, 835.
3. Nathenson, S. G., and Cullen, S. E., *Biochim. Biophys. Acta* (1974) *344*, 1.
4. Hirst, G. K., *J. Exp. Med.* (1948) *87*, 301.
5. Cook, G. M. W., Heard, D. H., and Seaman, F., *Nature* (Lond.) (1960) *188*, 1011.
6. Ambrose, E. J., James, A. M., and Lowick, J. H. B., *Nature* (Lond.) (1956) *177*, 576.
7. Abercrombie, M., and Ambrose, E. J., *Cancer Res.* (1962) *22*, 525.

8. Aub, J. C., Tieslau, C., and Lankester, A., Proc. Natl. Acad. Sci. U.S.A. (1963) 50, 613.
9. Burger, M. M., and Goldberg, A. R., Proc. Natl. Acad. Sci. U.S.A. (1967) 57, 359.

RECEIVED March 31, 1978.

21

Role of Glycoconjugates in Expression of the Transformed Phenotype

S. STEINER, D. VIA, M. KLINGER, G. LARRIBA, and S. SRAMEK
Department of Virology, Baylor College of Medicine, Houston, TX 77030

R. LAINE
Department of Biochemistry, University of Kentucky, Lexington, KY 40506

Recent studies have implicated changes in glycoproteins and glycolipids as being of importance in many of the altered properties associated with oncogenic transformation, e.g., altered antigenicity, decreased cellular adhesiveness, and altered lectin agglutinability. Moreover, in human tumors the blood group isoantigens, which are fucosyl glycolipids, are decreased (Hakomori and Jeanloz, 1964), as are the glycosyl transferases involved in the synthesis of these fucosyl glycolipids (Stellner et al., 1973). In contrast, the level of a simple and novel fucolipid is enhanced (Watanabe et al., 1976). Changes in fucoproteins in human tumors (Simmons and Pearlman, 1973) and in trypsin-sensitive surface fucopeptides in a variety of transformed cells (e.g., Buck et al. 1970; Pietropaolo et al., 1977) have also been observed. Furthermore, alterations in the serum level of fucosyl transferase in patients with tumors has been reported (Bauer et al., 1977; Kessel et al., 1977). These observations and related ones which suggested possible differences in the metabolism of fucosylated components in cancer cells as compared to normal cells prompted studies in this laboratory of changes in fucose metabolism in virus-transformed cells as compared to normal cells. These studies initially focused on the small molecular weight, organic-extractable fucosylated components.

Murine sarcoma virus-transformed rat (MSV-NRK) cells and their nontransformed counterparts (NRK) were grown in medium supplemented with either [^{3}H]- or [^{14}C]fucose. The cells were extracted either with $CHCl_3$-CH_3OH (2:1, v/v) or 60% ethanol (v/v), and the extract chromatographed on AG50, H+ form, followed by silicic acid thin-layer chromatography (TLC). The distribution of radioactivity on the thin-layer chromatogram is shown in Figure 1. The normal cells have two prominent peaks of radioactivity, i.e., component FL4 and FL3. In contrast, in the MSV-transformed rat cells there is a marked decrease in the incorporation of labeled fucose into FL4. This decreased labeling of FL4 is approximately five-fold and is evident when either [^{14}C]- or [^{3}H]fucose is used. Moreover, all normal mammalian cells thus far examined

0-8412-0452-7/78/47-080-378$06.50/0

(human, mouse, rat, baboon, monkey and hamster) contain FL3 and FL4. In addition, many different transformed cells, e.g., mouse mammary carcinoma, simian virus 40-transformed cells, herpes simplex virus-transformed cells, manifest a decrease in labeling of FL4 (Steiner et al., 1973, 1974). The decreased incorporation of label into FL4 appears to be more pronounced in highly tumorigenic (as compared to weakly or nontumorigenic) simian virus 40-transformed cells, herpes simplex virus-transformed cells (Steiner and Steiner, 1975/76), or mouse mammary carcinoma cells (unpublished observations).

Based on these observations, studies were undertaken to examine, in a temporal manner, the relationship between decreased incorporation of labeled fucose into FL4 and alterations in the metabolism of other glycoconjugates as a function of the expression of the transformed cell phenotype. We have observed that growth of MSV-NRK cells in medium supplemented with 2 mM sodium butyrate results in cellular flattening (within 96 hr) such that the MSV-NRK cells resemble the nontransformed NRK cells (Figure 2). Moreover, the process is fully reversible upon shift of cells to standard medium. There is also an elaboration of the cytoskeletal elements, i.e., microfilaments and microtubules, in these treated cells (Altenburg et al., 1976). Along with the morphological changes, there is a three-fold increase in the level of [^{3}H]-fucose incorporated into FL4 (Table I). The increased labeling is observed in MSV-NRK cells grown in butyrate-supplemented medium for short (1 passage) and long (30 passages) periods of time. Furthermore, shift of butyrate-treated MSV-NRK cells to control medium resulted in reversal to the level of control cells.

We also examined the effects of sodium butyrate on the ganglioside composition of the MSV-NRK cells. As indicated in Table II, growth of MSV-NRK cells in sodium butyrate-supplemented medium results in an increase in both hematoside GM_3 and in the disialoganglioside GD_{1A}. The increased levels of hematoside and GD_{1A} were observed within a single passage in sodium butyrate-supplemented medium, although the maximal level of hematoside was not observed for several passages. Another butyrate-induced effect is the marked increase in GD_{1A} in normal (NRK) cells (Table II). The elevated level of GD_{1A} in NRK cells (which are already flattened) was paralleled by a change in the organization of the microfilaments as examined by indirect immunofluorescent microscopy with anti-actin. The microfilaments shifted from a random distribution to a parallel distribution (Via and Steiner, unpublished data).

An increase in the average size distribution of trypsin-sensitive surface fucopeptides has been observed in a wide range of transformed cells as compared to normal cells. This increase is evident in the MSV-NRK cells as compared to NRK cells (Figure 3A). There is little detectable shift in the molecular weight distribution of these components in the first-passage sodium butyrate-treated MSV-NRK cells. By passage 2 in sodium butyrate-supplemen-

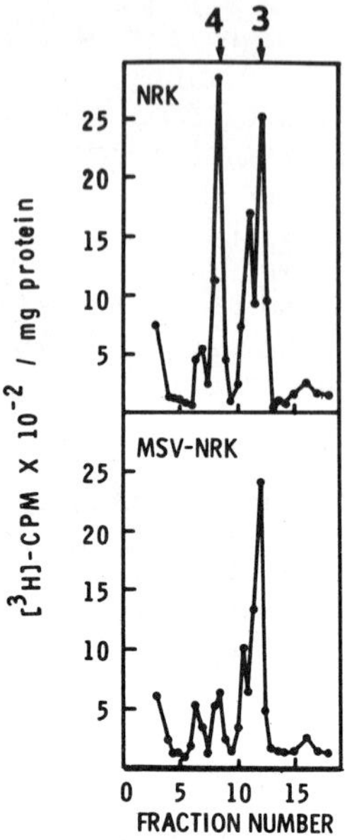

Figure 1. Aminoacyl fucosides of NRK and MSV–NRK cells. (Top) *NRK cells;* (bottom) *MSV–NRK cells; 3 = FL3; 4 = FL4. Cells were grown in [^{3}H]fucose-supplemented medium for 72 hr, harvested by scraping into 60% ethanol (v/v), and extracted thrice in 60% ethanol at 100°C. The pooled extracts were subjected to column chromatography on AG50 H^+ and the bound fraction eluted with 0.5N NH_4OH. The NH_4OH eluant was dried and chromatographed on silicic acid thin-layer plates in $CHCl_3$–CH_3OH–NH_4OH (40:80:25, by volume). The distribution of radioactivity was determined by scraping 0.5-cm fractions from origin to solvent front. Radioactivity was quantitated with a scintillation spectrometer.*

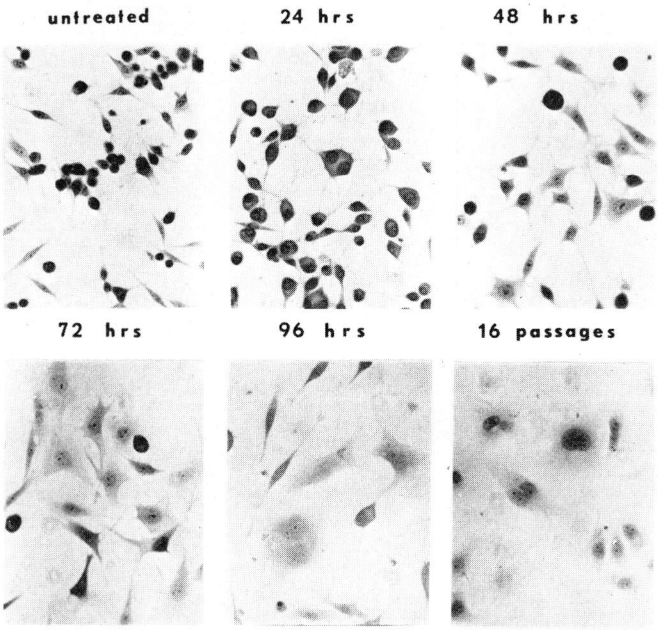

Figure 2. Morphological changes induced by sodium butyrate. MSV–NRK cells were cultured in control medium, i.e., untreated, and in 2mM sodium butyrate-supplemented medium for 24, 48, 72, and 96 hr and for 16 passages.

Table I. Effect of Sodium Butyrate on the Amincacyl Fucosides of MSV-NRK Cells

Cell line	No. of passages in sodium butyrate	FL4 cpm/ mg protein*	FL3 cpm/ mg protein*	FL4/FL3
NRK	--	5320	3276	1.6
MSV-NRK	0	1006	3062	0.3
MSV-NRK	1	3732	4334	0.9
MSV-NRK	30	3690	4167	0.9
↓ reversal, 3 passages no sodium butyrate				
MSV-NRK		999	3342	0.3

For details of [^{3}H]fucose labeling, harvesting and analysis of FL3a and FL4a, see Figure 1.

* Average of two separate experiments done in duplicate.

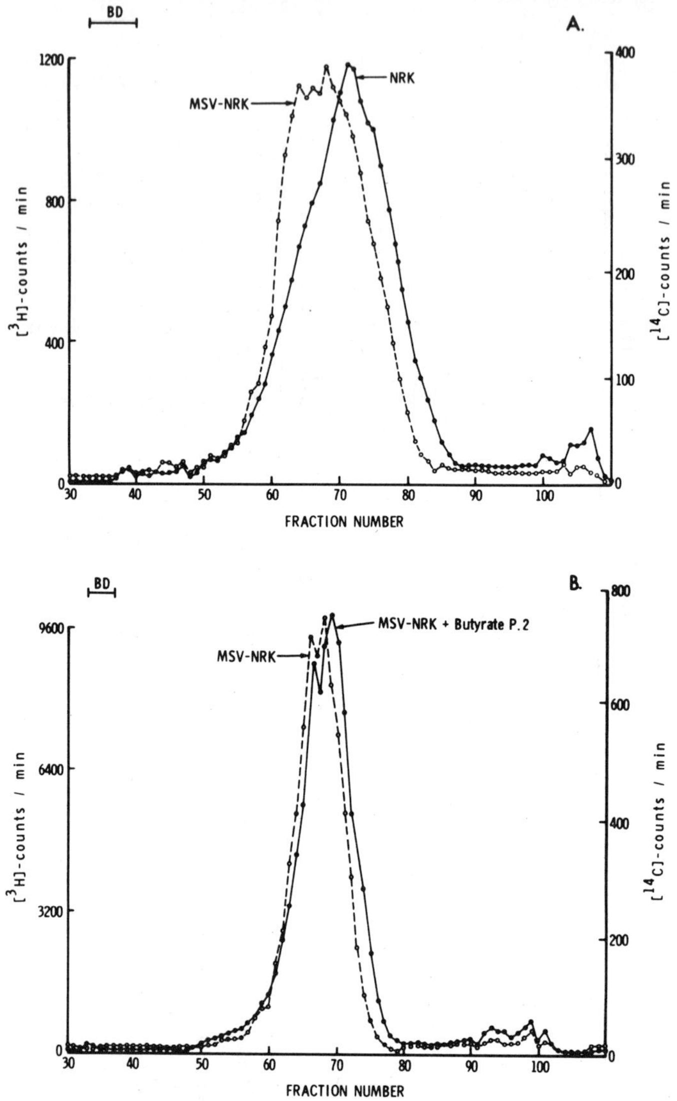

Figure 3. Sephadex G-50 fucopeptide profile of 2mM butyrate-treated MSV–NRK cells. Exponentially growing cells labeled with radioactive fucose for 48 hr were used in all experiments. Fucopeptides were prepared and analyzed by the general method described by Buck and co-workers (1970).

(A) ^{3}H-labeled (●) NRK cells vs. ^{14}C-labeled (○) control MSV–NRK cells. (B) ^{3}H-labeled (●) MSV–NRK cells passaged twice in butyrate medium vs. ^{14}C-labeled (○) control MSV–NRK cells.

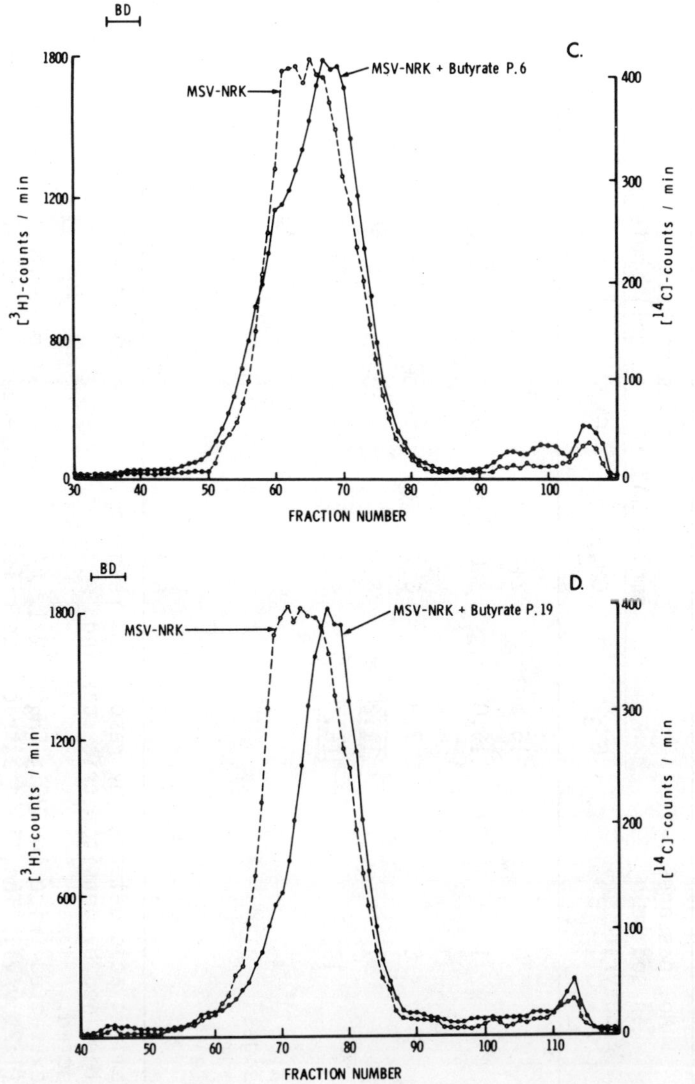

(C) ³H-labeled (●) MSV–NRK cells passaged six times in butyrate medium vs. ¹⁴C-labeled (○) control MSV–NRK cells. (D) ³H-labeled (●) MSV–NRK cells passaged 19 times in butyrate medium vs. ¹⁴C-labeled (○) control MSV–NRK cells. The phenol red dye marker eluted at fractions 136–146 in column A, fractions 123–133 in column B, fractions 122–132 in column C, and fractions 125–135 in column D. Radioactivity was not detected in fractions beyond those shown in the figure. Labeling experiments with reversed radioisotopic forms of fucose, i.e., ³H and ¹⁴C, gave comparable results.

Table II. Effect of Sodium Butyrate on the Ganglioside Composition of Murine Sarcoma Virus-Transformed NRK Cells as Compared with Normal NRK Cells

Cell line	No. passages in sodium butyrate	Ganglioside nmoles NANA[1]/10 mg cell protein				
		GM_3	GM_2	GM_1	GD_{1A}	GD_{1B}
NRK	0	16.8	2.4	ND[2]	ND	ND
NRK + SB	1	16.0	3.0	ND	[5.8]	ND
MSV-NRK	0	7.8	6.7	trace	1.2	0.4
MSV-NRK + SB	1	[11.2]	6.5	trace	[2.2]	0.3
MSV-NRK + SB	19	[16.1]	6.2	trace	[2.4]	0.2
↓ reversal (3 passages no sodium butyrate)						
MSV-NRK		7.4	6.3	trace	1.4	0.3

1 N-acetylneuraminic acid detected by resorcinol reagent, relative amount of NANA in each component determined by densitometric scanning.

2 ND = Not detected.

3 Lipids were extracted twice in $CHCl_3$-CH_3OH (2:1, v/v) followed by two extractions in $CHCl_3$-CH_3OH (1:2, v/v). The lipid extract was dried, resuspended in $CHCl_3$-CH_3OH (2:1, v/v) and partitioned according to the method of Folch and coworkers (1957). The Folch upper phase was dried, resuspended in a small volume of H_2O and dialyzed. The sample was chromatographed on silicic acid thin-layer plates (Q5, Quantum Industries, Fairfield, N.J.) in $CHCl_3$-CH_3OH-H_2O (60:35:8, by vol). The individual gangliosides were identified by comparison with authentic standards.

ted medium, a small shift can be observed (Figure 3B). By passage 6 (Figure 3C) in sodium butyrate-supplemented medium, there is a marked increase in the molecular weight distribution of the trypsin-sensitive surface fucopeptides to a distribution comparable to that of normal cells (Figure 3A). Thus, the shift in the size distribution of the trypsin-sensitive surface fucopeptides in butyrate-treated MSV-NRK cells not only appears to follow cellular flattening but also follows the increased incorporation of [^{3}H]-fucose into FL4 and the increased levels of the gangliosides GM_3 (hematoside) and GD_{1A}.

In order to understand the biochemical basis for the reduced level of FL4 in transformed cells, it was necessary to know the structures of FL3 and FL4. The initial examination involved indirect analyses with radioisotopically labeled FL3 and FL4 from cultured cells. These analyses indicated that these compounds had a positive charge, could be extracted with chloroform-methanol and might possess a long chain base (Skelly et al., 1976). However, because of the difficulties in unequivocally determining structure by indirect means, an isolation method for the purification of sufficient quantities for direct chemical characterization was developed (Figure 4). Four- to six-week old rats were injected intraperitoneally with [^{14}C]fucose. The liver, brain, intestine and kidneys were removed and extracted thrice with 60% ethanol. [^{3}H]FL3 and [^{3}H]FL4, purified from cell culture, were added to the extract to monitor for the fractions of interest during the purification and to evaluate the final recovery of FL3 and FL4. The ethanol-soluble fraction was chromatographed on Sephadex G-25, followed by high-voltage electrophoresis and descending paper chromatography. This last step separates FL3 from FL4, which in turn were chromatographed on silicic acid thin-layer plates and were each observed to be separable into two components, i.e., FL3a and FL3b, FL4a and FL4b. The purity of each component was examined in several TLC systems. Gas-liquid chromatography and an amino acid analyzer were used to identify the constituents of each component. The compounds were subjected to HCl-methanolysis, silylated and the carbohydrates analyzed by gas-liquid (GC) chromatography (Figure 5). The GC results indicated the presence of fucose in the FL3 series (as illustrated in Figure 5C for FL3b) and both fucose and glucose in the FL4 series (as illustrated in Figure 5D for FL4a). No other carbohydrate residues were observed (Figure 5A) nor did we observe the presence of long chain base (Figure 5B) or fatty acid (Larriba et al., 1977). Since these compounds contain a positively charged moiety, the possibility that there is an amino acid constituent was considered. Each component was subjected to strong acid hydrolysis and analyzed with an amino acid analyzer (Table III). FL3a contained threonine; FL3b, serine; FL4a, threonine; and FL4b, serine. The molar ratios of the amino acid and carbohydrate are indicated in Table IV. As can be seen, FL3a contains fucose and threonine; FL3b, fucose and serine; FL4a, glucose, fucose and threonine; and

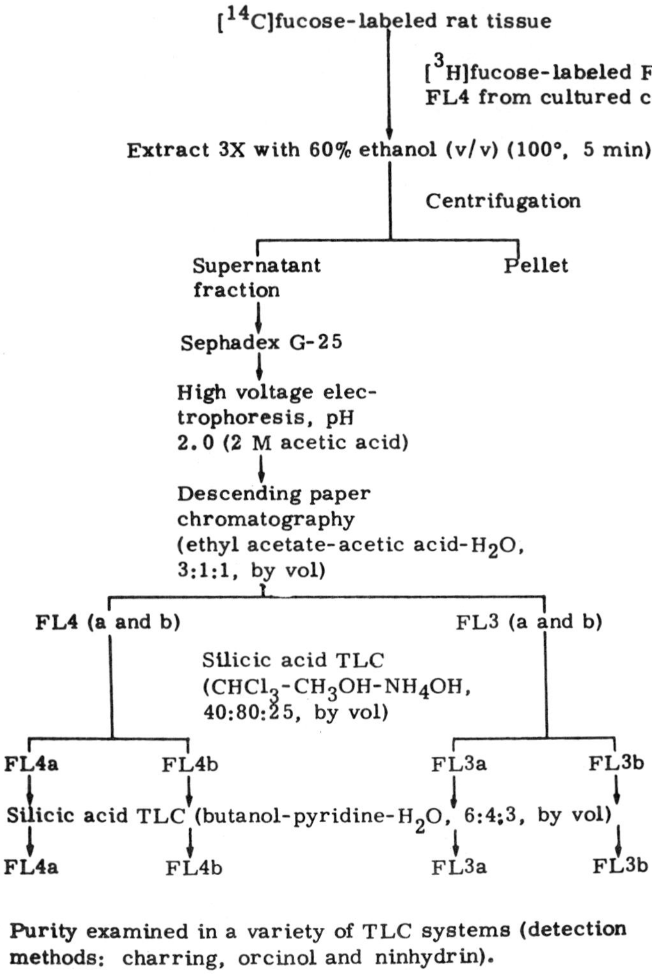

Figure 4. Purification of aminoacyl fucosides from rat tissue

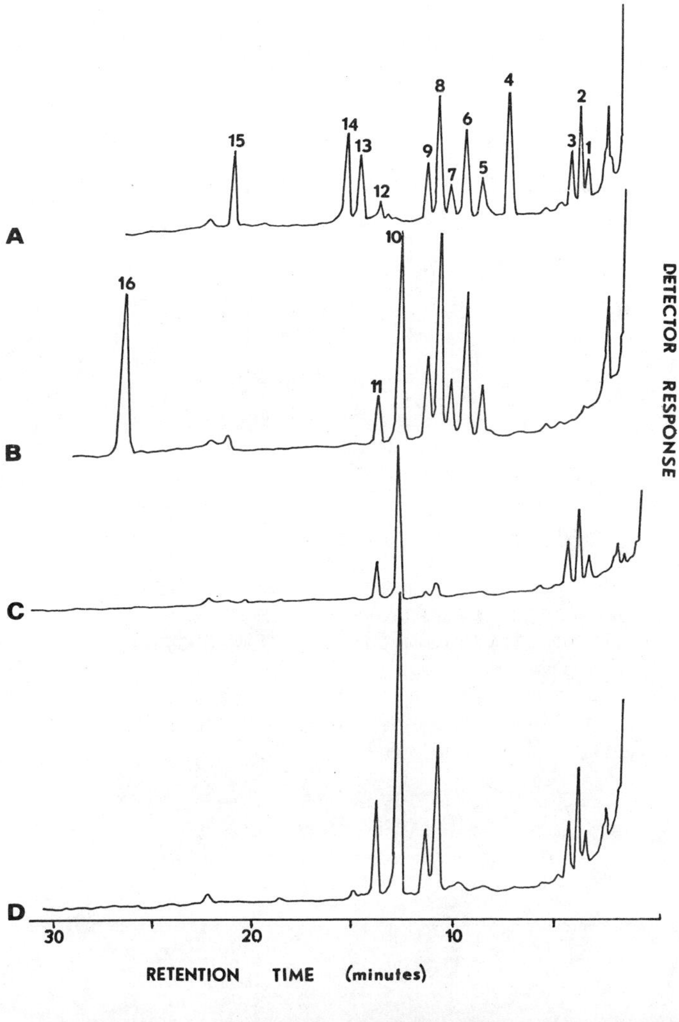

Biochemical and Biophysical Research Communications

Figure 5. Gas–liquid chromatography of FL3b and FL4a.

(A) Top profile: equimolar standard mixture of monosaccharides. (B) Second profile from top: lactosylceramide. (C) Third profile from top: FL3b. (D) Bottom profile: FL4a. Peaks correspond to the following compounds: 1, 2, 3, fucose; 4, arabinotol; 5, 6, 7, galactose; 8, 9, glucose; 10, 11, mannitol; 12, 13, N-*acetylgalactosamine; 14,* N-*acetylglucosamine; 15,* N-*acetylneuraminic acid; and 16, dihydrosphingosine. Samples were methanolyzed,* N-*acetylated and silylated (Larriba et al., 1977). 1–2 μL of each sample was then injected onto six-foot columns packed with 3% OV-1 on 80/100 Supelcoport in a Packard Becker model 419 dual column gas chromatograph with flame ionization detector.* N_2 *flow rate (effluent), 30 mL/min. Injector and detector temperatures, 270°C. Temperature program: temperature maintained at 160°C for 5 min, then raised 5°/min to 250°C, and maintained at 250°C for 15 min. Peak areas were integrated using a Spectra-Physics Autolab System I computing integrator.*

Table III. Amino Acid Analysis of Aminoacyl Fucosides

Compound[1]	Retention time (minutes)	
	Before hydrolysis	After hydrolysis
FL3a	21.9	43.0
FL3b	23.6	46.2
FL4a	16.0	42.9
FL4b	ND[2]	46.3
Serine[3]	--	46.8
Threonine[3]	--	43.3

[1] Samples were hydrolyzed in 6.0 N HCl at 110 C for 24 hr, concentrated by flash evaporation, resuspended in a small volume (0.05-0.1 ml) of 0.11 M sodium citrate, pH 3.2, and chromatographed on a Beckman 121 amino acid analyzer using a single column, physiological fluid program.

[2] ND = Not done.

[3] The amino acids with closest retention times to serine and threonine are aspartic acid (40.6) and proline (65.9).

Biochemical and Biophysical Research Communications

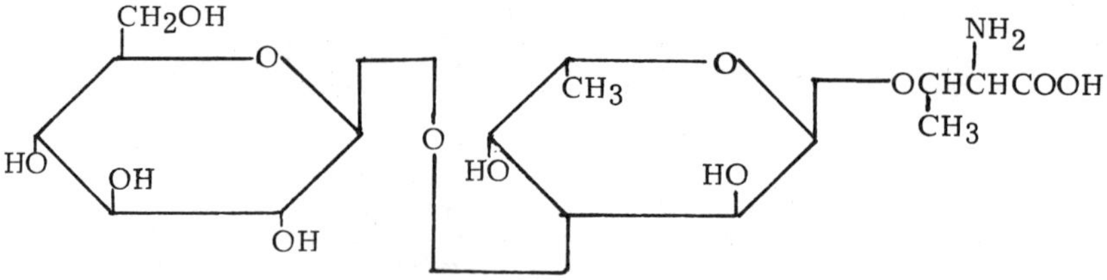

O-β-D-glucopyranosyl-(1→3)-O-α-L-fucopyranosyl-L-threonine

Figure 6. Structure of O-β-D-glycopyranosyl(1→3)-O-α-L-fucopyranosyl-L-threonine

Table IV. Molar Ratios of Carbohydrate and Amino Acids of FL3a, FL3b, FL4a and FL4b[1]

Compound	nmoles[2]				Fuc / Glc / Thr / Ser			
	Fucose	Glucose	Threonine[3]	Serine[3]				
FL3a	91	0	84	0	1.1	0	1.0	0
FL3b	137	0	0	120	1.1	0	0	1.0
FL4a	53	56	46	0	1.2	1.2	1.0	0
FL4b	9	11	0	8	1.1	1.4	0	1.0

1 For details of analysis of carbohydrate and amino acids, see Figure 5 and Table III, respectively.

2 The nmoles of amino acid and carbohydrate represent the total purified material recovered from the rat tissues. As judged by exogenously added FL3a and FL4a from cultured cells, recovery of FL3a from rat tissue was approximately 16.0% and recovery of FL4a was approximately 10.0%.

3 The amount of threonine and serine is not corrected for breakdown (∼10%) due to strong acid hydrolysis.

Biochemical and Biophysical Research Communications

FL4b, glucose, fucose and serine; all in an approximate ratio of 1:1. Like FL4a, which is the fucose-labeled component designated as FL4 in the earlier labeling studies, a urinary fucoside, which contains fucose, glucose and threonine, has recently been described (Hallgren et al., 1975). This urinary fucoside has the structure shown in Figure 6, including the novel feature of an internal fucose O-glycosidically linked to amino acid.

GC/mass spectrometry studies were undertaken to determine the carbohydrate sequence and linkage of FL4a. Synthetic fucoside, of the same structure as the urinary fucoside, was most generously supplied by Dr. Svensson (University Hospital, Lund, Sweden; synthesized by Drs. Garegg and Norberg, 1976) and was used as a standard. FL4a and the synthetic fucoside were permethylated and subjected to GC/mass spectrometric analyses. A GC/mass fragmentogram indicates that four fragments (m/e 393, 329, 219 and 187) are obtained from both permethylated synthetic fucoside (Figure 7A) and permethylated FL4a from rat tissue. The permethylated FL4a and synthetic fucoside had identical GC retention times (Table V). Moreover, the fragments m/e 219 and m/e 187 are consistent with the presence of terminal hexose and m/e 393 and m/e 329 are consistent with a disaccharide in which 6-deoxyhexose (fucose) is internal and hexose is terminal. Neither component had an observable fragment, m/e 189, typical of terminal fucose. The structure of the permethylated compounds was further examined following methanolysis and acetylation by analysis of the partially methylated alditol acetates. Each compound gave two peaks with major fragment m/e 262 and m/e 264, respectively. As indicated in Table VI, the m/e 262 peak from derivatized FL4a and synthetic fucoside had the same retention times. Similarly, the peaks which gave a major fragment m/e 264 from derivatized FL4a and synthetic fucoside had the same retention times. The mass spectra of the partially methylated alditol acetates from FL4a and synthetic fucoside were compared. The components from FL4a and the synthetic fucoside with a major peak m/e 262 also had fragments m/e 290, 118 and 131 (Figure 8). These fragments are consistent with an internal 6-deoxyhexose (Figure 9). The mass spectra of the components from FL4a and synthetic fucoside with a major fragment m/e 264 also yielded fragments m/e 292, 118, 161 and 162, which is consistent with a terminal hexose (Figure 10). The combined GC, amino acid analysis, and GC/mass spectrometry data indicate that FL4a has a similar structure, i.e., glucosyl(1⟶3)fucosyl⟶ threonine, to that of the urinary fucoside.

The novel structure of this component, i.e., with a fucose moiety which appears to be both internal and O-glycosidically linked to amino acid, raises a question of its biosynthetic origin. If FL4a is derived from fucoprotein which has fucose O-glycosidically linked to amino acid, then it should be possible to release a disaccharide which contains glucosylfucitol by alkaline borohydride treatment of that fucoprotein(s). An ethanol-insoluble fraction of fucose-labeled cells was subjected to alkaline boro-

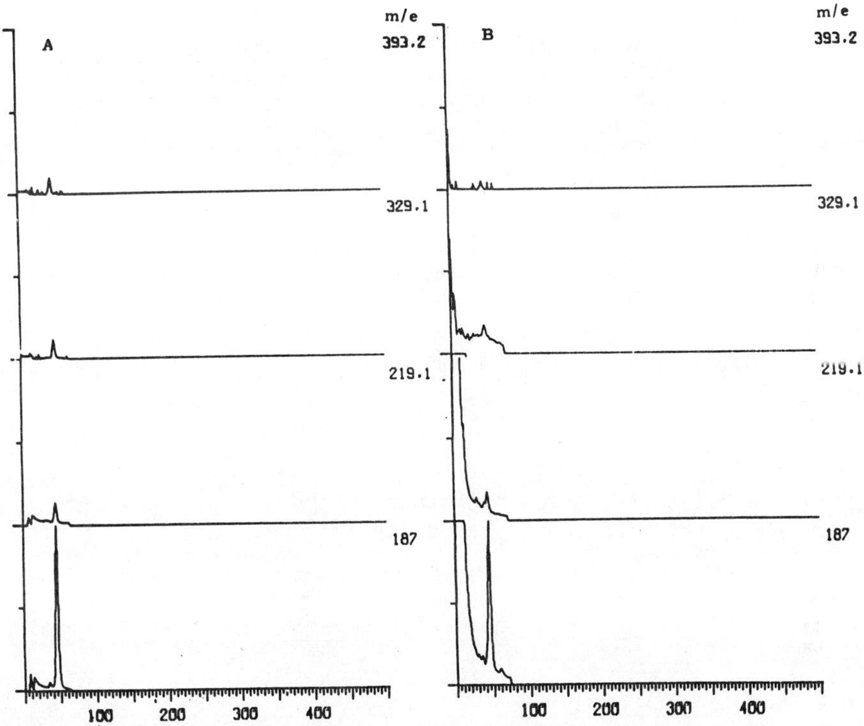

Figure 7. Mass fragmentogram of permethylated synthetic glucosyl fucosyl threonine and FL4a. (A) Authentic glucosyl fucosyl threonine. (B) FL4a. FL4a and synthetic glucosyl fucosyl threonine were permethylated (Hakomori, 1964) and analyzed by GC/mass spectrometry. A preliminary total mass scan was used to detect the major ions of interest. GC parameters: OV-101 column at 175°C, N_2 carrier gas. Mass parameters: Finnagen chemical ionization; cell volts, 34.2; lens volts, −0031; EM volts, −2318; lost mass, −0640; first mass, +0099; emission mass, +00.59; collective mass, +00.02; ion voltage mass, +00.58; elect. energy, +0150; ion energy, +011.9, −014.6; extr. volts, +008.6.

Table V. Gas-Liquid Chromatographic Retention Times of Permethylated Compounds

Compound	Time (min)
Synthetic O-β-D-glucopyranosyl-(1$\rightarrow$3)-O-α-L-fucopyranosyl	3.83
FL4a	3.83

FL4a and synthetic glucosyl fucosyl threonine were derivatized and analyzed as described in Figure 7. The permethylated compounds were detected by scanning for the fragments m/e 393.2, 329.1, 219.1, and 187, which may be derived as indicated below.

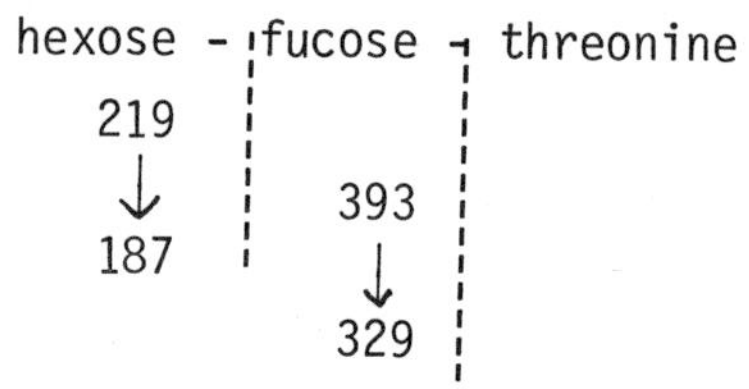

Table VI. Gas-Liquid Chromatography Retention Times of Partially Methylated Alditol Acetates

Parent compound	Fragment used for detection (m/e)	Retention time (min)
FL4a	262	3.92
Synthetic glc-fuc-thr	262	3.92
FL4a	264	4.42
Synthetic glc-fuc-thr	264	4.42

Partially methylated alditol acetates of FL4a and synthetic glucosyl fucosyl threonine were prepared by acidic methanolysis of the methylated compound followed by treatment with trifluoroacetic acid, reduction with sodium borodeuteride and acetylation with acetic anhydride. The partially methylated alditol acetates were analyzed by GC/mass spectrometry. The major ion of each partially methylated alditol acetate was determined in a preliminary mass scan and used to detect the respective partially methylated alditol acetates. GC parameters: OV-101 column at 120 C. Mass parameters: as in Figure 7.

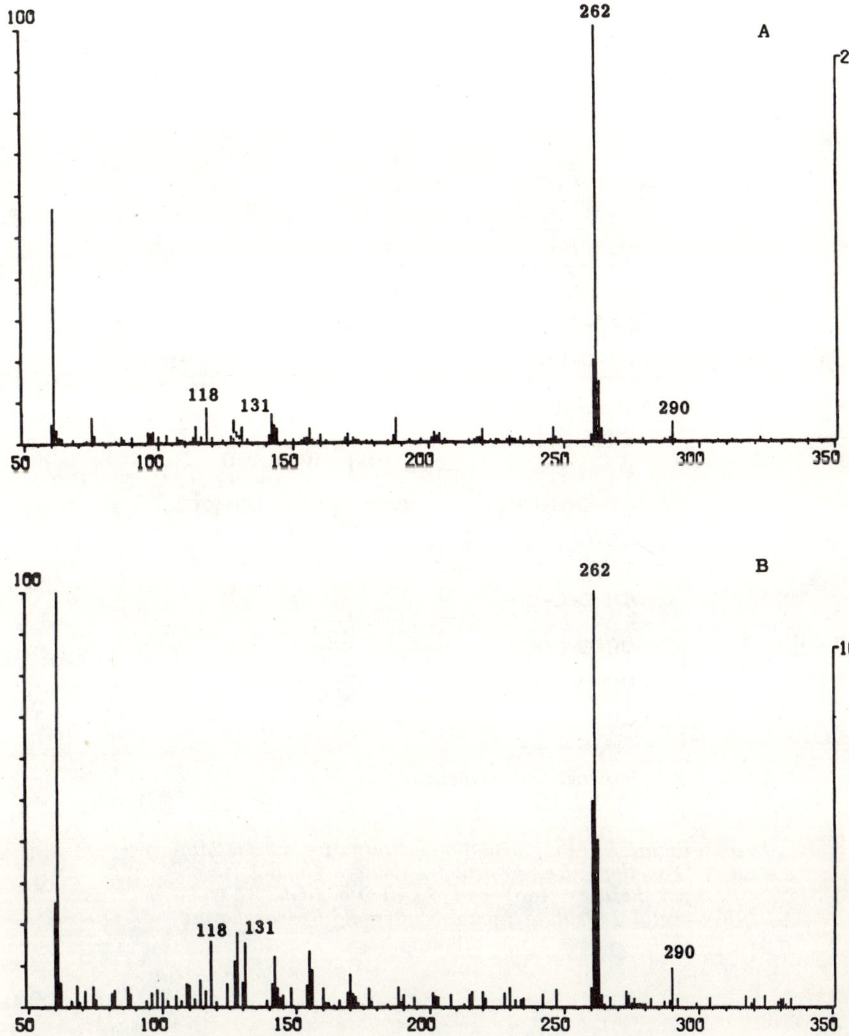

Figure 8. Total mass spectra of partially methylated alditol acetates obtained from glucosyl fucosyl threonine and from FL4a. (A) Total mass spectrum of partially methylated alditol acetate, with a major ion m/e *262 from glucosyl fucosyl threonine. (B) Total mass spectrum of partially methylated alditol acetate, with the same retention time as in (A), from FL4a. For experimental details see Table VI.*

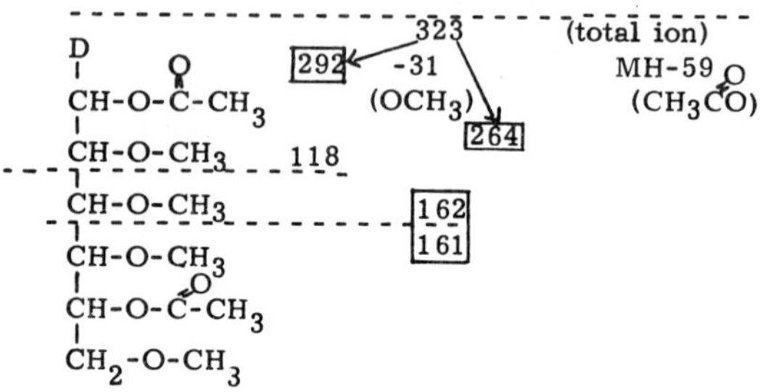

Terminal hexose

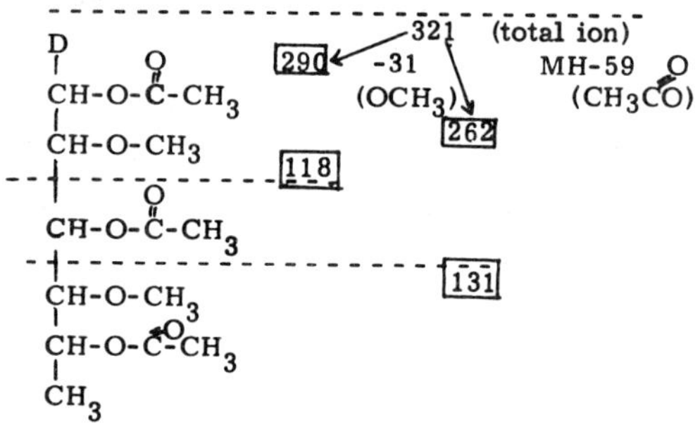

Internal 6-deoxyhexose

Figure 9. Fragmentation patterns for 2,3,4,6-tetra-O-methyl-1,5-di-O-acetyl-1-deutero-glucitol (terminal hexose) (top) *and 2,4-di-O-methyl-1,3,5-tri-O-acetyl-1-deutero-fucitol (internal 6-deoxyhexose)* (bottom)

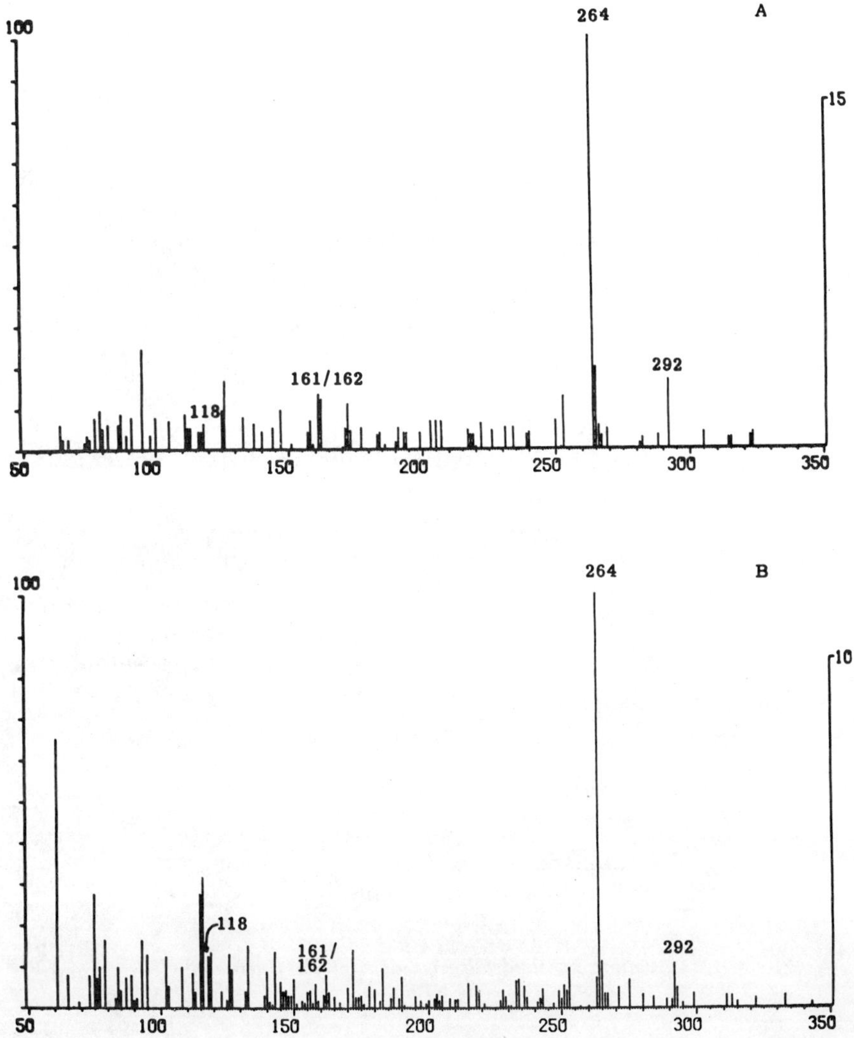

Figure 10. Total mass spectra of partially methylated alditol acetates obtained from glucosyl fucosyl threonine and from FL4a. (A) Total mass spectrum of partially methylated alditol acetate with a major ion m/e *264 from glucosyl fucosyl threonine. (B) Total mass spectrum of partially methylated alditol acetate, with the same retention time as in (A), from FL4a. For experimental details see Table VI.*

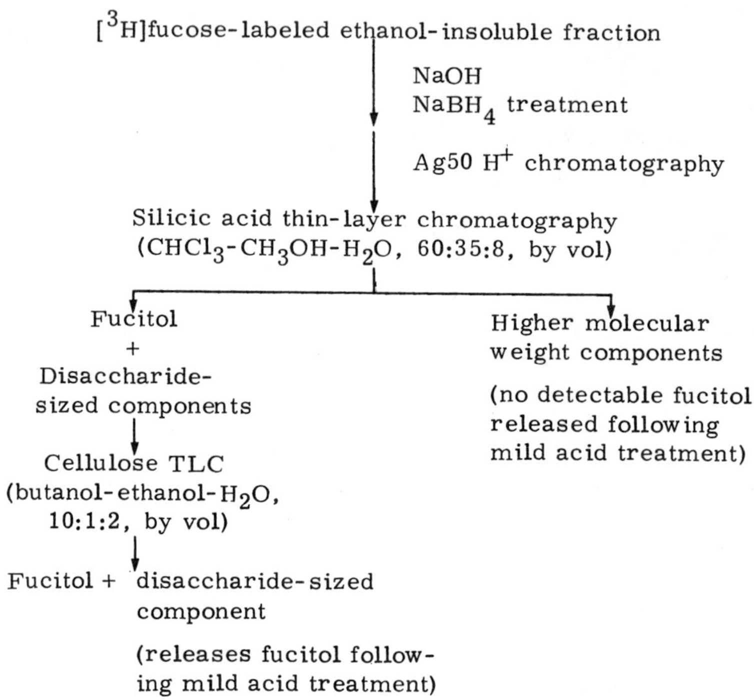

Figure 11. Alkaline borohydride treatment of [^{3}H]fucose-labeled glycoprotein fraction

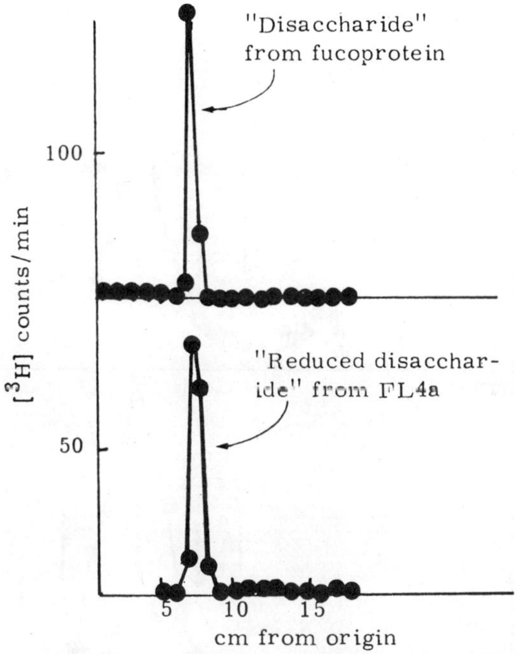

Figure 12. Chromatographic comparison of "disaccharide" from alkaline borohydride-treated glycoprotein and "glucosyl fucitol" from FL4a. The disaccharide component was from alkaline borohydride-treated [^{3}H]fucose-labeled protein. The glucosyl fucitol was obtained by mild acid hydrolysis of FL4a (0.1N HCl, 100°C, 1 hr) followed by reduction with 0.5M $NaBH_4$ (37°C, 2 hr). These samples were chromatographed on silicic acid thin-layer plates in solvent I ($CHCl_3$–CH_3OH–H_2O, 60:35:8, by volume), eluted, and chromatographed in solvent II (butanol–ethanol–H_2O, 10:1:2, by volume), eluted, and chromatographed in solvent III (butanol–propionic acid–H_2O, 6:3:4, by volume). Chromatography of reduced disaccharide (bottom) from FL4a and disaccharide from fucoprotein (top) in solvent III is shown. The two components had comparable chromatographic mobilities in all three solvents.

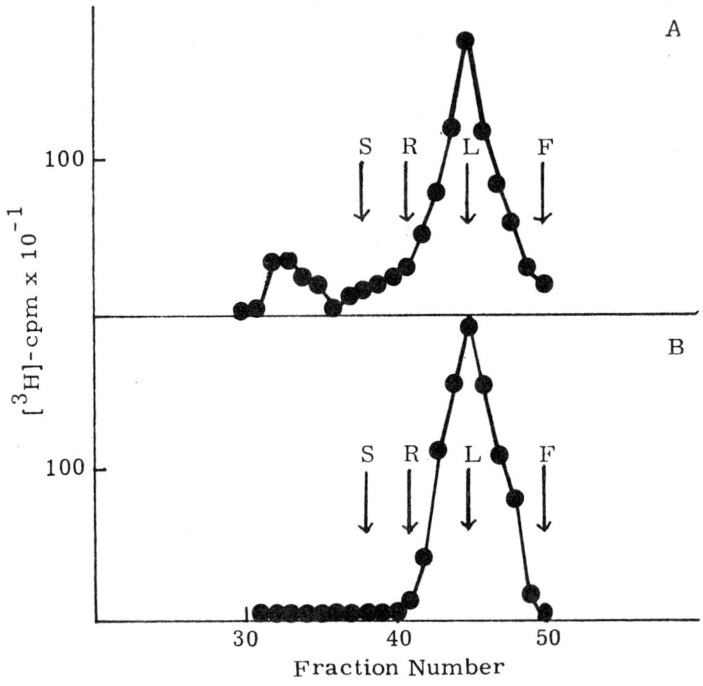

Figure 13. Gel filtration analysis of "disaccharide" material from glycoprotein and from FL4a. (A) Sephadex G-10 chromatography of reduced disaccharide from [^{3}H]fucose-labeled FL4a (see Figure 12). (B) Sephadex G-10 chromatography of disaccharide obtained from sodium borohydride-treated, [^{3}H]fucose-labeled protein. The arrows indicate the elution volumes of the standards: S, stachyose; R, raffinose; L, lactose; and F, fucose. Sephadex G-10 chromatography was carried out on 0.8 × 70 cm columns in 0.2M ammonium acetate at pH = 7.0 and 0.8-mL fractions were collected. The position of the authentic sugars was detected by spotting a portion of each fraction on silicic acid plates followed by chromatography of the samples in butanol–pyridine–H_2O (6:4:3, by volume) and estimation of the position and relative amount of each sugar with orcinol. Radioactivity was used to locate the disaccharide components.

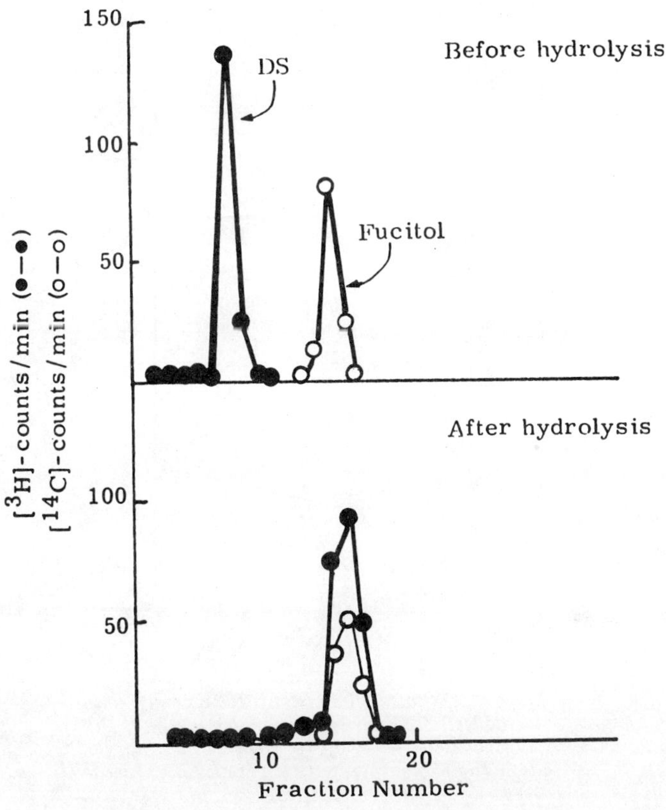

Figure 14. Acid hydrolysis of [³H]-labeled disaccharide component. Before hydrolysis: (●—●) purified [³H]-labeled disaccharide-sized material from [³H]fucose-labeled protein (see Figure 8); (○—○) authentic ¹⁴C-fucitol. After hydrolysis: (●—●) disaccharide-sized material following acid hydrolysis (1N HCl, 100°C, 1 hr); (○—○) authentic ¹⁴C-fucitol added to hydrolysis mixture. Chromatography in butanol–ethanol–H_2O (10:1:2, by volume).

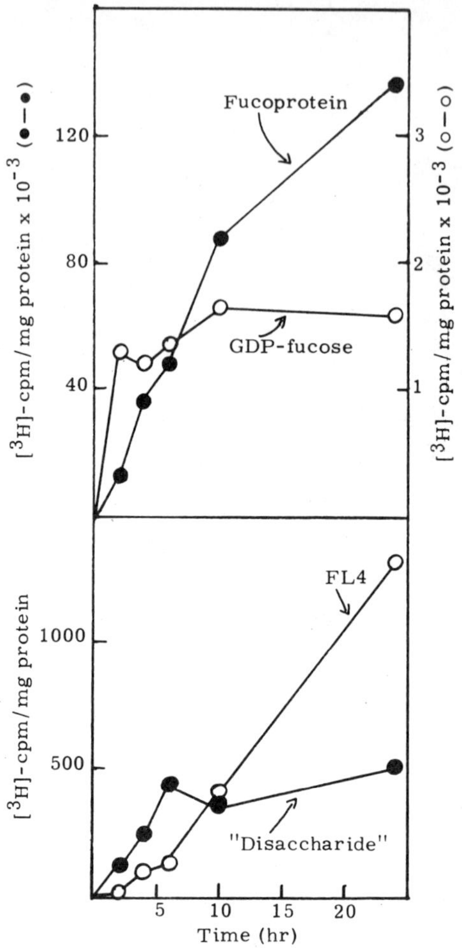

Figure 15. Pulse-labeling of NRK cells with [^{3}H]fucose. NRK cells were pulse-labeled with [^{3}H]fucose for 2, 4, 6, 10, and 24 hr and the radioactivity incorporated into (top) *GDP-fucose (○—○) and fucoprotein (●—●) and* (bottom) *FL4 (○—○) and "disaccharide" from alkaline borohydride-treated protein fraction (●—●) measured. Cells were extracted in 60% ethanol (v/v). The ethanol extract was used for the analysis of FL4 (Figure 11) and GDP-fucose (descending paper chromatography, ethanol–1M ammonium acetate, pH = 7, 7:3, v/v). The radioactivity in the ethanol-insoluble fraction was used as a measure of labeling of fucoprotein. The disaccharide was obtained by alkaline borohydride treatment* (see *Figure 12*) *of a portion of the fucoprotein fraction.*

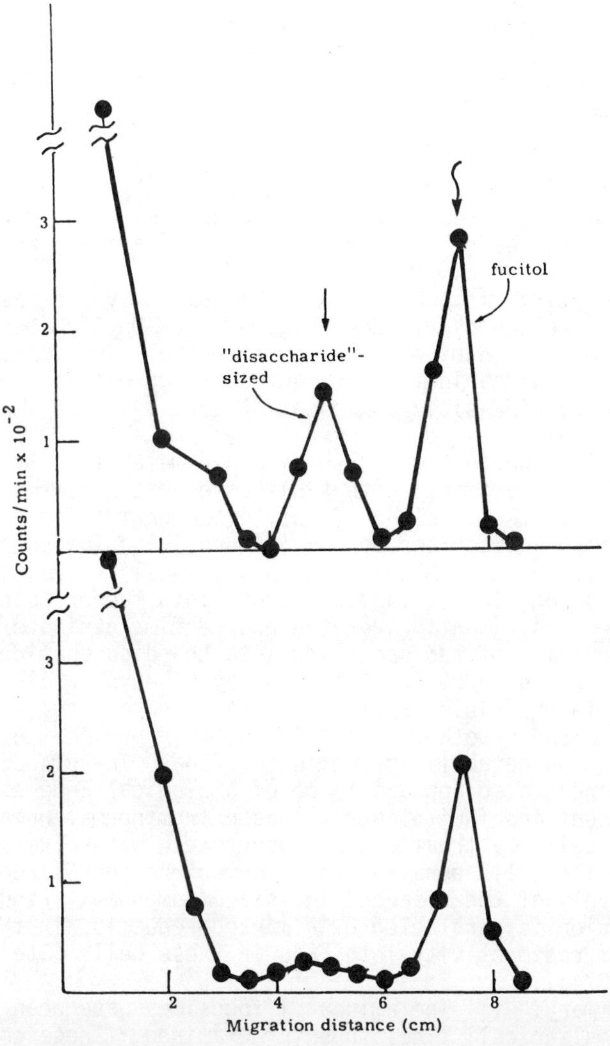

Figure 16. Alkaline borohydride treatment of [^{3}H]fucose-labeled proteins from normal (human embryonic lung) and transformed (A-204; human rhabdomyosarcoma) cells. Cells were labeled with [^{3}H]fucose for 72 hr, harvested, extracted, and protein-treated with alkaline borohydride as described in Figure 12. Samples were chromatographed in solvent I. The arrow (↓) indicates the position of standard glucosyl fucitol obtained from FL4a (Figure 12). The other arrow indicates the position of standard fucitol. Samples and standards also co-chromatographed in solvents II and III (Figure 12). The radioactivity in the peak labeled "disaccharide" quantitatively yielded fucitol upon mild acid hydrolysis as described in Figure 14.

hydride treatment, i.e., β-elimination reaction (Figure 11). The products were analyzed, after AG50 H^+ column chromatography, by silicic acid TLC and cellulose TLC. Both fucitol and a putative disaccharide component were observed. The [^{3}H]fucitol from alkaline borohydride-treated ethanol-insoluble fraction comigrated with standard fucitol in the three TLC systems employed. The higher molecular weight fucosylated components, which comprise the majority of the radioactivity, did not release detectable levels of fucitol; the radioactivity was quantitatively released as fucose.

The putative disaccharide had the same Rf values as reduced disaccharide released from FL4 (Figure 12). On Sephadex G-10, the putative disaccharide had a comparable elution volume to that of standard disaccharide, i.e., lactose (Figure 13). Hydrolysis of the disaccharide-sized component (Figure 14) quantitatively yielded fucitol.

The "disaccharide" was also examined metabolically. Specifically, pulse experiments with [^{3}H]fucose were performed in order to determine if the labeling of the "disaccharide" is consistent with its being a precursor of FL4 (Figure 15). The incorporation of radioactivity into GDP-fucose rapidly equilibrates, i.e., by 2 hours. Incorporation of labeled fucose into fucoprotein appears to increase significantly over the entire time period examined. The incorporation of radioactivity into the disaccharide-sized component from fucoprotein is maximal by 6 hours, while there is a 6-hour lag in the rapid labeling of FL4. These results are consistent with the hypothesis, but by no means prove, that "disaccharide" is the metabolic precursor of FL4a. The disaccharide-sized component also appears to be of biological interest. Both this component and fucitol are released from normal human embryonic lung cells by treatment of fucoprotein with alkaline borohydride, while cells derived from a human rhabdomyosarcoma have reduced levels of the disaccharide-sized component (Figure 16). This reduction is paralleled by a marked reduction in the incorporation of radioactivity into FL4a in these cells (Steiner and Melnick, 1974).

In summary: (1) The aminoacyl fucosides have been observed in all mammalian cell lines thus far examined. These components, FL3a (threonine and fucose), FL3b (serine and fucose), FL4a (threonine, fucose and glucose) and FL4b (serine, fucose and glucose), appear to be a novel series of compounds with fucose O-glycosidically linked to amino acid. Furthermore, glucose appears to be the terminal sugar in FL4a. (2) There is a marked decrease in the incorporation of radioisotopically labeled fucose into FL4a in a variety of oncogenically transformed cells. The level of FL4a also appears to vary as a function of the expression of the transformed morphological phenotype. (3) There is a glycoprotein(s) in which fucose appears to be O-glycosidically linked to amino acid. The results of pulse studies with [^{3}H]fucose are consistent with the fucoprotein(s) being a precursor of FL4a.

Acknowledgments. We wish to thank Dr. A. Beaudet and Robert Graham for running the samples on a Beckman 121 autoanalyzer, Dr. M. Steiner for helpful suggestions, and Dr. J.L. Melnick for his support during the course of this work. This work was supported by research grant CA 16,311 from the National Cancer Institute, DHEW, and by Faculty Research Award FRA-131 from the American Cancer Society, Inc. G. Larriba is a Fellow of the Program of Cultural Cooperation between Spain and the United States of America. S. Sramek is a recipient of a National Research Service Award (#1 F32 CA 05706) from the National Cancer Institute, DHEW.

LITERATURE CITED

Altenburg, B.A., Via, D.P. and Steiner, S. 1976. Exp. Cell Res. 102:223.

Bauer, C., Kottgen, E. and Reutter, W. 1977. Biochem. Biophys. Res. Comm. 76:488.

Buck, C.A., Glick, M.C. and Warren, L. 1970. Biochemistry 9: 4567.

Folch, K., Lees, M. and Sloane-Stanley, G.H. 1957. J. Biol. Chem. 26:497.

Garegg, P.J. and Norberg, T. 1976. Carbohyd. Res. 52:235.

Hakomori, S.I. 1964. J. Biochem. (Tokyo) 55:205.

Hakomori, S.I. and Jeanloz, R.W. 1964. J. Biol. Chem. 239:3606.

Hallgren, P., Lundblad, A. and Svensson, S. 1975. J. Biol. Chem. 250:5312.

Kessel, D., Sykes, E. and Henderson, M. 1977. J. Natl. Cancer Inst. 59:29.

Larriba, G., Klinger, M., Sramek, S. and Steiner, S. 1977. Biochem. Biophys. Res. Comm. 77:79.

Pietropaolo, C., Yamaguchi, N., Weinstein, I.B. and Glick, M.C. 1977. Int. J. Cancer 20:738.

Simmons, D.A.R. and Perlmann, P. 1973. Cancer Res. 33:313.

Skelly, J., Gacto, M., Steiner, M.R. and Steiner, S. 1976. Biochem. Biophys. Res. Comm. 68:442.

Steiner, S.,Brennan, P. and Melnick, J.L. 1973. Nature New Biol. 245:19.

Steiner, S. and Melnick, J.L. 1974. Nature 251:717.

Steiner, S., Melnick, J.L., Kit, S. and Somers, K. 1974. Nature 248:682.

Steiner, S. and Steiner, M. 1975/76. Intervirology 6:32.

Stellner, K., Hakomori, S.I. and Warner, G. 1973. Biochem. Biophys. Res. Comm. 55:439.

Watanabe, I., Matsubara, T. and Hakomori, S.I. 1976. J. Biol. Chem. 251:2385.

RECEIVED March 31, 1978.

22

Fucosylation—A Role in Cell Function

MARY CATHERINE GLICK

Department of Pediatrics, University of Pennsylvania Medical School,
Children's Hospital of Philadelphia, Philadelphia, PA 19104

The original observation that virus transformation is accompanied by an alteration of membrane glycoproteins (1) has been extended in a number of laboratories to many virus transformed and tumor cells, including human tumors (see ref. 2 and 3 for reviews). The appearance of specific glycopeptides has been directly correlated with tumorigenesis (4) and has been suggested as a diagnostic procedure for leukemia (5). The studies have progressed to the point of elucidating the partial structure of one of the major glycopeptides predominant on the surface of virus transformed cells (6). This glycopeptide is triantennary and appears to be more highly branched than the major fucose-containing glycopeptide isolated from the nontransformed counterpart. These initial results are compatible with the suggestion that the key to the alteration which leads to the formation of more highly branched oligosaccharides after virus transformation may reside around the β-mannose core. The partial structures were assembled from data obtained by enzyme degradation of the isolated glycopeptides with purified exoglycosidases and recovery of the released monosaccharides by gas liquid chromatography.

Even though these glycopeptide differences were detected using radioactive L-fucose to mark the glycoproteins, no difference has been seen thus far between the transformed and nontransformed cell surfaces in fucose *per se*. In order to accent this point a variety of data from a number of cell types have been summarized (Table 1) and used to suggest that the fucosylation of membrane glycoproteins may have a special role in the functioning of the cell.

The data which suggest a role for fucosylation fall into several categories: (a) more directly correlated to cell function; (b) the molecule of fucose; and (c) ancillary data.

More Directly Correlated to Cell Function

The first three observations listed in Table 1 fall into a category which may be more directly suggestive of a role for fucose in membrane glycoproteins. The first, the fact that most membrane glycoproteins contain fucose, has been shown with many

0-8412-0452-7/78/47-080-404$05.00/0

cell types using double labeling experiments with radioactive L-fucose and D-glucosamine or D-glucose. Similar membrane glycoproteins are labeled with either radioactive L-fucose or the more general labels, D-glucosamine or D-glucose when examined by polyacrylamide gel electrophoresis (7). An example of this is shown in Figure 1 which depicts the surface membrane glycoproteins of transformed baby hamster kidney fibroblasts marked with radioactive fucose (Figure 1a), and the same clone and the normal counterpart marked with radioactive glucosamine (Figure 1b). The only reproducible difference between the precursors was in the high molecular weight region of the gel (Fractions 5-15). Another example is seen in the partially purified membrane glycopeptides from hamster transformed cells or their normal counterparts. These cells were metabolically labeled with both D-[^{14}C] glucose and L-[^{3}H] fucose (6). Again the similarity of the radioactive patterns of the glycopeptides suggests that all of these membrane glycopeptides were fucosylated.

The second point (Table 1), that fucose is positioned in the core region of all of the membrane glycoproteins thus far examined, can be shown by the total recovery of the monosaccharide units after sequential enzyme degradation of the radioactive glycopeptides (6). The radioactivity was recovered as fucose only after treatment with rat testis α-L-fucosidase and was removed subsequent to the release of all the other monosaccharides with the exception of two residues of N-acetylglucosamine from the core region. If fucose was positioned in another location, the enzymatic degradation of the oligosaccharide unit would not have been complete.

The third point (Table 1) may be a phenomenon related to the glycoprotein structures observed. The activity of α-L-fucosidase was not elevated after virus transformation to the extent that the other glycosidases were elevated. The acid hydrolase activities of the virus transformed cells are expressed as percentage of the normal in Figure 2. Acid phosphatase, which represented other lysosomal enzymes, was similar in both cell types, as were several exoglycosidases. The activities of β-galactosidase, β-N-acetylglucosaminidase, α-mannosidase, and α,β-N-acetylgalactosaminidase were all elevated 150 to 300%. In contrast, the activity of α-L-fucosidase was elevated only 120%. In other experiments using 4-methylumbelliferyl-fucopyranoside as substrate, the activity of both cell types was even more similar. Three of the four glycosylhydrolases with elevated activities are directly concerned with the degradation of the oligosaccharide units which compose the branches of the membrane glycoproteins. The more branched glycoproteins are those which are more predominant on the transformed cell surface (6). The fact that α-L-fucosidase activity is similar in both transformed and nontransformed cells suggests that even though the branching of the glycoproteins is altered on transformation, the fucose residues and the core region may remain unchanged.

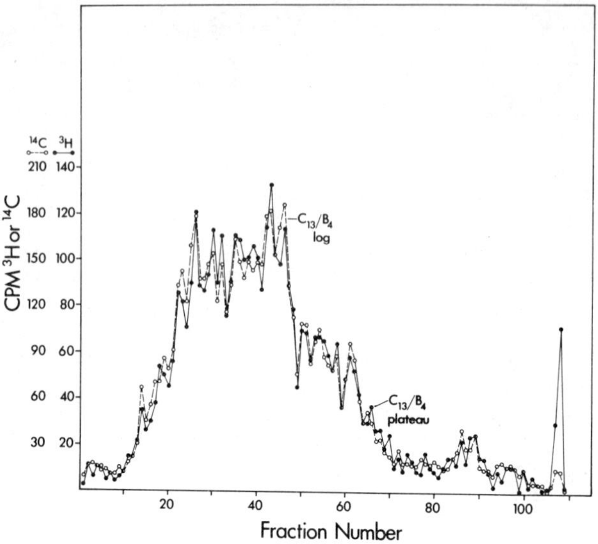

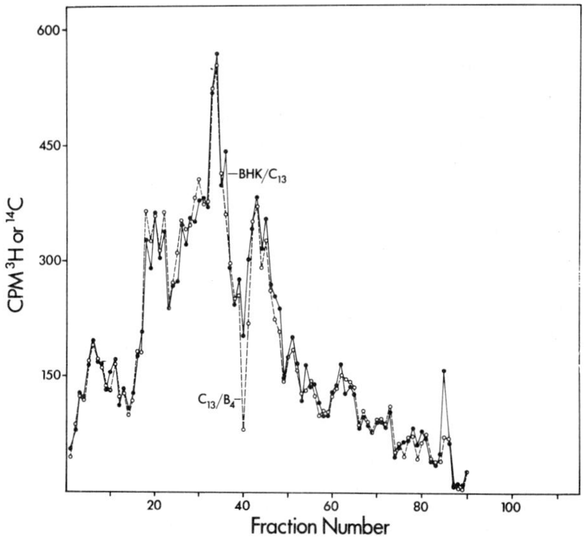

Figure 1. Polyacrylamide gel electrophoresis of surface membranes. Surface membranes were prepared from cells metabolically labeled with (a) L-*[³H] or -[¹⁴C] fucose and (b)* D-*[³H] or -[¹⁴C] glucosamine and examined by polyacrylamide gel electrophoresis. All details have been described (21). (a) C_{13}/B_4 cells logarithmically growing (○- - -○) or plateu phase (●—●) and (b) C_{13}/B_4 (○- - -○) and (BHK_{21}/C_{13} (●—●).*

TABLE 1

Observations which suggest a role for fucosylation in membrane functions of mammalian cells.

1. Most membrane glycoproteins are fucosylated.
2. Fucose is positioned near the glycopeptide core.
3. α-L-Fucosidase activity is not significantly elevated after virus transformation.
4. Fucose is the only monosaccharide which occurs as an L-isomer and a deoxy sugar but never as the N-acetyl derivative.
5. Fucose is not metabolized like other monosaccharides.

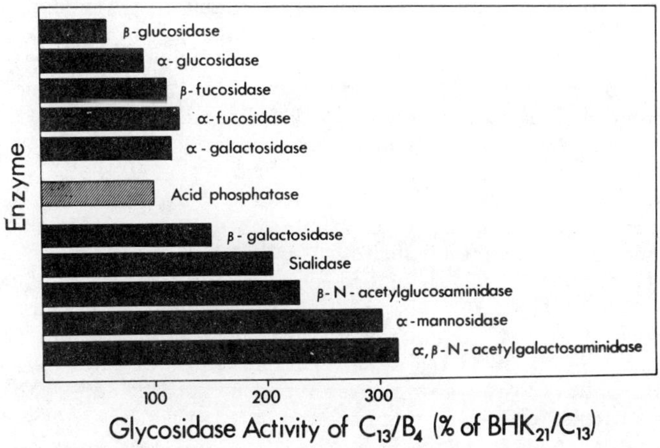

Figure 2. Acid hydrolase activities of transformed (C_{13}/B_4) and nontransformed (BHK_{21}/C_{13}) fibroblasts. The cells, harvested when growing exponentially, were homogenized in 0.1% Triton ×100. The enzyme activities were determined on the appropriate p-nitro*phenyl derivatives in 3mM citrate buffer, pH = 4.5, with the exception of sialidase where fetuin in 50mM acetate buffer, pH = 4.5, served as substrate. The enzyme activities (per mg protein) of C_{13}/B_4 cells are expressed as the percentage of the particular activity obtained with BHK_{21}/C_{13} cells. The values represent the mean obtained from three different cultures of each cell type. Acid phosphatase, which is not a glycosidase, is included for comparison.*

Uniqueness of Fucose in Mammalian Cells

Point 4 in Table 1 suggests that due to its structure alone fucose may be in a unique position. Thus far, fucose has been described only as the L-isomer, in contrast to other monosaccharides which occur as D-isomers. It is the only deoxy sugar present in mammalian glycoproteins and thus far has not been described as the N-acetyl derivative. The fact that it is not subject to the normal metabolic pathways as the other monosaccharides is also suggestive of a unique role. This is shown by the fact that radioactive fucose is recovered only as fucose not only in cells in culture (8), but also in animals (9). Subsequent studies may not support all of these points, and in fact, recent reports suggest that, contrary to the current dogma, fucose is not always found in a terminal position (10,11).

Ancillary Data

Additional data supporting the hypothesis of the uniqueness of fucose in mammalian cell membranes can be summarized from a number of other sources. Perhaps the most compelling is the presence of relatively large amounts of two unusual components associated with the surface membrane containing fucose. One of these is the major high molecular weight glycoprotein exported by human fibroblasts in culture (12). Although this glycoprotein has been shown by others to be a component of the fibroblast matrix, and for this reason called Fibronectin (13), little attention has been given to the fact that it is readily labeled with radioactive L-fucose.

The second unusual fucose-containing molecule is a low molecular weight component of the cell surface (14). Chromatography on precalibrated Fractogel PGM 2000, as well as other criteria, show that this fucose-containing molecule is approximately 500 daltons (Figure 3). This surface component is not the same as the low molecular weight component of NRK cells reported by Steiner (10), since fucose was released by purified α-L-fucosidase. This enzyme was free of other detectable exoglycosidases using synthetic and natural substrates (15), so fucose had to be in a terminal position.

Additional data concerning the presence of fucose in membrane glycoproteins is that in many different cell types examined, only 30-40% of the fucose content of the total cell was found in the membrane (7,15). This was in contrast to the sialic acid content of the membrane, which was as high as 70-80% of the total cell content (16).

In addition, fucose found in the membrane glycoproteins showed two different rates of acid hydrolysis and two different specific activities (8). Only 40% of the fucose from the membrane glycoproteins of transformed hamster fibroblasts (C_{13}/B_4) was removed after 2 hours with 0.1 N H_2SO_4 at 100°. This material showed a higher specific activity than the remaining fucose which was subsequently removed by α-L-fucosidase. The reasons for these differences are not apparent at this time, but the two

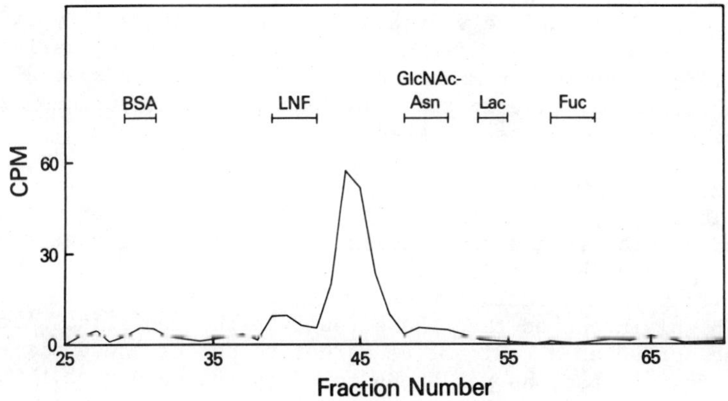

Figure 3. Low molecular weight fucose-containing components associated with the surface membrane (14). Radioactivity extracted from the surface of a human neuroblastoma cell line (IMR-32), metabolically labeled with L-[^{3}H] *fucose, was purified and chromographed on Fractogel PGM 2000. The column was precalibrated with BSA, bovine serum albumin; LNF, lacto-N-fucopentaose; GlcNAc-Asn, 2-acetamido-1-(*L-β-*aspartamide)-1,2-di-deoxy-β-*D-*glucose; Lac, lactose; Fuc, fucose.*

specific activities suggest the presence of at least two fucosyltransferases.

Role of Fucosylation

All of the above facts can be summarized to suggest that fucose as it occurs in membrane glycoproteins has a number of unusual characteristics. It is hypothesized that because of these, fucose performs a special function for the cell membrane. It can be postulated that fucose may serve to bind proteins or polypeptides for reinsertion into membranes, or transport through the membrane, thus its position near the glycopeptide core would be an advantage. The exact role it may play, if any, remains to be shown. Cell surface variants of a number of cell types have been reported (17-20). Perhaps through use of these variants which are defective in membrane glycoproteins, the special role which fucose may have in the relationship of the surface membrane to the cell will be ascertained.

SUMMARY

The fucose-containing glycopeptides of the membranes of virus transformed and tumor cells are clearly different from those of their normal counterparts. This difference is that the virus transformed cells express on their surfaces glycopeptides more highly branched than those found in the controls. Although fucose served as a radioactive marker to detect these differences, no difference has, as yet, been found in fucose *per se*. The similarities of fucose among several cell types and some unique properties of fucose in mammalian glycoproteins have been summarized and used to suggest that fucose may play a special role in the relationship of the surface membrane to the cell.

ACKNOWLEDGEMENTS

Supported by U.S.P.H.S. Grants CA 14037, CA 14489, and HD 08536.

Literature Cited

1. Buck, C.A., Glick, M.C., and Warren, L. (1970) Biochemistry 9, 4567-4576.
2. Glick, M.C. (1976) In *Fundamental Aspects of Metastasis*, L. Weiss, ed. North Holland Publishing Co., Amsterdam, pp. 9-23.
3. Poste, G., and Nicholson, G.L., eds. (1977) Cell Surface Reviews, Vol. 4. Elsevier-North Holland Inc., Amsterdam.
4. Glick, M.C., Rabinowitz, Z., and Sachs, L. (1974) J. Virol. 13, 967-974.
5. Van Beek, W.P., Emmelot, P., and Homburg, C. (1977) Br. J. Cancer 36, 157-165.
6. Glick, M.C., and Santer, U.V. (1977) In *Cell Surface Carbohydrate Chemistry*, Adv. Chem. Series, R.E. Harmon, ed., Academic Press, New York, pp. 13-26.

7. Scanlin, T.F., and Glick, M.C. VII International Cystic Fibrosis Conference. In press.
8. Fischer, A., and Glick, M.C. Manuscript in preparation.
9. Bocci, V., and Winzler, R.J. (1969) Am. J. Physiol. 216, 1337-1342.
10. Steiner, S., in these proceedings.
11. Hallgren, P., Lundblod, A., and Svensson, S. (1975) J. Biol. Chem. 250, 5312-5314.
12. Scanlin, T.F., Voynow, J.A., and Glick, M.C. Unpublished observations.
13. Vuento, M., Wrann, M., and Ruoslahti, E. (1977) FEBS Lett. 82, 227-231.
14. Mihalik, G.M., and Glick, M.C. Unpublished observations.
15. Santer, U.V., and Glick, M.C. Manuscript in preparation.
16. Glick, M.C. (1976) In Mammalian Cell Membranes, Vol. 1, G.A. Jamieson and D.M. Robinson, eds. Butterworths, London, pp. 45-77.
17. Stanley, P., Narasimhan, S., Siminovitch, L., and Schachter, H. (1975) Proc. Natl. Acad. Sci. USA 72, 3323-3327.
18. Meager, A., Ungkitchanukit, A., and Hughes, R.C. (1976) Biochem. J. 154, 113-124.
19. Briles, E.B., Li, E., and Kornfeld, S. (1977) J. Biol. Chem. 252, 1107-1116.
20. Pouysségur, J., and Pastan, I. (1977) J. Biol. Chem. 252, 1639-1646.
21. Greenberg, C.S., and Glick, M.C. (1972) Biochemistry 11, 3680-3685.

RECEIVED April 17, 1978.

23

Cell-Surface Glycoproteins of Normal and Malignant Rat Liver Cells

JAMES J. STARLING[1], DOUGLAS C. HIXSON, SYLVIA C. CAPETILLO[2], EDWARD M. DAVIS[3], GIOVANNI NERI[4], and EARL F. WALBORG, JR.

Research Division, University of Texas, System Cancer Science Park, Smithville, TX 78957

Malignant transformation is accompanied by cell-surface alterations which are directly involved in many of the aberrant properties displayed by the transformed cell (1). Lectins have been used extensively as probes to study these membrane alterations because, in several in vitro cell systems, non-transformed cells are not agglutinated at lectin concentrations sufficient to agglutinate transformed or protease-treated non-transformed cells (2). Despite extensive investigation, the molecular mechanism(s) responsible for lectin-induced cytoagglutination is (are) still not clear, although several hypotheses have been proposed: exposure of cryptic lectin-binding sites (3,4); clustering and lateral mobility of lectin receptors (5-8); reduction of surface charge repulsive forces (2); increased membrane deformability (2); increased density of lectin-binding sites due to altered surface morphology (9); and structural alterations of lectin receptors (10,11). Two such lectins, concanavalin A (Con A) and wheat germ agglutinin (WGA), have been used to probe the membrane structure of normal and malignant rat liver cells. Novikoff ascites hepatoma cells were agglutinated at low concentrations of Con A and WGA (12), whereas AS-30D ascites hepatoma cells (13) were agglutinable by WGA but required high Con A concentrations for agglutination (14). Macrosialoglycopeptides isolated from the surfaces of these cells possessed potent Con A and/or WGA receptor activity suggesting that these glycopeptides may be involved in the cytoagglutination properties of the intact cells (11). Isolated adult

*current address: Dept. of Biochem., Univ. of Florida, Gainesville, FL 32610

**current address: Dept. of Pathology, Univ. of Texas System Cancer Center, Houston, TX 77030

***current address: Dept. of Biology, Yale Univ., New Haven, CT 06520

****current address: Catholic Univ. School of Medicine Rome 01168, Italy

0-8412-0452-7/78/47-080-412$05.00/0

rat liver cells, on the other hand, were not agglutinable by high concentrations of Con A or WGA (11,15). The papain-labile sialoglycopeptides isolated from normal hepatocytes exhibited a Sephadex G-50 elution profile which was strikingly different from that of cell-surface sialoglycopeptides isolated from Novikoff and AS-30D hepatoma cells and possessed no detectable Con A or WGA receptor activity (11). These data suggested major differences in membrane structure between normal and malignant rat liver cells. However, the methods employed in the above studies to obtain single cell suspensions from adult liver involved mechanical dispersal and/or collagenase digestion which did not employ perfusion techniques. The structural and metabolic superiority of hepatocyte suspensions obtained by collagenase perfusion over that obtained by other isolation methods has been well documented (16). It is possible, therefore, that the reported low agglutinability of normal liver cells and/or the Sephadex G-50 elution profile of cell-surface glycopeptides isolated from hepatocytes may not reflect the in vivo properties of these cells but may instead be due to the methods used for their isolation.

The research reported herein describes the Con A-and WGA-induced agglutination properties of mechanically and enzymatically dispersed hepatocytes (refers to hepatocytes isolated by the technique of Jacob and Bhargava (17) and by collagenase perfusion of the intact liver, respectively) and compares these properties to those exhibited by Novikoff ascites hepatoma cells. The Sephadex G-50 elution profiles and lectin receptor activity of papain-labile glycopeptides isolated from the surface of enzymatically dispersed liver cells and Novikoff ascites hepatoma cells are also described and an attempt is made to delineate those factors primarily responsible for the increased Con A-induced agglutinability of rat hepatocytes following papain digestion.

Materials and Methods

Hepatoma Cells. Novikoff rat ascites hepatoma cells, maintained in 6-9-week-old female Sprague-Dawley rats (ARS/Sprague-Dawley, Madison, WI) were collected in Medium 199 with L-glutamine, obtained from Schwarz Mann, Orangeburg, NY), pH 7.4, containing 0.1g neomycin sulfate per liter (M199); resuspended in the same medium, and incubated 90 min in a shaking water bath at 37° in stoppered flasks gassed with 5% CO_2 in air.

Rat Hepatocytes. Suspensions of enzymatically dispersed hepatocytes were prepared by collagenase perfusion of the intact liver according to the method of Bonney (18) as described by Starling, et al (19). Suspensions of mechanically dispersed rat hepatocytes were prepared by the method of Jacob and Bhargava (17) as described by Walborg et al (11) except that the cells were incubated for only 90 minutes in M199 instead of 20 h.

Determination of cell viability and ultrastructural integrity. Cell viability was determined by trypan blue staining (20) and ultrastructural integrity was ascertained by morphological examination of the electron microscope. Samples were fixed in 3%

glutaraldehyde, post fixed in 2% OsO_4, and processed by standard methods for electron microscopy.

Papain digestion of hepatoma cells. Novikoff ascites hepatoma cells, washed twice in buffer 1 (21) and once in buffer 2 (21) at 50g for 5 minutes were incubated with papain (2 times crystallized, 666 U/ml Worthington Biochemical Corp.) for 40 minutes as described previously (11). Proteolytic digestion was performed using 3 U of papain per ml of cell suspension, containing 1.5 x 10^7 cells/ml. These incubation conditions approximate the mildest conditions which yield maximal release of peptide-bound sialic acid from the surface of hepatoma cells (14). Cell counts were performed using a Model ZBI Coulter Counter (Coulter Electronics, Hialeah, FL).

Papain digestion of liver cells. Mechanically dispersed liver cells were digested with papain as described previously (11). Enzymatically dispersed liver cells, incubated in M199 + 3% fetal calf serum, were washed twice in buffer 1 (21), once in buffer 2 (21) at 20g for 2 minutes, and resuspended in 30 ml of buffer 2 (21) at a concentration of 0.5 to 1 x 10^7 cells/ml. Ten ml aliquots of the cell suspension were placed in separate 250 ml Erlenmyer flasks containing 10 U of papain per flask and incubated for 30 minutes in a shaking water bath at 37° while being gassed continuously with a 95% O_2/5% CO_2. Cell counts were performed using a hemacytometer.

Assay of lectin-induced cytoagglutination. Lectin purification and preparation of serial dilutions was performed as previously described (11,12). Quantitative assessment of cytoagglutination properties of intact, incubated control, or papain-treated hepatoma and liver cells was accomplished using the particle counter agglutination assay of Davis et al (22) as described earlier (19). Microscopic evaluation of lectin-induced agglutination of hepatoma cells and hepatocytes was performed using a modified version (19) of the method of Wray and Walborg (23). Con A-induced agglutinability of female Sprague-Dawley rat erythrocytes, washed 3 times in Ca^{+2}-free Hanks' medium (24), pH 7.4 (CFH) at 210 g for 10 minutes, was determined using the test tube hemagglutination assay of Leseney et al (25).

Preparation of glycopeptides cleaved from the surface of normal and malignant rat liver cells by papain. A papain-labile, cell-surface sialoglycopeptide fraction (C-SGP) was prepared from Novikoff cells using the procedure of Neri et al (12), while a papain-labile, cell-surface glycopeptide fraction (C-GP) was prepared from mechanically and enzymatically dispersed liver cells by the method of Walborg et al (11). Glycogen was removed as previously described (19).

Analysis of Sephadex G-50 column effluents. Sialic acid was determined by the method of Aminoff (26) after hydrolysis in 0.1N H_2SO_4 for 1 h at 80°. N-acetylneuraminic acid (NANA) was used as a standard. Neutral sugar was determined by the method of Dubois et al (27) using D-galactose as a standard. Hexosamine was assayed according to Gatt and Berman (28) after hydrolysis in 2N HCl for 8 h at 100 degrees using 2-amino-2-deoxy-D-glucose. HCl as a standard.

Assay of Con A and WGA receptor activity. The lectin receptor activities of the cell-surface glycopeptide fractions were assayed by their ability to inhibit Con A-or WGA-induced agglutination of guinea pig erythrocytes (14). Erythrocytes used in the determination of the lectin receptor activity of the Novikoff cell-surface glycopeptide fractions were suspended in Ca^{+2}-and Mg^{+2}-free phosphate buffered saline, pH 7.4 (CMF-PBS), containing 0.5% BSA.

Preparation and purification of [^{125}I] Con A. Iodination was performed according to the procedure of Sonoda and Schlamowitz (29). After iodination, the [^{125}I] Con A was purified by affinity chromatography at 22 degrees on Sephadex G-100 (30). Concentration of the [^{125}I] Con A and preparation of serial dilutions was performed as previously described (31).

[^{125}I] Con A binding assay. Quantitation of [^{125}I] Con A binding to the enzymatically dispersed hepatocytes was accomplished using the lectin binding tube assembly (LBTA) assay (32) as described earlier (31).

Preparation and purification of ferritin-conjugated Con A (Fer-Con A). Fer-Con A was prepared by slowly adding 0.15 ml of 0.25% glutaraldehyde to CMF-PBS containing 4.5 mg Con A, 54 mg methyl α-D-glucopyranoside (Sigma Chem. Co.) and 30 mg of recrystallized horse spleen ferritin (Pentex, Miles Laboratories, Kankakee, Ill.). The conjugation reaction was allowed to proceed for 6 h at 22° and was then terminated by the addition of lysine to a final concentration of 2 mg/ml. The conjugated lectin was purified on a 15-30% linear sucrose gradient. The Fer-Con A was dialyzed against CMF-PBS, concentrated to approximately 3 ml by dialysis against polyethylene glycol (average molecular weight 6000 D), and further dialyzed against CMF-PBS. This stock solution of Fer-Con A (approx. 4 mg/ml) agglutinated rabbit erythrocytes up to a dilution of 1:128 using the hemagglutination assay described by Smith et al (14).

Hepatocyte labeling by Fer-Con A. Binding of Fer-Con A to rat hepatocytes and preparation of sections for electron microscopy was performed as described earlier (31).

Results and Discussion

Viability and cytostructural integrity of hepatocytes isolated by enzymatic and mechanical procedures. Accurate assessment of the surface properties of hepatocytes requires the use of cells which closely approximate their in vivo condition. Previous reports (11) on lectin-induced agglutination of adult hepatocytes utilized cells which were isolated by procedures which yield cells possessing low viability and poor cytostructural integrity. We now report the use of hepatocytes isolated from livers perfused with collagenase, a technique demonstrated to yield cells exhibiting high viability and excellent ultrastructural integrity. Hepatocytes isolated by this enzymatic technique were obtained in a yield of 2-4 ml of packed (210 g, 10 min) cells/liver prior to incubation in M199. Less than 20% of the cells were lost during the incubation in M199 + 3% FCS. Those cells surviving the short-term incubation were 78 ± 9.5 S.D. (n=75) percent viable as judged by their exclusion of vital stain. Some of the hepatocytes were present as small aggregates of 2-6 cells (33); e.g., in seven different experiments an average of 72% of the cells were present as single cells, 21% in aggregates of two cells and 7% in aggregates of 3-6 cells. As shown in fig. 1B these hepatocytes exhibited excellent ultrastructural morphology. They possessed intact plasma membranes, mitochondria with cristae evident and a well-defined endoplasmic reticulum.

For comparative purposes, hepatocytes were also isolated using perfusion with citrate followed by mechanical dispersal. This procedure yielded an average of 1.2 ml of packed (210 g, 10 min) cells/liver prior to incubation in M199. Of these, 70% were recovered after incubating 90 minutes in M199. Vital dye staining revealed that 100% of these cells were non-viable. As reported by others (16), the ultrastructural integrity of the mechanically dispersed hepatocytes was inferior to that of the enzymatically dispersed cells (fig. 1). The mechanically dispersed hepatocytes exhibited disrupted plasma membranes, swollen mitochondria devoid of critae, and no visible endoplasmic reticulum (figure 1A).

Lectin-induced agglutination of hepatocytes. Lectin-induced agglutination was quantitated using two complementary assays which measured the degree of agglutination as a function of the size of the cell aggregates (microtest II plate) or the disappearance of single cells (electronic particle counter). Enzymatically dispersed hepatocytes were agglutinated by low concentrations of Con A or WGA, i.e. lectin concentrations comparable to those required for the agglutination of Novikoff hepatoma cells (table I). Precise correlation of two cell types is not possible because of differences in cell size and the presence of some hepatocyte cell aggregates in the absence of lectin (table II). (The mean volumes of rat hepatocytes and Novikoff hepatoma cells, determined using the Model ZBI Coulter Counter, were 4850 and 1560 μm, respectively. Calibration of the particle counter was performed using micropolystyrene spheres (Lot no. 1001, Coulter Electronics) of 18.04 μm diameter. Assuming the cells to be spherical, this indicates that hepatocytes and Novikoff cells have diameters of 21.0 μm and 14.4 μm, respectively). The presence of small aggregates of hepatocytes in the absence of lectin desensitizes the electronic particle counter assay since their participation in the agglutination reaction escapes measurement. No aggregation of Novikoff cells was present in the absence of lectin. The Microtest II plate assay gave results similar to those obtained by the particle counter assay (table I) indicating that the decrease in single cell number measured by the electronic particle counter at increasing lectin concentrations was due to increased aggregate

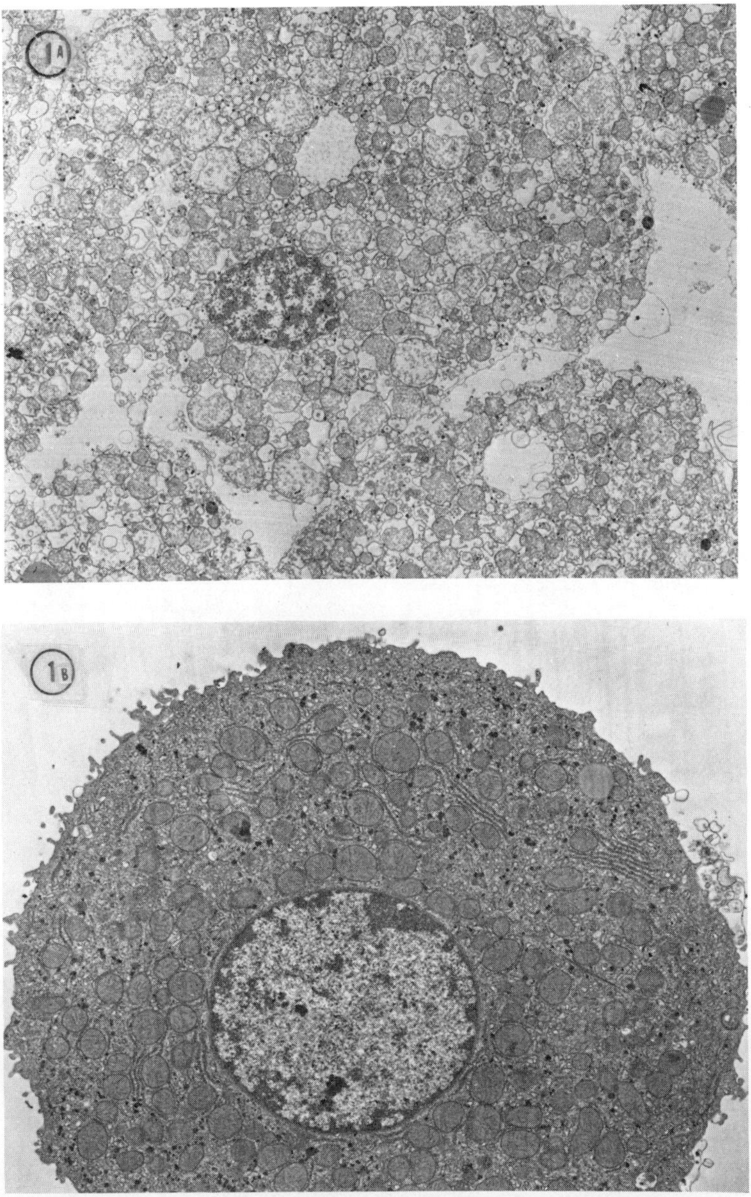

Experimental Cell Research

Figure 1. Morphological comparison of mechanically and enzymatically dispersed hepatocytes. Electron micrographs of glutaraldehyde-fixed cells postfixed with OsO_4. Sections were stained with uranyl acetate and lead citrate. (A) Mechanically dispersed hepatocytes after 90 min incubation in M199 $\times$3700; (B) enzymatically dispersed hepatocyte after papain digestion $\times$3000. Untreated hepatocytes showed comparable cytostructural integrity.

TABLE I

AGGLUTINABILITY OF NORMAL AND MALIGNANT RAT LIVER CELLS BY CON A AND WGA

Cell Type and Treatment	Assay System	Lectin concentratin (μg/ml) necessary for cytoagglutination of normal and malignant rat liver cells[a]	
		Con A	WGA
		Threshold	Threshold
Enzymatically Dispersed Hepatocytes			
Intact	Microtest II plate	1.9	1.5
Intact	Particle counter	2.5	4.8
Incubated control[b]	Particle counter	3.4	6.4
Papain-treated	Particle counter	0.78	0.78
Novikoff Hepatoma Cells			
Intact	Microtest II plate	3.2	0.56
Intact	Particle counter	0.97	0.51
Incubated control[b]	Particle counter	0.93	0.63
Papain-treated	Particle counter	1.7	0.63

[a] Concentrations as defined by Davis _et al_ (22) or Neri _et al_ (12).
[b] Incubated under the conditions of papain digestion, but without enzyme.

TABLE II

EFFECT OF INCUBATION TIME ON CON A-INDUCED AGGLUTINABILITY AND REAGGREGATION OF ENZYMATICALLY DISPERSED HEPATOCYTES

Incubation time	Con A conc. (µg/ml) required for cytoagglutination of enzymatically dispersed hepatocytes[a]		Cell aggregate size % of total[b]			% Viable cells
	Threshold[c]	1/2 Maximal[c]	1	2	3-6	
30 min	3.4	23	72	21	7	77
4 h	5.6	96	54	31	15	70
8 h	2.5	56	56	35	9	71

[a] Determined using the particle counter assay of Davis *et al* (22).

[b] Expressed as the percent of total cells in the indicated aggregate size.

[c] Concentrations as defined by Davis *et al* (22).

size and that the agglutinability of the enzymatically dispersed hepatocytes was exhibited at widely different cell concentrations, e.g. 6.7×10^5 cells/ml (Microtest II plate assay) and 1.3×10^7 cells/ml (particle counter assay). Lectin-induced agglutination of the intact hepatocytes was saccharide-specific since inclusion of 2.1 mM methyl α-D-mannopyrannoside in the agglutination reaction (particle counter assay) resulted in a 1.6-fold increase in the Con A concentration necessary for threshold and half-maximal agglutination or inclusion of 4.2 mM 2-acetamido-2-deoxy-D-glucose in the agglutination reaction resulted in a 2-fold increase in the WGA concentration necessary for threshold and half-maximal agglutination. The saccharide specificity of Con A-or WGA-induced agglutination of Novikoff cells has been demonstrated previously (12).

In agreement with previous studies (11), mechanically dispersed hepatocytes were not agglutinated by concentrations of Con A or WGA as high as 270 μg/ml (Microtest II plate assay). The lectin-induced agglutinability of the enzymatically dispersed hepatocytes at low concentrations of WGA or Con A (table I) is probably due to their increased viability and superior morphological integrity. Our observations differ from those of Becker (15) who reported that adult rat hepatocytes, also obtained by dispersal with collagenase, were not agglutinated by high concentrations of Con A. However, Becker (15) utilized an isolation technique which did not employ perfusion of the liver in situ with collagenase, and such techniques have been demonstrated to yield cells of low viability (16).

Investigation of the cell-surface properties of enzymatically dispersed hepatocytes is complicated by the possible alteration of cell-surface components by contaminating proteases present in preparations of crude collagenase. The procedure for isolation of hepatocytes described herein utilized 0.5% BSA in the perfusion medium. A control experiment indicated that this concentration of BSA was sufficient to minimize surface alterations by contaminating proteases. Rat erythrocytes offer a sensitive system for investigation of the effect of protease on Con A-induced cytoagglutination (25). Rat erythrocytes, incubated at 37° for 30 min in CFH, were not agglutinated by Con A concentrations as high as 133 μg/ml in the test tube agglutination assay of Leseney et al (25). When incubated at 37° for 30 min in the presence of 0.05% collagenase in CFH or 0.01% papain in CFH they were agglutinated by 22 and 4 μg Con A/ml, respectively. Rat erythrocytes, incubated in the presence of 0.05% collagenase and 0.5% BSA in CFH were not agglutinated at Con A concentrations up to 133 μg/ml, indicating that the BSA serves as substrate for the protease present in the crude collagenase preparation and prevents their action on the surface of the rat erythrocytes.

Since plasma membrane components undergo metabolic turnover (34), proteolytic or other damage to the plasma membrane is subject to repair processes. If the enzymatically dispersed liver cells used in this study were rendered agglutinable by alterations resulting from the collagenase perfusion, then incubation of viable liver cells should result in decreasing lectin-induced agglutinability as a function of surface repair. The data in table II show the effect of increasing incubation time on the Con A-induced agglutinability of enzymatically dispersed liver cells. There is a 2-to-4 fold decrease in the Con A-induced agglutinability of hepatocytes incubated for 4 h compared with hepatocytes incubated for 30 min; however, reaggregation of the hepatocytes during the 4 h incubation may be partially responsible for this decreased agglutinability since aggregate formation will desensitize the particle counter assay as previously discussed. After 8 h of incubation, the Con A-induced agglutinability of enzymatically dispersed hepatocytes is 2-fold greater than hepatocytes incubated for 4 h even though the extent of cell reaggregation was the same for hepatocytes incubated 4 or 8 h (table II). These data indicate that the non-agglutinability of adult hepatocytes by Con A reported by Becker (15) is not a function of repair processes which are operative during short-term incubation in M199 + 3% FCS.

Effect of papain digestion on Con A-or WGA-induced agglutination of hepatocytes. Papain treatment markedly altered the lectin-induced agglutinability of enzymatically dispersed hepatocytes. Papain treatment was accompanied by a 4-fold or 8-fold reduction in the lectin concentration required for threshold agglutination by Con A or WGA, respectively (table I). Papain digestion did not alter the number or distribution of small aggregates of hepatocytes which were present in the absence of lectin, consequently the increased lectin-induced agglutinability of papapin-treated cells cannot be attributed to an increase in the percentage of single cells in the cell suspension. By contrast, mechanically dispersed hepatocytes remained non-agglutinable by concentrations of Con A or WGA as high as 270 µg/ml, even after papain digestion, an observation consistent with a previous investigation (11). In comparison to the enzymatically dispersed hepatocytes, Novikoff cells exhibited only minor alterations of their Con A-or WGA-induced agglutinability as a result of papain digestion (table I).

Gel filtration of glycopeptides cleaved from the surface of hepatocytes and hepatoma cells by papain. Digestion of normal and malignant liver cells with papain degrades the cell-surface glycoproteins, releasing glycopeptides into the extracellular fluid (11). Papain treatment of enzymatically dispersed hepatocytes resulted in the release of 0.08 µmole sialic acid per ml of packed (210 g, 10 min) cells. A similar quantity of sialic acid, 0.09 µmole/ml of packed cells, was released from the surface of Novikoff cells by digestion with papain. The non-dialyzable glycopeptides cleaved from the surface of enzymatically dispersed hepatocytes and Novikoff hepatoma cells were submitted to gel filtration on Sephadex G-50 (fig. 2). The enzymatically dis-

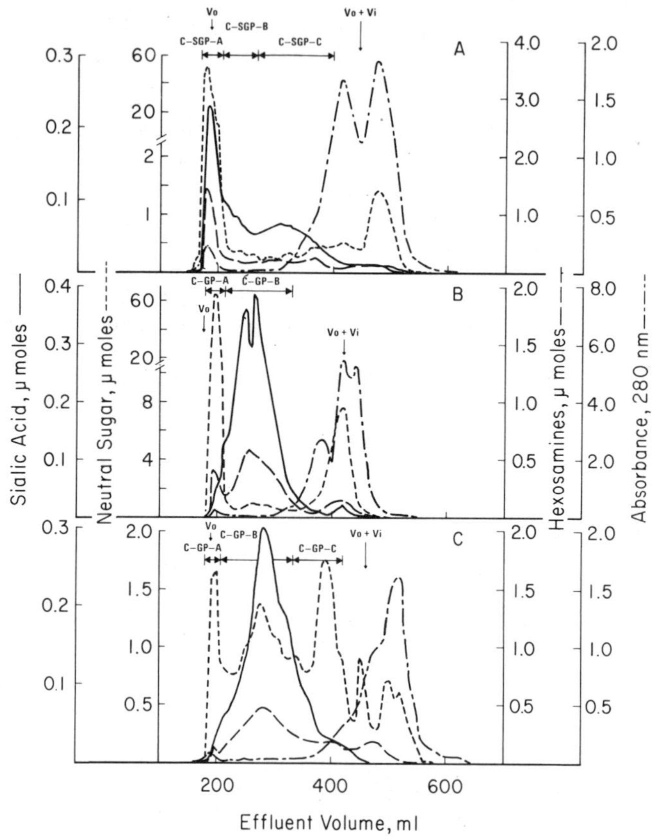

Experimental Cell Research

*Figure 2. Gel filtration of the cell-surface glycopeptides from (A) Novikoff hepatoma cells; (B) enzymatically dispersed hepatocytes; and (C) mechanically dispersed liver cells. The column (2.5 × 90 cm) of Sephadex G-50 was equilibrated and eluted with 0.1*N *acetic acid at a flow rate of 20 mL/hr at 22°C. Analysis of the column effluents is described in the Materials and Methods section. Sugar concentrations are expressed as μmol per fraction.*

persed hepatocyte cell-surface glycopeptides (fig. 2B) were resolved into two fractions: C-GP-A, a component excluded from the gel and C-GP-B, a prominent sialoglycopeptide fraction which was partially accessible to the gel and which contained over 90% of the sialic acid released from the cell-surface by papain. C-GP-A isolated from enzymatically dispersed hepatocytes contained a large quantity of neutral sugar which was due to glycogen released during the papain digestion, presumably resulting from the lysis of a few hepatocytes rich in glycogen. The glycogen present in C-GP-A was removed by precipitation with 50% ethanol.

Comparison of the gel filtration profiles of the glycopeptides released from the surface of enzymatically dispersed hepatocytes and Novikoff hepatoma cells suggests major qualitative differences in their plasma membrane glycoproteins (fig. 2). In contrast to the cell-surface sialoglycopeptides from hepatocytes, a major portion of the sialoglycopeptides from Novikoff cells (C-SGP-A) was excluded from the gel (fig. 2A). This fraction contained 31% of the total sialic acid released from the cell surface of papain.

Lectin receptor activities of cell-surface glycopeptides from hepatocytes and hepatoma cells. A comparison of the Con A and WGA receptor activites of the cell-surface glycopeptides from hepatocytes and Novikoff hepatoma cells provided additional evidence for qualitative differences between the cell-surface glycoproteins of normal and malignant rat liver cells (table III). Since lectins bind to specific saccharide determinants, differences in the lectin receptor activites suggest that these qualitative differences, are, in part, structural. The glycopeptides isolated from the surface of enzymatically dispersed hepatocytes possessed no detectable Con A or WGA receptor activity indicating that the cell surface components responsible for the agglutination of rat hepatocytes are not released by digestion with papain. On the other hand, the cell-surface glycopeptide fractions from Novikoff cells possessed potent Con A and/or WGA receptor activity. The major portion of the Con A and WGA receptor activity resided in C-SGP-A as reported previously (12). The loss of Con A-binding sites from the surface of Novikoff cells following papain treatment has been demonstrated by binding studies using Con A labeled with [^{125}I] (35). These observations indicate a differential lability of the cell-surface lectin receptors of normal and malignant rat liver cells to degradation by papain.

Papain-labile, cell-surface glycopeptides isolated from Novikoff hepatoma cells reflected the cytoagglutination properties of the intact cells from which they were derived while glycopeptides isolated from enzymatically dispersed cells did not. These data suggest that, contrary to Novikoff ascites hepatoma cells, the papain-labile, cell-surface glycopeptides isolated from enzymatically dispersed hepatocytes do not function in the agglutination of the intact cells. This finding emphasizes the importance

TABLE III

LECTIN RECEPTOR ACTIVITIES OF CELL-SURFACE GLYCOPEPTIDES ISOLATED FROM NORMAL AND MALIGNANT RAT LIVER CELLS[a]

Glycopeptide Fractions	Yield mg/100 mg	Specific Lectin Receptor Activity HAIU/mg[b]		Total Lectin Receptor Activity HAIU/Fraction $x10^{-2}$	
		Con A	WGA	Con A	WGA
Enzymatically dispersed hepatocytes					
C-GP-A[c]	17	<70	<70	--	--
C-GP-B	83	<70	<70	--	--
Incubated Novikoff hepatoma cells					
C-SGP-A[c]	26	2500	620	650	160
C-GSP-B[c]	29	710	170	210	49
C-GSP-C[c]	45	250	<70	110	--
Ovalbumin glycopeptide[d]		2200 ± 200 S.D.			
Ovomucoid[e]			250 ± 60 S.D.		

a Determined using assay of Smith *et al* (14).
b Hemagglutination inhibitory units defined by Smith *et al* (14).
c Ethanol treated prior to lectin receptor activity assay.
d Ovalbumin glycopeptide prepared as described by Montgomery and Lee (42).
e Ovomucoid obtained from Worthington Biochemical Corp.

of the papain-stable lectin receptor sites in determining the lectin-induced agglutination properties of normal liver cells and indicates that the increased agglutinability of hepatocytes by Con A and WGA after papain digestion may be due to several factors: exposure of cryptic lectin receptor sites; topographical redistribution of lectin receptor sites; and/or a change in membrane morphology.

<u>[^{125}I] Con A binding to rat hepatocytes</u>. Since brief proteolysis of non-transformed cells rendered them as agglutinable by WGA as transformed cells, Burger (3) proposed that the increased agglutinability of transformed cells was due to the exposure of cryptic cell-surface lectin-binding sites during transformation. Since this proposal was formulated, however, several investigators (36,37) have reported that normal and transformed cells bind equivalent amounts of radioactive lectin suggesting that the increased lectin-induced agglutinability of transformed cells was not due to exposure of cryptic lectin-binding sites. More recently, Noonan and Burger (38) reported that trypsinized non-transformed and transformed 3T3 cells bound 2.5-3 times more [3H] Con A than non-transformed 3T3 cells under carefully controlled conditions which limited endocytosis of the radioactive lectin. These experiments again suggested the possibility that increased lectin-induced agglutinability was brought about by exposure of cryptic lectin-binding sites. As shown in table IV, incubated control and papain-digested hepatocytes bind equivalent amounts of [^{125}I] Con A and exhibit similar apparent binding affinity constant. The similarity of [^{125}I] Con A binding to incubated control and papain-digested hepatocytes is not due to endocytosis of the radioactive lectin because the binding assay utilized in these determinations measures only the [^{125}I] Con A binding that is specifically displaced by methyl α-D-mannopyranoside. This is confirmed further by the low amount of non-specific binding (less than 15%) of [^{125}I] Con A in the binding assay (table IV). As discussed earlier, hepatocyte suspensions obtained by collagenase perfusion possessed 28% of the cells in aggregate of 2-6 cells. The amount of [^{125}I] Con A nbound to papain-digested hepatocytes is not enhanced by an increase in the number of single cells in the hepatocyte suspension following papain digestion since the aggregate size of hepatocytes obtained by collangease perfusion is not altered by papain digestion. These data indicate that rat hepatocytes do not exhibit increased Con A binding following papain digestion even though the hepatocytes become significantly more agglutinable (table I). Thus, these data do not support the hypothesis of increased Con A-induced agglutinability of papain-digested hepatocytes due to expression of cryptic lectin binding sites following papain digestion. The unaltered lectin binding of the hepatocytes following papain digestion is consistent with the finding that cell-surface, papain-labile glycopeptides from enzymatically dispersed hepatocytes possess no detectable Con A receptor activity, indica-

TABLE IV

BINDING OF CON A TO RAT HEPATOCYTES[a]

Conditions or treatment	Specific binding[b] Molecules bound/cell $\times 10^{-8}$	Specific binding[b] Ka Liter/moles $\times 10^{-6}$	Non-specific binding[b] Molecules/cell $\times 10^{-8}$
Intact	1.8	0.80	0.13
Incubated control	1.1	1.1	0.16
Papain-digested	0.96	1.3	0.13

a Determined by binding assay of Davis et al (32).
b Data plotted according to the method of Steck and Wallach (41).

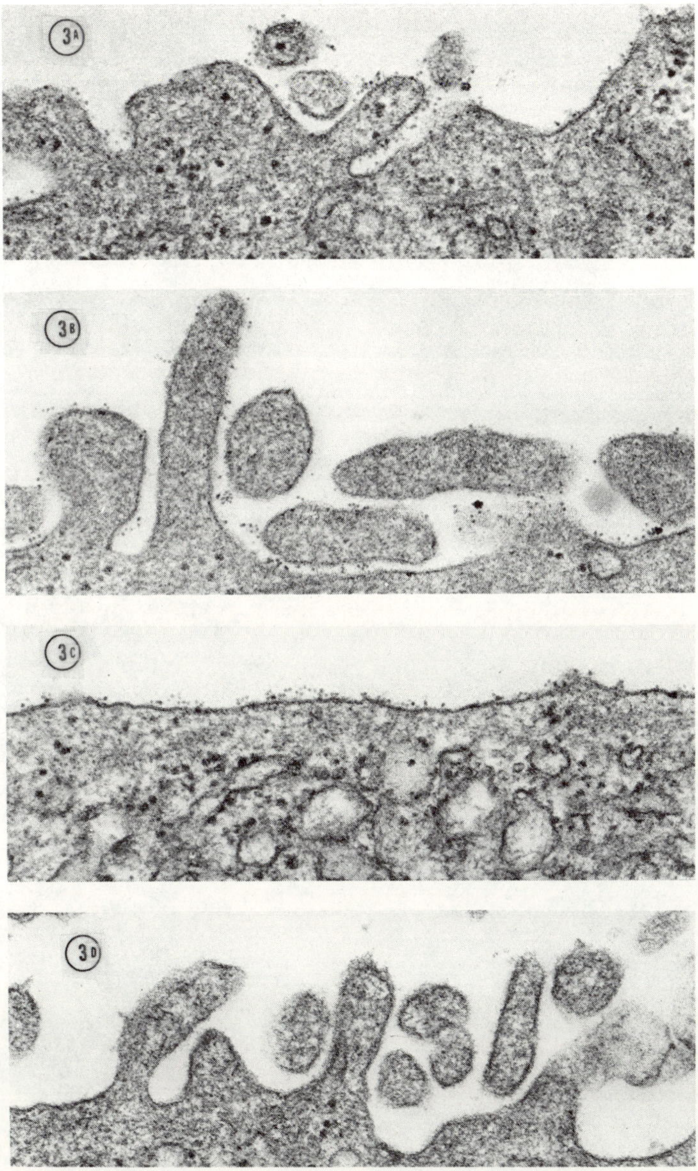

Experimental Cell Research

*Figure 3. Fer-Con A topographical distribution of rat hepatocytes. Electron micrograph of rat hepatocytes labeled with Fer-Con A as described in the Materials and Methods section. (A) Intact + Fer-Con A; (B) incubated control + Fer-Con A; (C) papain-digested hepatocytes + Fer-Con A; (D) intact hepatocytes + 0.17*M *methyl* α-D-*mannopyranoside + Fer-Con-A.* ×*55,000.*

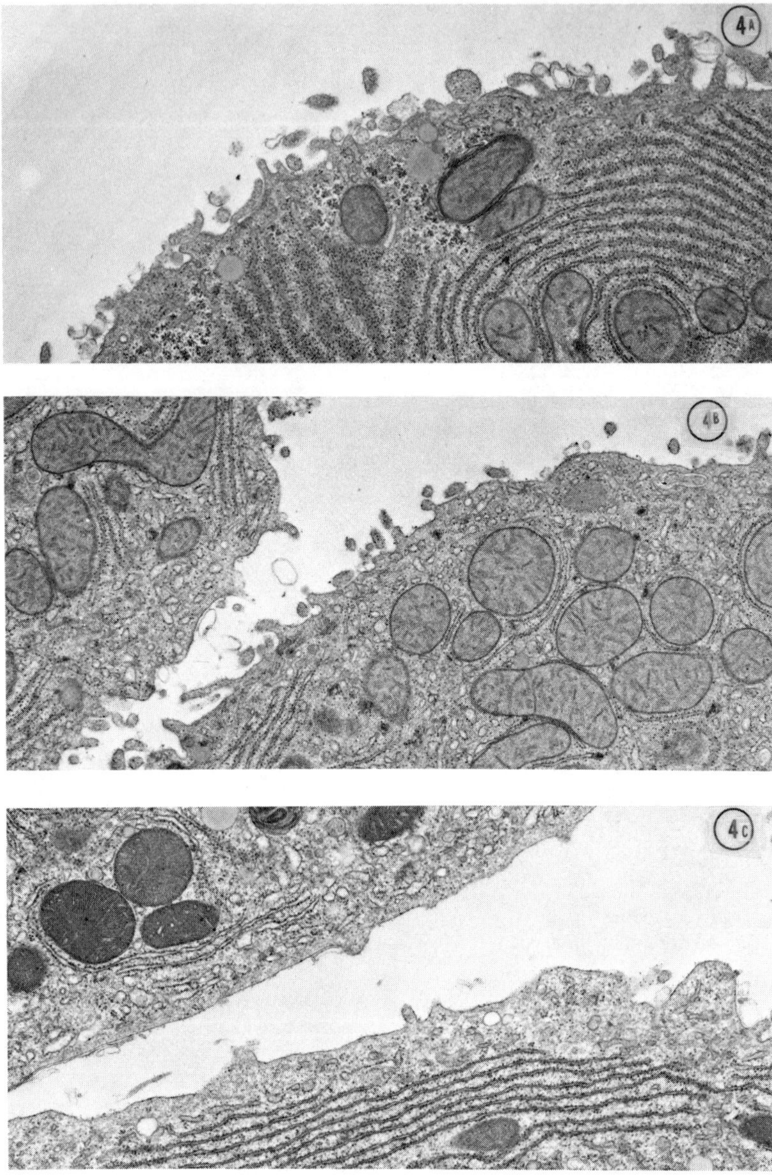

Experimental Cell Research

Figure 4. Cell-surface morphology of rat hepatocytes. Electron micrograph of glutaraldehyde-fixed cells post-fixed with OsO_4*. Sections were stained with uranyl acetate and lead citrate. (A) Intact; (B) incubated control; (C) papain-digested hepatocytes.* $\times$*11,800.*

ting that these glycopeptides are not involved in the Con A-induced agglutination of the intact cells (table III).

Con A receptor topography. Nicolson (39) observed that Fer-Con A receptors on transformed and trypsinized 3T3 cells were in a clustered topographical distribution compared with non-transformed 3T3 cells. Based on this observation, Nicolson postulated that the increased agglutinability of trypsinized non-transformed and transformed 3T3 cells was due to the clustered distribution of Con A receptors on the surface of these cells which enable agglutination to occur between cell-surface areas of high lectin receptor density. The topographical distributions of hepatocytes labeled with Fer-Con A are shown in Fig. 3. Altered distribution of Con A receptor sites after papain digestion does not appear to be responsible for the increased agglutinability of papain-digested hepatocytes since intact, incubated control, and papain-digested hepatocytes all exhibit a clustered distribution of Fer-Con A molecules (fig. 3). Since several investigators (5,6,7) have reported that clustered distributions of lectin molecules may reflect the lateral mobility of the lectin receptors rather than their inherent topographical distribution, these data do not rule out possible differences which may exist in the inherent topographical distributions of hepatocyte Con A receptors, although such differences would not explain the increased Con A-induced agglutinability of papain-digested hepatocytes compared with intact and incubated control hepatocytes.

Cell-surface morphology. A plausible explanation for the increased Con A-induced agglutinability of papain digested hepatocytes is derived from the change in surface morphology from a villous to a smooth membrane as a result of papain digestion (fig. 4). This membrane alteration may allow greater areas of contact between agglutinated cells thereby permitting more extensive lectin bridging to occur. This morphological change will also result in a higher surface density of Con A bound receptors on papain-digested hepatocytes as compared to incubated control hepatocytes although this increased receptor density may not account entirely for the increased agglutinability of papain digested hepatocytes (31). The above conclusions are further supported by the observation that high Con A-and WGA-induced agglutination of human lymphoblasts and normal human peripheral lymphocytes was correlated with a smooth cell surface whereas low Con A and WGA agglutinability was correlated with a villous surface morphology (40).

The above data suggest that there are significant structural differences in the cell-surface glycoproteins of normal and malignant rat liver cells and that viable hepatocyte suspensions offer a valuable in vivo system for studying cell-surface properties associated with normal cells.

ACKNOWLEDGEMENTS

The authors express appreciation to Dr. Max Schlamowitz and Dr. Kay Hillman for their assistance in the Con A iodination procedure. This work was supported by grants from NCI (CA-11710), the Robert A. Welch Foundation (G-354), the Paul and Mary Haas Foundation and the George and Mary Josephine Hamman Foundation. J.J.S. was recipient of a Rosalie B. Hite predoctoral fellowship in cancer research. Portions of the tables and figures from references 19 and 31 are reproduced by permission of Academic Press.

Literature Cited

1. Emmelot, P. (1973) Eur. J. Cancer 9, 319.
2. Nicolson, G. L. (1974) Int. Rev. Cytol. 39, 89.
3. Burger, M. M. (1969) Proc. Natl. Acad. Sci. US 62, 944.
4. Burger, M. M. (1970) Permeability and Function of Biological Membranes (ed. L. Bolis et al.), p. 107. North-Holland, Amsterdam.
5. Nicolson, G. L. (1973) Nature New Biol. 243, 218.
6. Nicolson, G. L. (1974) Control of Proliferation in Animal Cells (ed. B. Clarkson and R. Baserga), p. 251. Cold Spring Harbor, New York.
7. Rosenblith, J. A., Ukena, T. E., Yin, H. H., Berlin, R. D., and Karnovsky, M. J. (1973) Proc. Natl. Acad. Sci. US 70, 1625.
8. Rutishauser, U. and Sachs, L. (1975) J. Cell Biol. 66, 76.
9. Collard, J. G. and Temmink, J. H. M. (1975) J. Cell Sci. 19, 21.
10. Jansons, V. K., Sakamoto, C. K., and Burger, M. M. (1973). Biochim. Biophys. Acta 291, 136.
11. Walborg, E. F., Jr., Wray, V. P., Smith, D. F., and Neri, G. (1975) Immunological Aspects of Neoplasia, p. 123. Williams and Wilkins Co., Baltimore, Md.
12. Neri, G., Smith, D. F., Gilliam, E. B., and Walborg, E. F., Jr. (1974) Arch. Biochem. Biophys. 165, 323.
13. Smith, D. F., Walborg, E. F., Jr., and Chang, J. P. (1970) Cancer Res. 30, 2306.
14. Smith, D. F., Neri, G., and Walborg, E. F., Jr. (1973) Biochem. 12, 2111.
15. Becker, F. F. (1974) Proc. Natl. Acad. Sci. US 71, 4307.
16. Schreiber, G. and Schreiber, M. (1972) Sub-cell. Biochem. 2, 321.
17. Jacob, S. T. and Bhargava, P. M. (1962) Exp. Cell Res. 27, 453.
18. Bonney, R. J. (1974) In Vitro 10, 130.
19. Starling, J. J., Capetillo, S. C., Neri, G., and Walborg, E. F., Jr. (1977) Exp. Cell Res. 104, 177.
20. Phillips, H. J. (1973) Tissue Culture: Methods and Applications (ed. P. F. Kruse and M. K. Patterson) p. 406. Academic Press, New York.
21. Walborg, E. F., Jr. Lantz, R. S., and Wray, V. P. (1969) Cancer Res. 29, 2034.

22. Davis, E. M., Starling, J. J., and Walborg, E. F., Jr. (1976) Exp. Cell Res. 99, 37.
23. Wray, V. P. and Walborg, E. F.,Jr. (1971) Cancer Res. 31, 2079.
24. Hanks, J. H. and Wallace, R. E. (1949) Proc. Soc. Exp. Biol. Med. 71, 196.
25. Leseney, A. M., Thomas, M. W., O'Neil, P. A., and Walborg, E. F., Jr. (1974) Méthodologie de la Structure et du Métabolisme des Glycoconjugues. Actes du Colloque Internationaux n 221 due Centre National de la Recherche Scientifique, vol. 2, p. 859. Editiones du CNRS, Paris.
26. Aminoff, D. (1961) Biochem. J. 81, 384.
27. Dubois, M., Gilles, K. A., Hamilton, J. K., Rebers, P. A., and Smith, F. (1956) Anal. Chem. 28, 350.
28. Gatt, R. and Berman, E. R. (1966) Anal. Biochem. 15, 167.
29. Sonoda, S. and Schlamowitz, M. (1970) Immunochem. 7, 885.
30. Agrawal, B. B. L. and Goldstein, I. J. (1965) Biochem. J. 96, 23C.
31. Starling, J. J., Hixson, D. C., Davis, E. M., and Walborg, E. F., Jr. (1977) Exp. Cell Res. 104, 165.
32. Davis, E. M., Tsay, D. D., Schlamowitz, M., and Walborg, E. F., Jr. (1977) Anal. Biochem. 80, 416.
33. Seglen, P. O. (1972) Exp. Cell Res. 74, 450.
34. Warren, L. (1969) Current Topics in Developmental Biology (ed. A. A. Moscona and A. Monroy) vol. 4, p. 197. Academic Press, New York.
35. Walborg, E. F., Jr., Davis, E. M., Gilliam, E. B., Smith, D. F., and Neri, G. (1975) Cellular Membranes and Tumor Cell Behavior, p. 337. Williams and Wilkins, Baltimore, Md.
36. Cline, M. J. and Livingston, D. C. (1971) Nature New Biol. 232, 155.
37. Ozanne, B. and Sambrook, J. (1971) Nature New Biol. 232, 156.
38. Noonan, K. D. and Burger, M. M. (1973) J. Cell Biol. 59, 134.
39. Nicolson, G. L. (1971) Nature New Biol. 233, 244.
40. Temmink, J. H. M., Collard, J. G., Roosien, J. and van den Bosch, J. F. (1976) J. Cell Sci. 21, 563.
41. Steck, T. L. and Wallach, D. F. H. (1965) Biochim. Biophys. Acta 97, 510.
42. Montgomery, R. and Lee, Y. C. (1965) Biochem. 4, 566.

RECEIVED March 30, 1978.

24

Membrane Glycoproteins of Rat Mammary Gland and Its Metastasizing and Nonmetastasizing Tumors

KERMIT L. CARRAWAY, JOHN W. HUGGINS, ANNE P. SHERBLOM, ROBERT W. CHESNUT, ROBERT L. BUCK, SUSAN P. HOWARD, CHARLOTTE L. OWNBY, and CORALIE A. CAROTHERS CARRAWAY

Department of Biochemistry, Oklahoma State University, Stillwater, OK 74074

Glycoproteins are ubiquitous components of animal cell surfaces (1) and are generally believed to play important roles in neoplastic diseases (2). Glycoprotein differences between non-malignant and malignant cells have been demonstrated by such diverse techniques as lectin agglutination (3), glycopeptide analyses (4), gel electrophoresis (5) and enzymatic cell surface labeling (6). The origin of these differences and the roles of the glycoproteins in most cellular phenomena are still unknown.

One intriguing possibility is that cell surface glycoproteins may be involved in the mechanism by which tumor cells escape destruction by the immune system of the host. Two very different hypotheses have been presented to explain the role of glycoproteins in this escape from immune surveillance. The first involves the shedding of antigens from the cell surface. These antigens combine with circulating antibodies to form complexes which block antibody-mediated destruction of the tumor cells (7). Kim *et al.* (8) have extended this hypothesis to suggest that glycoprotein shedding in mammary tumors is related to their metastatic potential. Using a series of mouse mammary tumors selected for variations in ability to metastasize, they found an inverse correlation between metastasis and three parameters: 5'-nucleotidase activity, cell surface glycocalyx staining and immunogenicity of the tumors in rabbits. It was suggested that the loss of the cell surface constituents was due to shedding, but the mechanism was not specified.

A second means by which glycoproteins might be involved in "protection" of the tumor cells was suggested by studies on the TA3 mouse mammary tumor. Two sublines of this tumor which are different in their immunological properties have been isolated. The TA3-St subline grows only in the syngeneic host, but the TA3-Ha subline is able to cross histocompatibility barriers (9). This loss of strain specificity is accompanied by a greatly increased amount of a large sialoglycoprotein (10) named

0-8412-0452-7/78/47-080-432$05.00/0

epiglycanin. It was hypothesized that the glycoprotein in the TA3-Ha subline "covers" the histocompatibility antigens at the cell surface (11), preventing their expression and thereby preventing the recognition of these cells as foreign to the host.

Our own work has been concerned with the cell surface glycoproteins of rat mammary tumors and the relationship of these glycoproteins to the properties of the tumor cells. In this presentation we shall focus on three questions. 1) Is there a necessary relationship between the expression of cell surface glycoproteins and metastasis of mammary tumors as suggested by Kim *et al.* (8)? 2) Is there a relationship between the "form" or environment of the tumor and the cell surface glycoproteins it expresses? 3) How are the cell surface properties of tumors related to their biological activities as tumors?

Comparison of membrane glycoproteins from normal mammary gland, a nonmetastasizing tumor (R3230 AC) and a metastasizing tumor (13762).

To determine whether there is a necessary difference in expression of cell surface glycoprotein between nonmetastasizing and metastasizing mammary tumors, we have examined the R3230 AC and 13762 rat mammary adenocarcinomas, both grown in the Fischer 344 rat. The R3230 AC is a slow-growing tumor derived from a spontaneous carcinoma (12). It is hormonally responsive and exhibits little or no metastatic capacity under the conditions used in our studies. The 13762 rat mammary adenocarcinoma is a methylcholanthrene-induced solid tumor (13) which is hormonally responsive and metastasizes widely and heavily.

For comparison of cell surface glycoproteins it is necessary to purify plasma membranes from the tissue samples. Purification of plasma membrane and other subcellular tissue fractions is difficult for several reasons (14). Disruption of the cells in the tissue almost necessarily fragments the plasma membrane. Fragmentation is not uniform and may result in a heterogeneous distribution of membrane fragments, both in terms of size and of physical and functional attributes. This complicates the separation of particular morphological units because of the variation of the sizes, densities, compositions and other factors upon which separation of fragments is based. Identification of morphological units is also complicated by fragmentation, which destroys morphological characteristics and reduces membranes to sheets or vesicles. In addition to morphological criteria membrane functional activities (enzymes, antigens) must be used for identification of subcellular components. This combination of heterogeneity and loss of morphological identity makes the preparation of coherent, highly purified subcellular components very difficult. Most of the work in this area has been performed on liver. Plasma membranes have been purified by a number of methods, most of which tend to give membranes derived from the bile canalicular regions (15).

Mammary tissue presents its own particular problems.

Because of the presence of large amounts of connective tissue the elasticity of the tissue makes homogenization difficult. The procedure which gives the best results for normal lactating mammary tissue involves vigorous homogenization in a Sorvall Omni-mixer and isolation of a microsomal fraction by differential centrifugation (5). Microsomes are further fractionated by flotation through a sucrose density gradient (Fig. 1). 5'-Nucleotidase is used as a marker for preliminary assessment of plasma membrane purification, since this enzyme has been shown to be present predominantly in the plasma membrane of a number of cell types (14), including mammary gland and mammary tumors (16, 17). Significant purification of 5'-nucleotidase (>25-fold) was achieved in both of the lighter fractions (F_1 and F_2) of the discontinuous gradient, and there is a parallel purification of Na^+, K^+-ATPase (5), another plasma membrane marker. Succinic dehydrogenase (a mitochondrial enzyme) is markedly reduced, but NADPH-cytochrome c reductase (an endoplasmic reticulum marker enzyme) is still present, and galactosyltransferase is significantly concentrated. The last observation suggests Golgi apparatus contamination (16).

Further purification is achieved by treating F_1 and F_2 fractions separately with digitonin and subjecting these to reflotation on a similar discontinuous gradient containing digitonin. Digitonin complexes with cholesterol and would be expected to shift the cholesterol-enriched plasma membrane fraction to a higher density (18). Cholesterol-poor fractions such as Golgi fragments should be unchanged in density. By this procedure a portion of the digitonin-treated F_1 is shifted to F_1DF_3, which shows a 3-fold increase in nucleotidase activity and is devoid of galactosyltransferase and NADPH-cytochrome c reductase (19). Clearly it has the characteristics expected of a highly purified plasma membrane fraction.

The F_2 fraction behaves differently. Part of it reequilibrates to the F_1 position upon reflotation. When treated with digitonin, part of F_2 is shifted to a higher density (F_2DF_3). F_2DF_3 is enriched in 5'-nucleotidase, cholesterol, sialic acid and galactosyltransferase, properties suggesting that this subfraction comes from a plasma membrane containing galactosyltransferase.

These procedures have been used with similar results to purify plasma membranes from the solid tumors. As shown in Table I, the 5'-nucleotidase of the metastasizing 13762 tumor is not greatly diminished when compared to the nonmetastasizing R3230 AC. Sialic acid analyses (20) of membrane fractions from these two tumors also fail to show any significant depletion in the metastasizing tumor (Table II).

Sialoglycoproteins are analyzed by polyacrylamide gel electrophoresis in sodium dodecyl sulfate (SDS). Whereas normal mammary membranes show three major classes of glycoproteins by this technique (Fig. 2), membranes from the tumor cells have a

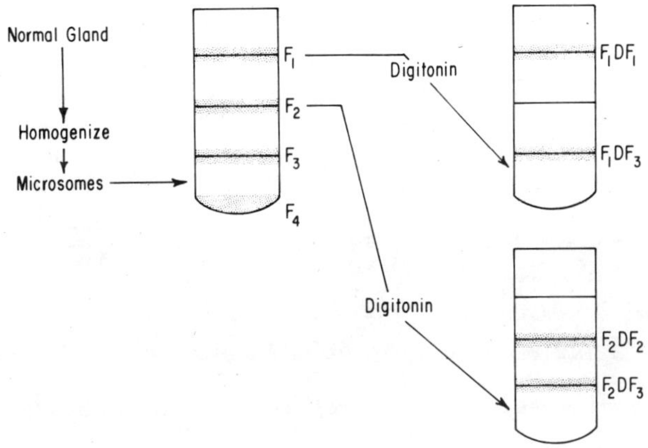

Figure 1. Schematic for membrane preparations from normal lactating mammary tissue and rat mammary tumor tissue samples

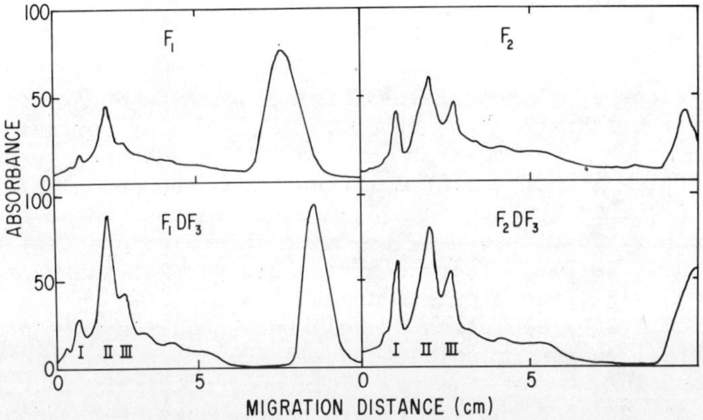

Figure 2. Membrane glycoproteins of normal lactating rat mammary tissue. Membrane fractions were prepared as previously described (5, 19) and analyzed by electrophoresis on polyacrylamide gels in the presence of SDS (5). Gels were stained by the periodate–Schiff procedure and scanned on a Gilford spectrophotometer with a linear transport device.

TABLE I

Nucleotidase Activities of Mammary Tumor Fractions

Fraction	R3230 AC	13762
	µmoles/hr·mg protein	
Homogenate	3.5	4.2
F_1DF_3	95	126
F_2DF_3	64	61

TABLE II

Sialic Acid Contents of Tumor Cell Membranes

	Homogenate	F_1	F_2
		µg/mg protein	
R3230 AC	0.7	13	11
13762	0.4	19	16

fourth class of glycoproteins (Fig. 3) with an apparently lower molecular weight than the other three (5). Glycoprotein patterns from the two tumors are very similar, though both have patterns more complex than those of membranes from the normal gland. It is not possible to ascertain at present whether these similar electrophoretic classes are the same molecular species, but it should be feasible to obtain at least a partial answer to that question using immunological techniques.

Clearly the depletion of cell surface glycoproteins is not a universal characteristic of plasma membranes isolated from metastasizing mammary tumors. This does not imply that these tumors do not shed their surface constituents, nor does it indicate that shedding is unimportant to metastasis. The concentration of any component in the cell reflects a balance between synthetic and degradative pathways. Since these need to be investigated more carefully, we have examined ascites cells derived from the metastatic 13762 tumor, because these cells are more easily manipulated.

Glycoproteins of the ascites form of the 13762 mammary adenocarcinoma.

The solid 13762 mammary adenocarcinoma has been converted into transplantable, metastatic ascites forms MAT-A, MAT-B and

MAT-C, which we obtained from the Mason Research Laboratory, Worchester, Mass. To serve as models for glycoprotein behavior in the solid tumor, these ascites tumors must meet two criteria. 1) The tumor sublines and their cell surface properties must be stable to in vivo passage over the period of experimentation. 2) The glycoproteins of the ascites tumor must be the same as those of the solid tumor.

The question of the stability of the sublines proved to be an interesting one. During passage of MAT-B and MAT-C sublines in vivo at weekly intervals over about six months, we noted that both were converted to variant forms. We shall term these MAT-B1 and MAT-C1 and will discuss their altered cell surface properties in a later section. We suspect that the MAT-A subline also undergoes changes with prolonged passage in vivo, but we do not have an appropriate marker to demonstrate these changes. We are maintaining the MAT-B1 and MAT-C1 tumors by transplantation and as frozen stock samples for future uses and comparisons.

To analyze cell surface glycoproteins of the ascites tumors, membranes were prepared by a modification of the method of Warren et al. (21), using zinc to "stabilize" the cell surface membranes. Polyacrylamide gel electrophoresis of SDS-solubilized membranes was used for analysis of sialoglycoproteins. All ascites tumor sublines showed a major band which barely penetrated the gels, illustrated in Fig. 4 for the MAT-B1 and MAT-C1 sublines. Although SDS electrophoresis is unreliable for molecular weight determinations on glycoproteins, gel filtration experiments also suggest that this sialoglycoprotein is a large molecule. This glycoprotein is observed as the major cell surface component when intact cells are labeled with lactoperoxidase and ^{125}I or with periodate and ^{3}H-borohydride.

These results indicate that glycoproteins in the solid tumor differ from those in ascites lines derived from it. However, it should be noted that the membrane preparations for solid and ascites tumors are different. Perhaps the more stringent conditions of cell disruption used in solid tumor membrane preparation cause breakdown of a large glycoprotein to give multiple smaller glycoproteins. Two types of experiments argue against this possibility. First, attempts to demonstrate a large glycoprotein in the solid 13762 tumor by periodate-borohydride labeling of the solubilized tissue followed by SDS electrophoresis were negative. Second, milk fat globule membranes contain the same three major sialoglycoprotein classes as the normal mammary gland and solid tumors. Since the milk fat globule has not been subjected to mechanical disruption, it is unlikely that the similar classes of glycoproteins of normal mammary and the solid tumors are products of the cell disruption.

What is the origin of this difference in glycoproteins between solid and ascites forms of the tumor? Probably it is a response to the altered environment of the ascites cells, but

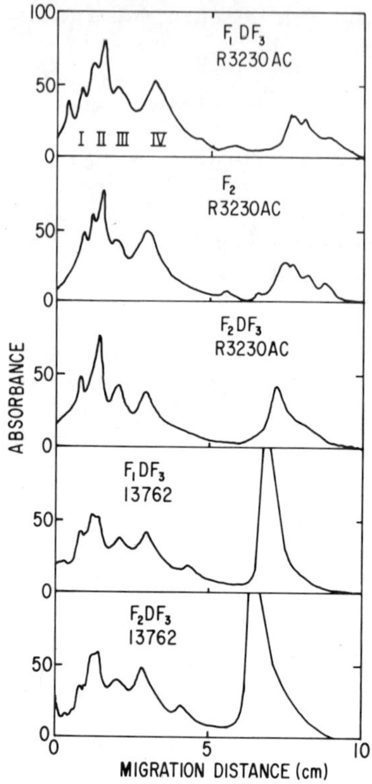

Figure 3. Membrane glycoproteins of rat mammary tumors. Membrane preparations and analyses were performed as described for Figure 2.

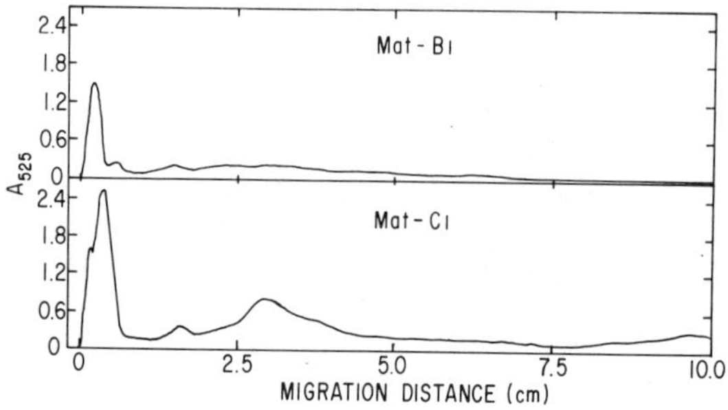

Figure 4. Membrane glycoproteins of the rat mammary adenocarcinoma MAT-B1 and MAT-C1 sublines. Membranes were prepared essentially as described previously (17) for MAT-A cells. Electrophoresis and staining were performed as indicated in Figure 2. About 50% more protein was applied to the gel for the MAT-B1 sample.

the mechanism of change is unclear. One possibility is an altered gene expression resulting in the synthesis of different proteins. Alternatively, the change could reflect an altered post-translational event such as an altered addition of carbohydrate or crosslinking of the proteins. Although we cannot answer these questions at present, it should be possible to obtain answers by biochemical and immunological comparisons of glycoproteins from solid and ascites tumors.

Surface properties of the 13762 ascites tumor cells.

The sublines of the 13762 ascites carcinoma differ not only from the solid tumor but also from each other. The difference in membrane sialoglycoprotein content between MAT-B1 and MAT-C1 cells is illustrated in Fig. 4. The actual difference between the two is greater than that shown because of the amounts of membranes loaded onto the gels. Cell surface sialic acid released by trypsin or neuraminidase is about 5-fold greater from the MAT-C1 cells than from MAT-B1 cells, even though the total cell sialic acid is only about 40% greater. Labeling studies with lactoperoxidase and ^{125}I or with periodate and ^{3}H-borohydride also show more major sialoglycoprotein in the MAT-C1 cells.

The most striking difference between these sublines is their surface morphology, MAT-C1 cells exhibiting a very unusual surface structure. Their surface is densely covered with microvilli extending from the cell body into highly branched structures (Fig. 5), whereas MAT-B1 cells have a more normal appearance with fewer microvilli, all apparently unbranched (Fig. 6).

MAT-B1 and C1 cells also have different cell surface properties (Table III). When treated with Concanavalin A (Con A), the MAT-B1 cells readily agglutinate. Under the appropriate conditions of temperature and lectin concentration these B1 cells also form patches and caps. The MAT-C1 cells do not respond to the lectin treatment under the same conditions, indicating a restriction of the mobility of the cell surface lectin receptor sites. It is also noteworthy that the MAT-B1 and MAT-C1 cells have different 5'-nucleotidase activities. This activity is essentially undetectable in B1 cells, even with a highly sensitive radiochemical assay.

In view of the large differences in cell surface properties one might expect significant biological differences between the B1 and C1 tumors. However, we have not found this to be true. The growth and killing rates of the two tumors are about the same. Tests for strain specificity of the tumors have not shown pronounced differences, and preliminary studies suggest that both sublines are metastatic. Thus it appears that differences in cell surface properties are not necessarily reflected in the gross properties of the tumors.

It is interesting that the properties of the MAT-B1 cells fit the criteria of Kim et al. (8) for a metastatic tumor under-

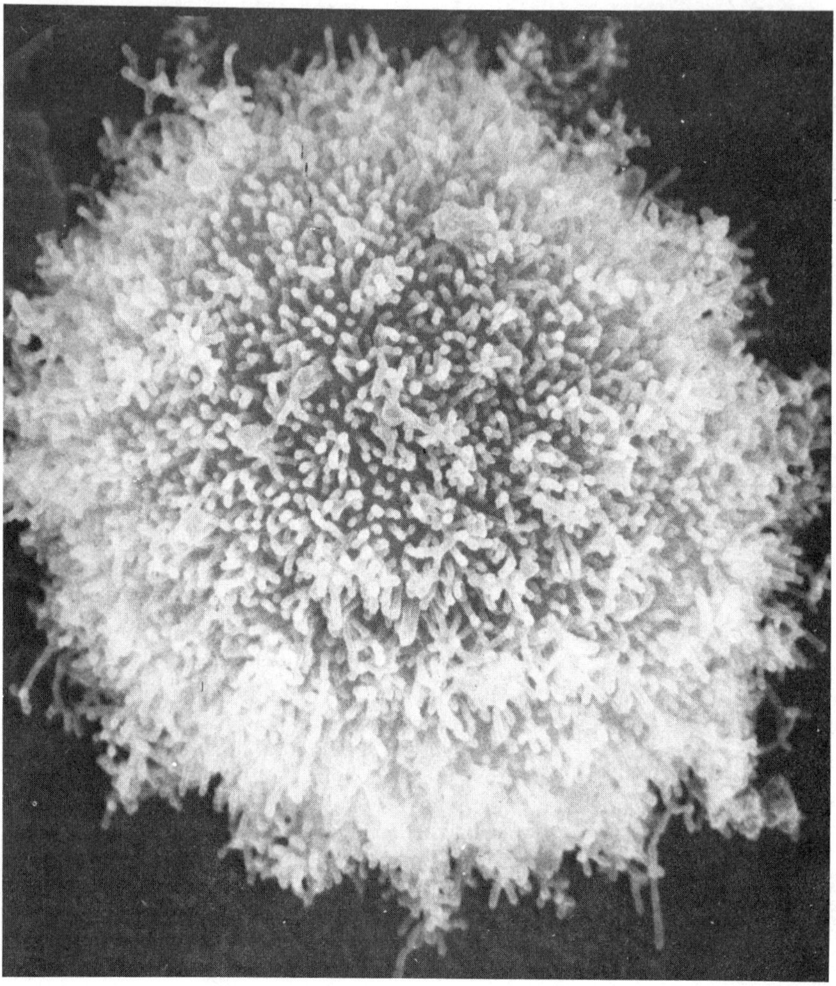

Figure 5. Scanning electron micrograph of MAT-C1 cells. Cells were removed from the animal, fixed in 3% glutaraldehyde in 0.1M sucrose–0.1M cacodylate, postfixed with 2% osmium tetroxide, dehydrated in ethanol, dried in a Polaron critical point dryer, coated with gold/palladium, and examined in a JEOL JSM-35 Scanning Electron Microscope. Note the extensive microvilli, whose branched structures can be readily observed at the periphery of the cell. Magnification, 8160.

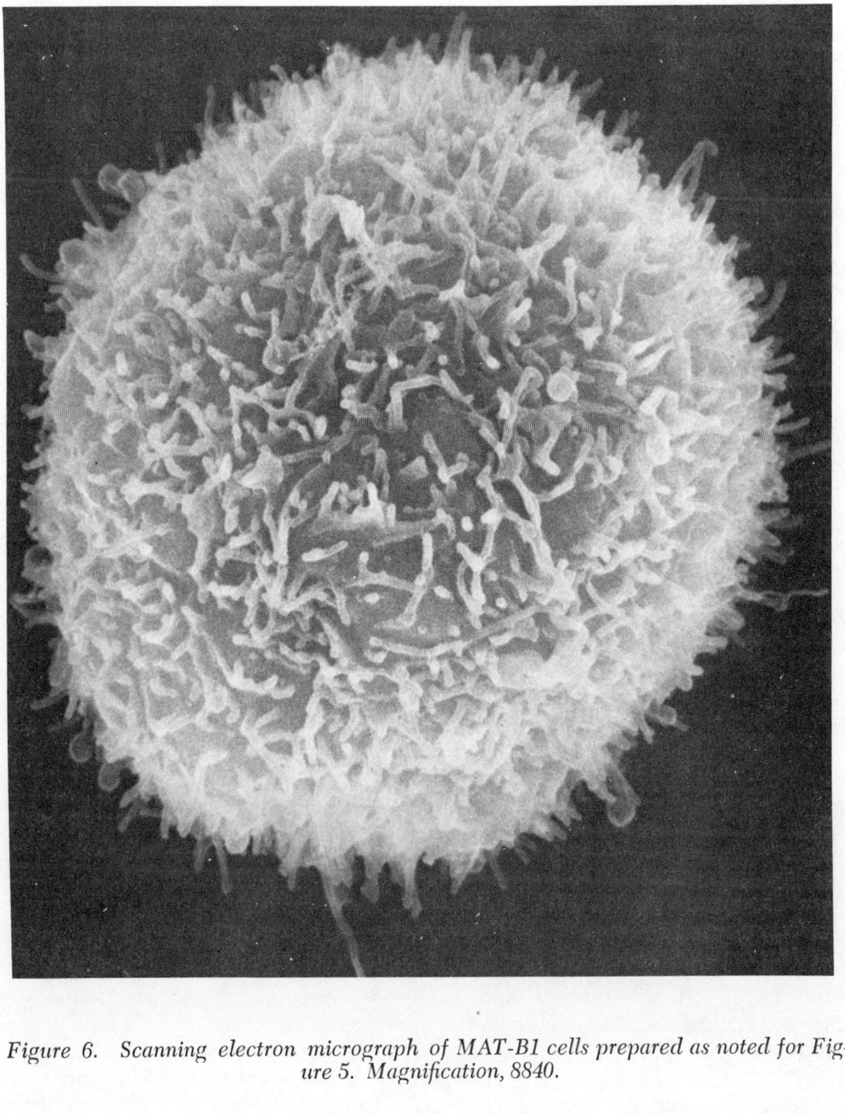

Figure 6. Scanning electron micrograph of MAT-B1 cells prepared as noted for Figure 5. Magnification, 8840.

TABLE III

Cell Surface Properties of MAT-B1 and MAT-C1
Mammary Ascites Cells

Property	MAT-B1	MAT-C1
Agglutination	++++	-
Patching and Capping	++++	-
5'-Nucleotidase, μmoles/hr•mg protein	0	2.0

going cell surface shedding. On the other hand the MAT-C1 cells show striking similarities to the TA3-Ha cells (11). Thus these two sublines, derived from the same solid tumor, may represent two different modes of escaping the immune system of the host.

Discussion

The results presented permit us to supply tentative answers to the questions raised at the beginning of this article. 1) There does not appear to be a necessary correlation in mammary tumors between metastasis and glycoprotein shedding. However, both glycoprotein expression and metastasis are very complex phenomena, so their relationships may also be complex. 2) Tumor environment apparently affects glycoprotein expression, as indicated by the differences between the ascites and solid tumors. These results point out the problems involved in extrapolating studies on ascites cells to conclusions about solid tumors. However, the studies on ascites cells have their own inherent value. In the process of metastasis individual tumor cells or small aggregates dissociate from the primary tumor into the body fluids. The similarity of these dissociated cells to ascites cells should be obvious. Since tumors lose many cells, the capacity of a tumor for metastasis must depend on the ability of the dissociated cells to survive in the circulation and become established at a new site. The ability of a tumor cell to alter its surface readily in response to a change in environment might be a useful attribute for the metastatic tumor. Since the 13762 tumor has this capability, further investigation of the relationships between the glycoproteins of the solid and ascites tumors may provide insight into the requirements for such changes. 3) The studies on the MAT-B1 and MAT-C1 cells clearly show that cells with very different cell surface properties can be very similar biologically. We suspect that these results reflect the fact that different tumor cells may use different ways of surviving the defense system of the host. However, it is also possible that the mechanisms of tumor survival are too subtle to be detected by the methodologies used in our studies.

The studies presented here have raised more questions than they have answered. The differences between the MAT-B1 and MAT-C1 cells are particularly intriguing. The abundance and complexity of the microvilli at the surface of the C1 cell indicate the presence of an extensive submembrane cytoskeletal network, since the microvilli contain microfilaments. One explanation for the extensive, branched microvilli is that there is an imbalance in the turnover of plasma membrane compared to cytoskeletal material. However, these cells also contain an excess of surface sialoglycoprotein. Is this a coincident or a result of some type of coupling between the submembrane cytoskeleton and extramembrane glycoprotein? This question assumes greater importance in view of the possible roles of cell surface component redistributions in various cellular responses. Our preliminary studies indicate that the sialoglycoprotein is not solely responsible for inhibition of Con A receptor movements in the C1 cells. There also appears to be a cytoskeletal involvement. Regardless of the exact mechanism by which receptor movement is repressed, the absence of mobility may be an important facet in escape from immune destruction. If mobility of the antigens of the target cell is a prerequisite for its destruction, the immobilization of the entire surface would be an effective defense. This might also be considered as an alternative mechanism for the lack of strain specificity of the TA3-HA cells.

The decreased nucleotidase and cell surface glycoproteins of the MAT-B1 cells are also intriguing. Can this really be explained by a "shedding" mechanism? Or has the capacity to synthesize these molecules been lost? It is necessary to examine the dynamics of the surfaces of these cells in order to understand their behavior. Such studies are complex but could provide insight into the relationship of cell structure and neoplastic behavior.

In attempting to understand tumor cell behavior in relationship to its cell surface properties, an important concept is that tumor cells appear highly adaptable. Perhaps this adaptability is the key to the success of the tumor in surviving a hostile environment and is the cause of much of the frustration in trying to establish definite relationships between cancer and particular cell properties. This adaptability is not limited to cell surface properties, as evidenced by the ability of tumors to develop drug resistance. However, the control of the cell surface would assist the cell in meeting the rigors of its environment. Studies on cell surface variations should help us understand the relationship of cell surface to tumor survival and the more general question of adaptability of tumor cells.

Acknowledgements

This is contribution P-490 of the Oklahoma Agricultural Experiment Station, Oklahoma State University, Stillwater, Oklahoma. This research was conducted in cooperation with the USDA, Agricultural Research Service, Southern Region and supported by the National Cancer Institute (NO-1-CB-33910 and CA 19985), the American Cancer Society (BC-246), a Presidential Challenge Grant from Oklahoma State University and the Oklahoma Agricultural Experiment Station.

Literature Cited

1. Hughes, R. C. (1976) in Membrane Glycoproteins, Butterworths, London.
2. Emmelot, P. (1973) Eur. J. Cancer 9, 319-333.
3. Nicolson, G. L. (1974) Int. Rev. Cytol. 39, 89-190.
4. Buck, C. A. Glick, M. C., and Warren, L. (1970) Biochemistry 9, 4567-4576.
5. Shin, B. C., Ebner, K. E., Hudson, B. G., and Carraway, K. L. (1975) Cancer Res. 35, 1135-1140.
6. Hynes, R. O. (1974) Cell 1, 147-156.
7. Hellström, K. E., and Helström, I. (1974) Adv. Immunol. 18, 209-277.
8. Kim, U., Baumler, A., Carruthers, C., and Bielat, K. (1975) Proc. Natl. Acad. Sci. U.S.A. 72, 1012-1016.
9. Friberg, S., Jr. (1972) J. Natl. Cancer Inst. 48, 1463-1476.
10. Codington, J. F., Sanford, B. H., and Jeanloz, R. W. (1973) J. Natl. Cancer Inst. 51, 585-591.
11. Miller, S. C., Hay, E. D., and Codington, J. F. (1977) J. Cell Biol. 72, 511-529.
12. Hilf, R., Michel, I., and Bell, C. (1967) Recent Prog. Hormone Res. 23, 229-289.
13. Segaloff, A. (1966) Recent Prog. Hormone Res. 22, 351-379.
14. DePierre, J. W., and Karnovsky, M. L. (1973) J. Cell Biol. 56, 275-303.
15. Steck, T. L., and Wallach, D. F. H. (1970) Methods in Cancer Res. 5, 93-153.
16. Keenan, T. W., Morré, D. J., and Huang, C. M. (1974) in Lactation: A Comprehensive Treatise, (B. L. Larson and U. R. Smith, eds.), Vol. 2, Academic Press, New York, pp. 191-233.
17. Carraway, K. L., Fogle, D. D., Chesnut, R. W. Huggins, J. W., and Carraway, C. A. C. (1976) J. Biol. Chem. 251, 6173-6178.
18. Beaufay, H., and Amar-Costesec, A. (1976) Methods in Membrane Biol. 6, 1-100.
19. Huggins, J. W., and Carraway, K. L. (1976) J. Supramol. Struct. 5, 59-63.

20. Warren, L. (1959) J. Biol. Chem. 234, 1971-1975.
21. Warren, L., Glick, M. C., and Nass, M. K. (1966) J. Cell. Physiol. 68, 269-287.

RECEIVED March 13, 1978.

AUTHOR INDEX

Names and numbers in bold are authors and first pages, respectively, of complete chapters. Numbers in parentheses are reference numbers which indicate that an author's work is referred to, followed by pages of that referral. Numbers in italics show the page on which the complete reference is listed.

C

D

E

F

G

H

M

N

T

U

V

W

Y

Z

SUBJECT INDEX

A

B

C

F

G

N

O